DATE DUE

BARK BEETLES IN NORTH AMERICAN CONIFERS

Number Six
The Corrie Herring Hooks Series

Bark Beetles in North American Conifers

A System for the Study of Evolutionary Biology

Edited by Jeffry B. Mitton and Kareen B. Sturgeon

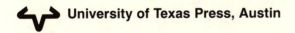 University of Texas Press, Austin

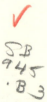

Contents

Contributors

Harvey Alexander
Department of Biology
University of New Mexico
Albuquerque, New Mexico 87131

Alan A. Berryman
Department of Entomology
Washington State University
Pullman, Washington 99163

John H. Borden
Department of Biological Sciences
Simon Fraser University
Burnaby, British Columbia
Canada V5A 1S6

Donald E. Bright, Jr.
Biosystematics Research Institute,
    Research Branch
Canada Department of Agriculture
Ottawa, Ontario
Canada K1A 0C6

Rex G. Cates
Department of Biology
University of New Mexico
Albuquerque, New Mexico 87131

Robert N. Coulson
Department of Entomology
Texas A&M University
College Station, Texas 77843

Donald L. Dahlsten
Division of Biological Control
Department of Entomological
    Sciences
University of California at Berkeley
Berkeley, California 94720

Jeffry B. Mitton
Department of Environmental,
    Population, and Organismic
    Biology
University of Colorado
Boulder, Colorado 80309

Ronald W. Stark
College of Forestry, Wildlife, and
    Range Sciences
University of Idaho
Moscow, Idaho 83843

Molly W. Stock
College of Forestry, Wildlife, and
    Range Sciences
University of Idaho
Moscow, Idaho 83843

Kareen B. Sturgeon
Department of Biology
Linfield College
McMinnville, Oregon 97128

H. Stuart Whitney
Department of the Environment
Canadian Forestry Service
Pacific Forest Research Centre
506 West Burnside Road
Victoria, British Columbia
Canada V8Z 1M5

Preface

Regrettably, biology is becoming more and more the province of narrowly trained specialists. Evidence of this can be seen in the growth of specialized departments of biology on college campuses, an increase in the kinds of professional societies, and a proliferation in the number of annual meetings attended by biologists. We are forced to increase our depth of knowledge in a specialty at the expense of our breadth of knowledge as biologists. The schism between applied and basic research seems to be getting wider and deeper just as we realize the necessity for exchanges of ideas and information.

Evolutionary biologists now recognize that there are several *modes* of speciation; evolution proceeds in a *variety* of ways. Population size, mating systems, the magnitude of gene flow, the diversity of habitats occupied, and life history characteristics may all influence population structure and the differentiation of populations. Thus, a reasonably comprehensive story of the evolution of any group may require knowledge of natural history, ecology, biochemistry, behavior, and genetics. Similarly, applied biologists, such as wildlife biologists, foresters, and agricultural biologists, now recognize the awesome complexity of protecting ecosystems, of controlling pest populations, and of managing such resources as natural forests or agricultural systems. Populations evolve, and species interact; nothing is constant except the challenge. Biological communities are complex, and simple attempts to modify communities often produce unexpected results.

Our interest in this project began with our attendance at separate meetings of evolutionary biologists, geneticists, forest entomologists, and forest geneticists. We perceived factions, semi-isolated groups of biologists publishing in different outlets, unaware of the resources provided by other groups. We anticipated advantages to be gained by bringing the different groups together. We have attempted to do that with this volume.

The story presented here is that of bark beetle communities. There are many facets to this story, and therein lies its strength or value. In fact, there are so many perspectives on this story, and so much information, that this story is perhaps too broad and intricate to be adequately presented by any one individual. Thus, we have collected original contributions on evolution, on the systematics of bark beetles, their life cycles and natural history, their pheromones, their symbionts, host resistance to attack by bark beetles, management of forests in the face of pressure from bark beetles, and the evolution of bark beetle communities. We hope that the integration of this information in one volume will be a service to naturalists and evolutionary biologists, as well as to the foresters and entomologists on the front lines.

J.B.M. and K.B.S.

Acknowledgments

The editors gratefully acknowledge the following: NSF grants DEB 77-162245 and DEB 7816 798 and a faculty fellowship from the C.R.C.W. at the University of Colorado to J.B.M.; American Association of University Women graduate research fellowships and Sigma Xi grant-in-aid of research to K.B.S. K.B.S. also wishes to acknowledge the Forest Research Laboratory at Oregon State University, a grant from the U.S.D.A. Forest Service Rocky Mountain Forest and Range Experiment Station, and grants from the vice-chancellor for research at the University of Colorado and from Linfield College. The book was reviewed, in whole or in part, by D. J. Futuyma, M. C. Grant, Y. B. Linhart, and T. Paine. A majority of the figures were drawn by Jan Logan and most of the manuscript was typed by Jeanie Cavanagh. The cover photograph of ponderosa pine bark was taken by Allan Doerksen and the cover illustration of *Dendroctonus ponderosae* was provided by the U.S.D.A. Forest Service.

Comments and reviews by Stephen L. Wood and Gerald N. Lanier are acknowledged for Chapter 3 and by Alan Berryman, Malcolm Shrimpton, Mary Alexander, Mike Freehling, Colin Henderson, Rick Redak, Janice Moore, Tim McMurray, and John Weins for Chapter 7. Research discussed in Chapter 7 was partially supported by NSF grant DEB 7927067 to R.G.C. The Canadian Forest Tree Service provided administrative and editorial support for the preparation of Chapter 6.

BARK BEETLES IN NORTH AMERICAN CONIFERS

1. Biotic Interactions and Evolutionary Change

J. B. MITTON and K. B. STURGEON

"...the most important of all causes of organic change...is the mutual relation of organism to organism..."

(Charles Darwin 1859)

The vast range and staggering diversity of extinct and extant species, as well as the wealth of variation within species, are the products of evolutionary genetical change. The adaptive characteristics of species which promote their survival and procreation in the face of physical hardship, predation, parasitism, and competition are forged, honed, and finely polished by natural selection, the differential reproduction of individuals with diverse hereditary characteristics. A few examples recall to mind the remarkable molding of living protoplasm to the most unsuitable of living spaces. Enzymes of the icefish, Trematomus, which lives in close proximity to Antarctic sea ice, function most efficiently near 0° Centigrade, while enzymes of several species of algae which live in hot springs have evolved so that optimal efficiency occurs at 45°C. Both animals and plants have evolved to the extreme osmotic stress of salinities which often exceed 40% in tidal pools and salt pans, and some arctic and alpine plants are able to withstand temperatures of -40°C, thriving in growing seasons lasting from a few days to several weeks. Finally, in the dark and cold Pacific trenches we find an abundance and diversity of living forms living under extreme pressures. Few places on earth support no life.

Yet the range and extremes of the physical environment may play a secondary role in the molding of adaptation. As

Darwin foresaw so clearly, the greatest range and assortment of adaptations have arisen, not in response to the rigors of the physical environment, but rather in response to pressures exerted by other species within a community. This book will focus on one community of organisms--bark beetles, their predators and parasites, the micro-organisms associated with them, and their host trees--and on the interrelationships among these organisms which have yielded a remarkable diversity of coadaptations. Largely because of the enormous economic impact of bark beetles on public and private lands and on the lumber industry in particular, we are fortunate to have a substantial amount of knowledge gathered over many years by researchers from several different disciplines. It is rare that an evolutionary story can be brought together as completely as can this one; yet this is not to say that we know the whole story. In the chapters to follow, we have brought together nine biologists whose interests include a broad range of topics, such as ornithology, botany, entomology, and plant pathology, and all of whose research subjects include one or more organisms in the bark beetle community. In each chapter, the author will present the current status of the knowledge in her or his area as well as suggest directions for future research. In this chapter, we hope to give the reader a perspective on the bark beetle story by considering the range and complexity of adaptations which have arisen in other plant and animal communities as a result of organisms interacting with one another.

NATURAL SELECTION

It is worthwhile, in consideration of forms of adaptation, to once again recall the mechanism of natural selection. The process of natural selection, first presented clearly by Charles Darwin (1859), can be described rather succinctly. A first tenet is that virtually all species have a wealth of heritable variation. Although individuals usually can be unambiguously identified to species and there are usually familial resemblances, there are always individual distinctions. Sexual reproduction and meiosis continually scramble the available genetic variation so that, although traits are passed from generation to generation, genetic differences exist between individuals. A second tenet is that populations are capable of growing geometrically, and thus are able to occupy all suitable space quickly. This idea and its consequences were first elaborated by Malthus (1798) in an essay that had a profound influence on Darwin. Given that a population regularly produces many more offspring than can be supported by the environment, only a small fraction of progeny will survive to reproduce. Many will suffer early mortality, and some, although living until reproductive age, will lack the vigor or competitive edge necessary to secure the energy, territory and mates necessary for reproduction. These tenets, working jointly, ineluctably produce the mechanism for evolutionary genetic change that Darwin called natural selection. Individuals genetically endowed to compete favorably with other genotypes in a particular environment will be more successful in reproducing and passing on their genetic material. The following generation will not be a random sample of the hereditary material of the previous generation, but will more closely reflect those individuals that enjoyed high reproductive success. Thus, natural selection is a consequence of heritable variation and differential reproduction in the context of a specific environment.

COEVOLUTION

In 1965, Ehrlich and Raven coined the phrase "coevolution" to describe those interactions between organisms of different species in which evolution in one member of an interacting pair elicits an evolutionary response from the other member. Although Ehrlich and Raven illustrated the phenomenon with examples of plant-herbivore interactions, coevolution is now recognized as a ubiquitous phenomenon.

The most widely accepted theory for the evolution of eucaryotic cells suggests that cell organelles such as mitochondria and chloroplasts originated as symbionts within larger procaryotic cells (Margulis 1974). Thus, a major evolutionary reorganization, second in importance only to the origin of life, may be the product of a mutually advantageous interaction between species. Among the myriad of coevolutionary adaptations between plants and their pollinators, two of the most remarkable are the associations between figs and their wasp pollinators (Ramirez 1970) and between yucca and their moth pollinators (Powell and Mackie 1966). The reliance of plant upon animal and animal upon plant in these systems is obligate; one cannot reproduce without the other. After pollination of the flower by the insect, the developing fruit serves as brood chamber for the developing larvae.

Interactions between predators and their prey have also resulted in many complex coadaptations. Papilio dardanus are palatable swallowtail butterflies which are found in parts of Africa where they are subject to heavy predation by birds (Sheppard 1962). Females of the species are highly polymorphic; it is estimated that there are probably 31 different morphs present. Apparently, predation on female swallowtail butterflies has resulted in disruptive selection for color morphs which mimic other sympatric species of unpalatable butterflies. In one species of model, Amauris echeria, there are three morphologically distinct subspecies, each of which is mimicked by a different morph of Papilio dardanus.

While we generally consider coevolution in the context of reciprocal interactions between two species, their joint evolution may have ramifications for other species in the community as well. A particularly clear instance of community coevolution is exemplified by the community of organisms associated with milkweed plants. Many species in the plant family Asclepiadaceae contain cardiac glycosides which are known to be potent heart toxins (Brower and Brower 1964). The poisons are believed to have evolved in response to predation pressures and now serve as anti-herbivore defenses for the plants. However, a small number of insects have developed the ability to circumvent this defense, and currently utilize the plant as a resource. Almost all of these species are aposematically colored, suggesting that they utilize the glycosides in their own defense. Monarch butterflies, (Danaus plexippus), which feed exclusively on milkweeds, sequester cardiac glycosides from the plants, thereby rendering themselves inedible to their vertebrate predators. Birds which attempt to consume either larvae or adult butterflies regurgitate their meal almost immediately, and subsequently learn to avoid monarchs.   The monarchs advertise their unpalatability with brightly colored markings which have subsequently served as the basis for evolutionary responses from yet another group of organisms.   A taxonomically unrelated species of butterfly, the viceroy, (Limenitus archippus), has phenotypically converged upon the warning color pattern of the monarch butterfly. Although viceroy butterflies do not feed on milkweeds and are not unpalatable, as are the monarchs, the viceroys are protected from predation by virtue of their resemblance to the unpalatable monarch. As expected, the community of predators that can safely prey upon these insects is small. Yet, in response to the morphological and chemical adaptations of the butterflies, birds have modified their behavior and have learned to avoid the most toxic monarchs. Birds are major predators of monarch butterflies in their overwintering congregations in Mexico and have learned to distinguish between individuals with low and high levels of glycosides, and to select those body parts which contain lower levels of the poisons (Calvert et al. 1979). In addition, two species of lygaeid bugs which live exclusively on milkweeds have striking red and black markings and secrete defensive compounds which contain cardiac glycosides. Although their coloration serves to warn potential predators of their toxicity, several species of parasitoids have evolved to be able to tolerate the presence of toxins and they now exploit their insect hosts without competition (Scudder and Duffey 1972).

ABUNDANT GENETIC VARIATION IN NATURAL POPULATIONS

At the heart of all reciprocal evolutionary interactions lies
a wide range of genetically based polymorphisms for those
characters involved in adaptation.
Although there were clear indications concerning the
amount of genetic variation in natural populations several
decades ago (Dobzhansky 1937), evolutionary biologists have
only recently received the information necessary to
appreciate the staggering abundance of genetic variation
maintained in natural populations. Genetic variation is
found in all aspects of the phenotypes of both animals and
plants.   Mayr (1963) documented extensive morphological
variation in populations of animals, and Ehrman and Parsons
(1976) summarized numerous examples indicating that many
aspects of behavior are genetically based and variable among
individuals. Advances in methods for staining of chromosomes
have greatly increased our knowledge of karyotypic variation
(e.g. chromosome number, inversions, translocations) in
natural populations (White 1978), and molecular biologists
are revealing more and more variation in repetitive DNA
(Tartof 1975) and in "multi-gene families" (Hood et al.
1975). Methods developed by molecular biologists (starch gel
and acrylamide electrophoresis) have revealed high levels of
genetic variation in the soluble proteins of both animals
(Selander 1976, Nevo 1978) and plants (Hamrick et al. 1979).
Plant secondary metabolites, in some cases known to be under
direct genetic control, have a diversity that has long
fascinated evolutionary biologists (Fraenkel 1959).   Perhaps
no aspect of plant or animal phenotypes will be found to be
invariant. Today, even pheromones (Averhoff and Richardson

1974, 1976; Lanier and Burkholder 1974) and hormones (Templeton and Rankin 1978), once believed to be too critical to survival and reproduction to maintain alternate forms, are known to possess considerable genetic variability.

Although the genetic bases of most coevolutionary interactions are not yet fully understood, a few are well documented.   Flor (1956), after studying complementary genetic systems in flax (<u>Linum</u> <u>usitatissimum</u>) and its parasitic rust (<u>Melampsora</u> <u>lini</u>), proposed the gene-for-gene concept of disease resistance.   Flor documented that two genes, each with two alleles, determine resistance of flax to rust.   In rust, there are two genes, each with two alleles, determining virulence on flax.   This matching of genes is probably the product of an evolutionary struggle between the two species.   Day (1974, Table 4.3, pp. 96-97) lists 27 examples, all involving domesticated plants and oligogenic inheritance, that seem to conform to the gene-for-gene concept.

Two cases from the literature adequately present the complexity of the genetic mechanisms behind specific resistance and hint at the dynamic evolutionary struggle between parasite and host.   The Hessian fly, <u>Mayetiola</u> <u>destructor</u>, lays its eggs on leaf surfaces of wheat seedlings, where the larvae will subsequently feed.   If the wheat is susceptible to the larvae, the leaves become stunted and dark and growth is inhibited.   Flies living on resistant seedlings do not permanently impair the growth of the seedling, and are unable to grow sufficiently to reach reproductive maturity.   Laboratory tests using different strains of wheat and flies from different localities indicate that flies have substantial genetic variation for virulence, and similarly, strains of wheat differ genetically in their patterns of susceptibility to flies with different genotypes (Hatchett and Gallun 1970). Patterns of susceptibility and virulence are consistent with the gene-for-gene hypothesis of disease resistance, and indicate that M. destructor carries at least five genes determining virulence on wheat.

A similar story with more genetic detail can be told for corn, Zea mays, and the common rust, Puccinia sorghi (Saxena and Hooker 1968). There are 15 alleles in corn on one arm of chromosome 10 which provide resistance to different forms of the common rust. Analysis of variability and pathogenicity in P. sorghi (Flangas and Dickson 1961) and inheritance of resistance of corn to rust (Hooker and Russell 1962) once again reveal a pattern of genetic architecture consistent with the gene-for-gene hypothesis of disease resistance. Fine structure analysis of this variation indicates that these are not 15 alleles segregating at a single locus, but that there are perhaps five tightly linked polymorphic loci determining resistance to common rust. Thus, this section of chromosome 10 may have evolved by tandem duplication to provide resistance to rusts. These systems of complex, polygenic control of disease resistance have been called multigene families by Hood et al. (1975) and in a discussion of the evolution of these systems, they point out the following common characteristics: multiplicity of gene loci, close linkage among genes, substantial sequence homology across loci, and related or overlapping phenotypic function. This similarity in genetic architecture between flax and corn may be much more than coincidence; the same pattern of structure and function is seen in the Rh antigen-antibody system and the HL-A histocompatibility system in man and in the major histocompatibility system of the mouse.

The two examples above are consistent with the gene-for-gene hypothesis, but one must wonder how general this sort of genetic architecture is. What proportion of the genome is designed by natural selection for disease resistance, and what proportion of that fraction of the genome is consistent with the gene-for-gene hypothesis? In the list of examples tabulated by Day (1974), one cannot help but notice that all of the plants are cultivated. Could these examples simply be an unrepresentative phenomenon resulting from artificial selection for high performance and genetic uniformity? This is probably not the case. Domestication and the consequent loss of genetic variability due to artificial selection probably makes the gene-for-gene relationship far easier to detect than it would be in a feral, highly variable population, but it is extremely unlikely that the genes for resistance to rust on chromosome 10 in corn evolved as a consequence of domestication. It is much more likely that these genes existed in the progenitors of corn, and domestication simply made them more evident.

If plant defenses are a significant evolutionary force for herbivorous insects, we might expect to see differentiation within polyphagous herbivores utilizing different hosts. A survey of protein variation in several sexual geometrid lepidopteran species revealed no significant genetic differentiation between individuals utilizing different hosts (Mitter and Futuyma 1979). Analyses of protein variation in a parthenogenetic race of the geometrid Alsophila pometaria, the fall cankerworm, uncovered some evidence for associations between genotype and host plant species. Stands with different dominant trees tended to be infested with different genotypes (Mitter et al. 1979). These investigators surmised that heterogeneity in the diet could create substantial diversifying selection pressures for an herbivore species, but host race formation might be frustrated by the reshuffling of genes in the sexual process and the migration pressure from diverse hosts. Another opportunity to study differentiation associated with hosts plants is provided by the treehopper, Enchenopa binotata, which utilizes seven different species of deciduous trees in eastern North America. Recent studies of treehoppers discovered sharp differentiation between individuals on different host species for color, chemical composition of the egg froth, time of mating, and time and site of oviposition. Females generally rejected males taken from different hosts, so this restriction in gene flow justified the designation of host races or perhaps incipient species (Wood 1980). Electrophoretic studies of these treehoppers from six host species in Ohio indicated that the differentiation had a broad genetic basis; the individuals utilizing different hosts are probably good species (Guttman et al. 1981, Wood 1980). Finally, the pine beetle, Dendroctonus ponderosae, commonly infests ponderosa pine (Pinus ponderosa), lodgepole pine, (Pinus contorta), and limber pine, (Pinus flexilis), in Colorado. Beetles taken from these three hosts at one locality exhibit both morphological and protein differentiation among the hosts (Sturgeon 1980; Chap. 10). Thus, herbivorous species may be forced to adapt in numerous ways to their different hosts, but we cannot presently predict the magnitude of the differentiation to be formed, nor can we accurately predict whether host races and ultimately species will evolve as a result of the diversifying selection.

THE RATE OF EVOLUTIONARY CHANGE

The term "evolution" often evokes images of subtle phenotypic changes over great gulfs of time in bizarre creatures seen only in museums; but natural selection can act very quickly, and can do so in virtually any species. Natural selection may act with startling rapidity in populations made predominantly homozygous through domestication, or when the environment is changed drastically by man or natural forces. Rapid responses to evolutionary pressures were illustrated nicely in a laboratory study by Pimentel and Bellotti (1976). They exposed house flies (Musca domestica) to various toxins (citric acid, copper sulfate, magnesium nitrate, sodium chloride, ammonium phosphate and potassium hydroxide) in sugar solutions in concentrations sufficient to cause heavy (80%) mortality. When flies were exposed to any of these toxins, they generally evolved resistance in seven to eight generations.

Studies in natural populations reveal that populations under heavy pressure from pests may maintain a greater diversity of defenses than those that are protected or not exposed to attack. Dolinger et al. (1973) studied a flower-feeding butterfly, Glaucopsyche lygdamus, and three of its host plants (Lupinus spp.) in southwestern Colorado. They found that the concentration of alkaloids in the plants was less important to the defense of the plant than was intraspecific variation in the types of alkaloids present. Populations suffering heavier attacks had a greater diversity of alkaloids than populations receiving little pressure from G. lygdamus. On the other hand, when populations are made genetically identical, and hence stripped of most of their defenses, mutation, recombination, and natural selection can quickly produce a pest or disease capable of circumventing the few remaining defenses. A common result is the recurrent pattern of epidemics seen in domesticated plants and animals (Day 1977).

Perhaps the finest demonstration of a broad based, rapid response to strong selection comes from observations on the evolution of resistance of insects to insecticides wielded by man. In 1972, Georghiou wrote:

> The experience gained from almost a quarter of a century on the evolution of resistance to synthetic organic insecticides inevitably leads the conclusion that, given adequate pressure, nearly every species of insect is capable of developing some tolerance to a given insecticide (Georghiou 1972).

At that time there was documentation for the evolution of insecticide resistance in 225 species of insects. Just five years later (Georghiou and Taylor 1977) the numbers of documented cases of evolution of resistance had climbed to 325. Pathways to the evolution of insecticide resistance included detoxification of the insecticide through modification or storage, decreased absorbance of the toxins through membranes, and evolution of behavioral modification to avoid the insecticide. Furthermore, individuals within a species, inhabiting different parts of the species' geographic range, may evolve resistance to a specific insecticide through different pathways. These observations suggest that the evolution of resistance to insecticides has a broad genetic base.

A case history in biological control may also be used to illustrate the speed at which populations can evolve (Fenner

1968). In 1859 rabbits were introduced into Australia, and, from a few propagules, the population grew and spread until drastic measures were required to control them. In 1950 a virus pest of South American rabbits was introduced. Initially, the virus was lethal in the vast majority of rabbits (98%), and it reduced populations to very low densities so quickly that it was no longer able to be transported by mosquitoes to uninfected individuals, so it, too, died out. Every year a few rabbits survived to reproduce again so that, in order to keep their populations under control, the virus had to be reintroduced each spring. This pattern of local extinction of the virus provided strong selection for a less toxic virus and a more resistant rabbit, i.e., the most virulent strains succumbed along with the most susceptible rabbits. Within three years new strains of virus had evolved and the mortality induced by these strains had dropped below 50%. But so also had the rabbits evolved a tolerance to the virus, so that when rabbits taken from the field were exposed to identical inoculations, mortality decreased steadily from 90% to 25% over a period of seven years. Thus, within a few years after man brought these species together, each had evolved to increase its own fitness and to coexist with the other.

EVOLUTION OF COEVOLUTIONARY RELATIONSHIPS

Several theoretical studies have addressed the question of how adaptations to interspecific pressures are developed and maintained. The studies all concur that apparent balance or stability in nature is dynamic, and is maintained by genetic variation responding to conflicting selective pressures. Pimentel stressed that competition between species would produce a genetic feedback which would result in the regulation of populations (Pimentel 1961, 1968). The assertion that this genetic response regulates populations has not stood unchallenged (Lomnicki 1974), but this model has heuristic value for the study of coevolution. Pimentel (1968) considers two species, one common, one rare, competing for a limiting resource. The abundant species contacts the other only occasionally, and hence evolves more to satisfy intraspecific rather than interspecific pressures. The rare species is constantly in competition with the common one, and hence receives strongest pressures to evolve to be a better competitor. As its competitive ability evolves, the population density of the rare species increases, and perhaps it supplants the other species as the more abundant. Now the roles have reversed, and once again the rarer species experiences greater pressure to enhance its competitive ability. This process may result in a series of dampened oscillations that ultimately stabilize.

Another perspective on the development of interspecific adaptations was formulated by Van Valen (1973). He envisioned each species on a treadmill, evolving to keep up with what it perceives to be a constantly deteriorating environment. The deterioration of the environment is caused by the evolution of other species; an increase in fitness of one species may produce decrements in the fitness of others. Those species that cannot evolve fast enough to keep pace with this continual challenge go extinct. He presented a massive amount of data from the paleontological literature to indicate that the probability of extinction for any group is a constant, i.e., not dependent upon the age of the group. Van Valen interpreted the constant probabilities of extinction as evidence suggesting that the majority of evolutionary challenges comes from competition between species.

Maynard Smith (1976) considered interactions between species and the rates of extinction produced by these interactions. He asserted that increases in fitness enjoyed by one species would not always be balanced by decrements in fitness suffered by others. If fitness increases enjoyed by one species are not balanced by decrements in others, the rate of extinction may decrease. Alternatively, if an advance of one species produces greater losses in fitness of others, the rate of extinction should increase.

A mathematical model was utilized by Rosenzweig (1973) to determine the conditions needed for the interactions between an exploiter and its victim to come to a dynamic equilibrium. He proposed, just as Fisher did years before (1930), that the closer a species came to a phenotypic optimum, the more slowly it would evolve. As a predator became more and more proficient, and approached its paragon, its rate of acquisition of improvements would decrease. The realized improvements would present a challenge to the prey, which would then evolve away from the slowly evolving predator. The result is a dynamic balance, with the exploiter evolving to be more efficient, and with the victim escaping extinction each time the rate of evolution of the exploiter declines.

Clearly, many models adequately describe some aspects of the evolution of adaptations to interspecific pressures. Presently, our knowledge of this subject will not permit us to accept or reject any of these abstractions. Past experience with evolutionary theory suggests that it is not likely that any model will be applicable to all situations. Yet all of these models warn us not to confuse an apparent equilibrium with constancy, or to interpret apparent stability as lack of competition.

COEVOLUTIONARY STUDIES

Studies of the interactions between species and the resulting adaptations hold great promise for substantially improving our knowledge of evolution. The detailed study of reciprocal evolutionary pressures, or coevolution, does not reach far into the past (Ehrlich and Raven 1965), but the number of studies is growing rapidly. Most of the earliest studies inferred from present day patterns that the organisms in question must have coevolved sometime in their evolutionary past. Ehrlich and Raven, for example, compiled an enormous amount of data to support their hypothesis that families of butterflies and of flowering plants had diversified in a series of adaptive radiations in response to one another. More recently, studies have attempted to show that coevolution is not only a phenomenon of past geologic eras but rather is a ubiquitous component of present day community interactions. Finally, the most ambitious studies have attempted to understand the evolution of communities. Reciprocal evolutionary responses between interacting species feed back on the system to produce changes in abundance, distribution, and diversity of the interacting species, and, therefore, to produce changes in the structure of the community as a whole.

This volume is intended to contribute to the literature on coevolution by assimilating theory and data on a single but extensively studied system. The system presented is that of bark beetles of the family Scolytidae. Different chapters consider the population dynamics of the beetles, geographic variation and taxonomy, resistance and susceptibility of their host trees, and evolutionary interactions with predators and parasites and with microorganisms symbiotically associated with them. Lastly, we present a chapter on management and control of these forest pests. Sound management practices depend upon our accurate understanding of the biology of these organisms. Perhaps the most important conclusion that may be drawn is that the system is not static; it is changing, i.e., it is evolving, and change in one part of the system will have ramifications for other parts. Perhaps it is this ever shifting approach to equilibrium which simultaneously frustrates the forest manager and which fascinates and challenges the evolutionary biologist. Although attempts are being made to predict the future course of this evolution through computer modeling and simulations, it must suffice for now to have an understanding of the community as it is today and of the forces that have helped to mold it.

2. Generalized Ecology and Life Cycle of Bark Beetles

R. W. STARK

Insects are a highly specialized group belonging to the large and diverse phylum of joint-legged animals, the Arthropods, which includes lobsters, crayfish, crabs, barnacles, shrimps, millipedes, centipedes, scorpions, spiders, mites and others. In terms of the number of species, it is the most successful phylum in the animal kingdom, and within Arthropoda, Insecta is the largest class. The evolutionary success of insects is, at least partly, attributable to several characteristics: flight, an external skeleton, metamorphosis, and specialized systems of reproduction.

The most obvious attribute of insects, shared only with birds and bats, is flight. Wings enable them to escape rapidly from unfavorable situations and to seek out favorable ones. Wings also enhance their ability to locate at a distance and forage widely for food. The external exoskeleton is composed of chitin, a substance that is flexible, lightweight, and tough, and is impervious to water, many chemicals, and most microorganisms. These characteristics have enabled insects to exploit habitats that soft-bodied forms would have to avoid. In addition, insects are able to regulate aspects of their reproduction in accordance with environmental cues. For example, some are capable of storing sperm and delaying fertilization until the proper food plants or conditions have been found, or delaying oviposition until the appropriate site is found or prepared.

The more primitive insects, such as grasshoppers, pass

through a gradual metamorphosis. The post-embryonic stages grow to adults with little change in appearance other than in body proportions, and the appearance of reproductive organs and wings. The immature stages of species that undergo gradual metamorphosis are known as nymphs and they exploit habitats similar or identical to those used by their parents. Complete metamorphosis consists of three distinct post-embryonic stages: an early form without wing pads called the larva, a quiescent form with wing pads called the pupa, and the adult. Members of the family Scolytidae, which contains bark beetles, undergo complete metamorphosis (Figure 2.1). In insects with complete metamorphosis, the egg may exploit one environment, the larva another, and the pupa and adult yet others. Further, metamorphosis can be accommodated to environmental conditions. Any one of the developmental stages may enter an arrested stage of development (diapause) that permits it to last out unfavorable periods. Some forms (e.g. multivoltine) have a facultative diapause, i.e., the onset is dependent upon environmental cues such as shorter day length. Others (e.g. immature stages of univoltine species) have an obligatory diapause, which is not influenced by environmental cues, i.e., it is genetically controlled. The most common obligatory diapause is winter diapause, which decreases mortality due to low temperatures, but summer diapause in response to drought or extremely high temperatures is not uncommon. In some species, developmental rate and time can be shortened or extended. For example, some species of bark beetles in the genus Dendroctonus are able to complete development at a rate that is primarily dependent on temperature and may complete from 4 to 7 generations in a single season.

The family Scolytidae (order Coleoptera) is one of the most successful in the class Insecta. Scolytids are subcortical feeding insects -- bark beetles and ambrosia or

Figure 2.1.   Stages in the life cycle of the western pine
beetle, Dendroctonus brevicomis.   (a) beetle galleries in
phloem before eggs hatch.   (b) eggs in niches in the parent
gallery.   (c) mature larvae.   (d) pupal chamber with mature
pupa surrounded by a mass of associated blue stain fungus.
(e) newly formed adults beneath the bark.   (f) exit holes
left in bark.   (g) mature adult.

timber beetles. They are distributed world-wide and there
are many cosmopolitan genera. The bark beetles, in
particular, have flourished in the coniferous forests of
North America. The most destructive species are in the genus
Dendroctonus (literally "killer of trees") whose hosts
include all conifers and most hardwoods. There are about 24
known species of Dendroctonus in the world; 23 of these are
found only in North America and one is found in Europe. Of
the North American species, 19 are found north of Mexico, the
remainder are in Mexico and Central America.

Scolytids occupy a broad range of niches on woody and
herbaceous plants. Members of this family attack almost all
forms of plant life, including coniferous and broad leaved
trees, coffee, tea, sugar cane, cotton, rice, dates, corn,
legumes, gourds, various fruits, cacti, and even corks of
wine bottles. While ambrosia beetles are restricted to boles
or large branches, bark beetles are not confined to any
particular portions of plants and one or more species may be
found attacking all plant parts. Primitive mouthparts,
adapted for biting and chewing, have been retained in both
the larval and adult stages and the well-developed mandibles
enable them to penetrate and chew almost any material. They
can tolerate a wide range of environmental conditions; some
species can endure extremely high temperatures as well as
those considerably below freezing. They can tolerate toxic
substances such as conifer resin or resin vapor for long
periods, and they can develop in oxygen-deficient
environments. Some can withstand long periods without
water. Scolytids possess highly sophisticated systems of
communication based primarily on chemistry (Chap. 4) and
sound and they have developed symbiotic (Chap. 6) and
communal relationships (Chap. 5, 10) with several types of

organisms. Each of these features has contributed to the evolutionary success of these organisms.

The literature on the family, particularly on those species of economic importance, is extensive but there are several good general references (e.g. Anderson 1960, Baker 1972, Bright and Stark 1973, Chamberlin 1939, 1958, Graham and Knight 1965, Graham 1963, Furniss and Carolin 1977, Rudinsky 1962, Thatcher 1960, Wilson 1977). The family contains about 73 genera and over 625 species with a potential for injury to plant life that is awesome. Fortunately, however, relatively few have been elevated to

Table 2.1.   Important Forest Scolytidae of North America north of Mexico[1]

Dendroctonus adjunctus	Roundheaded pine beetle
D. frontalis	Southern pine beetle
D. brevicomis	Western pine beetle
D. engelmanni	Engelmann spruce beetle
(= D. obesus)	
D. jeffreyi	Jeffrey pine beetle
D. ponderosae	Mountain pine beetle
D. pseudotsugae	Douglas-fir beetle
D. valens	Red turpentine beetle
Hylurgopinus rufipes	Native elm bark beetle
Ips plastographus	California pine engraver
(= I. integer)	
I. lecontei	Arizona five spined engraver
I. pini (= I. oregonis)	Pine engraver
Scolytus multistriatus	Smaller European elm beetle
Trypodendron lineatum	Striped ambrosia beetle

[1]From Davidson and Prentice 1967.

the status of pests of national importance (Table 2.1).
Eight species of Dendroctonus and only the ambrosia beetle,
Trypodendron lineatum, are considered serious pests and this
list has remained remarkably constant for many years. For
example, for as long as we have records from the Pacific
Northwest, the fir engraver beetle, Scolytus ventralis, has
been a significant pest in the true fir forests and in the
Southeastern U.S., the black turpentine beetle, Dendroctonus
terebrans, continues to cause extensive damage to pines,
particularly in turpentine orchards (Baker 1972).

GENERALIZED LIFE CYCLE OF SCOLYTIDS

The extreme variety of life cycles in the family Scolytidae
mandates some restrictions in coverage. The ambrosia or
wood-boring species deserve some mention but this chapter
will emphasize the tree killing bark beetles (Dendroctonus)
whose life cycle is passed largely in galleries completely or
partially excavated in the phloem of conifers.

Ambrosia Beetles

The ambrosia beetles are wood-boring species that utilize
either the bole or branches of their hosts, and almost all
species in North America infest only dead or dying trees.
Their name is derived from their symbiotic relationship with
ambrosial (describing the food of Greek and Roman gods) fungi
which the beetles introduce into breeding galleries and tend
and harvest for food. Adults of some species actually

collect the fungal spores and place them near the developing larvae. Both sexes participate in maintaining the galleries clear of boring dust and frass, although it has been claimed that some adult and larval excrement is retained to "fertilize" the fungal gardens.

There are both monogamous and polygamous species. Either the male or female selects the host, excavates an entrance tunnel, and is joined by its mate. They excavate egg galleries deep in the xylem (Figure 2.3), where they establish their fungus "gardens" and where the larvae develop. Eggs may be deposited more or less indiscriminately along the gallery walls or in individual cradles or burrows. The larvae feed on the ambrosial fungi and, as adults, they emerge from the parental entrance hole or bore directly out through the wood and bark.

Bark Beetles

For convenience the life cycle will be treated in three phases (Figure 2.2); Dispersal, which includes emergence from the host tree, flight to a new host tree, and selection of that host for establishment of the next generation; Colonization, the process of invading the tree and initiating the brood gallery; and production, which includes mating, egg gallery construction, egg deposition or oviposition, and brood development up to the time of emergence (Cobb et al 1968; Wood 1972).

Dispersal. Once the adult bark beetle has fully developed (Figures 3.1-3.5) it constructs an exit hole from the pupal cell by boring directly through the outer bark, leaving

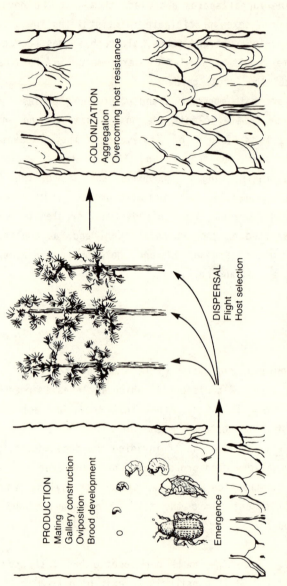

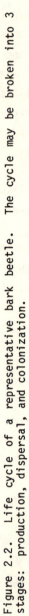

Figure 2.2. Life cycle of a representative bark beetle. The cycle may be broken into 3 stages: production, dispersal, and colonization.

a distinctive clean cut hole in the bark. The bark of abandoned trees is literally riddled with these holes. Emergence takes place over a period of several days or weeks. The sex that emerges first is usually the species that selects the host tree; these are generally females in the genus Dendroctonus. Environmental conditions, particularly temperature, significantly affect emergence and dispersal. For example, there is a typical temperature threshold below which beetles cannot fly, and an upper temperature limit that inhibits flight, and both emergence and development vary with altitude and latitude, probably also as a function of temperature.

There are many factors which influence the direction and duration of flight. In the Trypodendron species, for example, emerging beetles have a "gas bubble" in their gut that must be eliminated by flight exercise before the insect will settle on a substrate, thereby insuring dispersal (Graham 1959). Although bark beetles are capable of long distance flight, they generally attack trees in the vicinity of the brood tree from which they emerged.

There is a considerable mystery surrounding the ability of beetles to locate and select hosts. Some investigators believe that a "primary attraction" exists that results from odors emanating from physiographically weakened trees (Person 1931, Heikkenen 1977). Others believe that dispersing beetles land randomly on host and non-host trees (Chap. 9), perhaps guided initially by visual cues (Gara et al. 1965). Whether the beetles continue to the next phase, colonization, depends on their acceptance of the tree as an appropriate host (Borden 1974). The "host selection principle" asserts that insect species will attack the host species in which they developed as larvae (Walsh 1864, Craighead 1921). Others have shown that photic reactions are important in flight, dispersal, and orientation to the host (Rudinsky 1962, Schonherr 1976, Shepherd 1966), for some bark beetles respond to profiles of the tree crown, others to vertical (living tree) and horizontal (fallen tree) shapes. Initial attraction and host selection are still open questions and are, consequently, areas of active research.

Host preferences shown by scolytids are presented in Appendix 1 (Baker 1972, Chamberlin 1958, Bright and Stark 1973). Care must be taken when drawing conclusions from such host lists because they reflect the intensity and accuracy of collection records, and are subject to the idiosyncracies of taxonomists.

Colonization. The acceptance of a suitable host tree by a pioneer beetle begins the colonization process. Feeding in the host has been divided into three categories: (1) monophagy-feeding on host plants of one or more closely related species within a genus; (2) oligophagy-feeding on species in several genera within a family and (3) polyphagy-feeding on species in several families. There are many different classifications of coniferous trees that express various authors' opinions of their evolutionary or taxonomic relationships. This presentation relies upon the review of Harlow and Harrar (1941). Generally, it may be stated that a higher degree of monophagy is demonstrated by those species that attack living trees (e.g. Dendroctonus, Ips) whereas species with polyphagous habits are those attacking primarily dead or downed trees (e.g., the ambrosia

beetles, Trypodendron, Gnathotricus) (Table 2.2 and Appendix
1). With few exceptions, (e.g. certain Micracisella,
Xyleborus), species attacking conifers do not
attack hardwoods. For example, the genus Micracisella is
restricted mainly to hardwoods and six of the eight species

Table 2.2. Feeding Habits of Scolytidae[1]

Genus	No. spp. Considered	Monophagous[2] M-1	M-2	Oligophagous	Polyphagous
Scolytus	11	0	3	8	0
Crypturgus	4	0	0	3	1
Dolurgus	1	0	0	1	0
Polygraphus	2	1	0	1	0
Carphoborus	5	1	0	4	0
Phloeotribus	2	1	0	1	0
Dendroctonus	12	4	4	4	0
Phloeosinus	23	12	2	7	2
Xylechinus	2	0	0	1	1
Scierus	2	0	0	2	0
Pseudo-phylesinus	11	2	2	6	1
Hylastes	8	1	2	5	0
Hylurgops	5	0	3	2	0
*Trypodendron	4	0	0	2	2
Cryphalus	5	1	1	2	1
*Gnathotricus	4	0	0	1	3
Conophthorus	10	8	2	0	0

Table 2.2 (continued)

Genus	No. spp. Considered	Monophagous[2] M-1	M-2	Oligophagous	Polyphagous
Pity-					
ophthorus	52	15	25	12	1
Ips	22	6	12	4	0
Micracis	1	0	0	0	1
Pityogenes	7	1	3	3	0
Pityokteines	5	0	1	4	0
Orthotomicus	1	0	0	1	0
Pityokteines	1	0	0	1	0
*Xyleborus	9	0	1	0	7
Dryocoetes	6	1	3	2	0

*ambrosia beetles
[1]summarized from Appendix 1
[2]M-1, feeding on a single species in a genus
M-2, feeding on several species in a genus

of ambrosia beetles in the genus Xyleborus (listed in Appendix 1) feed primarily in hardwoods; the remaining two are confined to conifers.

There is a continuing controversy over whether scolytids are "primary" or "secondary" insects, i.e. whether they are able to overcome completely healthy trees in full vigor or only those trees of subnormal physiological condition (Rudinsky 1962, Amman 1978, Shrimpton 1978). Under endemic conditions the preponderance of species are termed "secondary", that is, they infest living trees of subnormal physiological condition, which have been temporarily or permanently weakened by drought, defoliation or disease, or have been fatally injured by cutting, fire, or windthrow. These authors agree that it is under epidemic or outbreak conditions that many species become primary pests and invade and kill trees in apparently normal health. The major areas of controversy center on whether epidemics are caused when large tracts of trees become weakened and susceptible to attack or when beetle populations experience genetically controlled behavioral changes causing them to become more aggressive and overcome normally resistant trees (see also Chap. 9).

Whatever the stimuli, successfully attacking beetles initiate a complex chain of events leading to colonization of the tree. Once a few beetles have successfully entered the bark of the selected host, aggregation of individuals of both sexes commences. This is a critical phase of the bark beetle life cycle, for the defenses of a living tree, even stressed trees, are formidable. Most conifers contain copious quantities of oleoresin with toxic properties and so a large number of beetles are needed to successfully attack a healthy tree (see Chap. 3, 9).

Aggregation on the host tree is regulated almost exclusively by pheromones, which are produced by the attacking beetles, in conjunction with host volatiles that synergize or activate the pheromones. This "secondary attraction" involving pheromones has been found in over 60 species of scolytids from some 20 genera (Borden et al. 1975). Because of their importance to the understanding of bark beetle life histories and in management, pheromones are discussed at length in Chapter 4.

The organs which detect pheromones (sensilla) are located on the antennae. There are literally hundreds of sensilla with thousands of receptor pores. Airborne molecules of the pheromones and host odors are collected and transmitted through the central nervous system, and elicit behavioral responses. The pheromones are released by one species to attract members of that same species but they also attract parasites, predators, and commensals of bark beetles (Chap. 5).

Spacing of gallery sites on the host tree and the density of attacks by bark beetles are controlled by a weakening or cessation of the attractant pheromone, by release of repellent pheromones, or by stridulation, which is the production of sound by friction. Stridulation is highly developed in scolytids. Barr (1969) reviewed the world literature and found that 77 species in 23 genera were known to possess stridulatory mechanisms. Possession of a stridulatory organ appears to be associated with the sex opposite to that which initiates the entrance tunnel. Apparently specific "signals" are associated with different kinds and timing of stridulations. Stridulation has been associated with stress, rivalry and aggression, "greeting", or with courtship in D. valens, D. ponderosae, D. pseudotsugae and D. frontalis (Rudinsky and Michael 1974, Ryker and Rudinsky 1976, Rudinsky and Ryker 1976). In D. ponderosae and D. brevicomis, male stridulation elicits release of pheromone by the female (Rudinsky et al. 1976).

Removal of stridulating organs of Ips confusus females resulted in their entrance to the gallery being denied or delayed by the males, although once they were admitted, there was no significant difference in egg gallery construction.

The state of our knowledge and understanding of sonic communication is embryonic and offers another area where much more research is warranted.

Microorganisms, such as the blue stain fungi of the genus Ceratocystis (Chap. 6), play important roles in the colonization phase (Francke-Grosmann 1963, Graham 1967, Safranyik et al. 1975, Shrimpton 1978). They probably influence aggregation, modify development and survival by providing nutrients and assist in overcoming the natural defense mechanisms of the tree (Chap. 7). For example, fungi, which are transported by the southern pine beetle in special structures called mycangia, influence the production of pheromones by the beetle (Brand et al. 1976, Brand and Barras 1977).

Production. Successful entry to the phloem in the bark-wood interface signals the beginning of the production phase of the life cycle. After attacking bark beetles have penetrated to the phloem layer, they begin construction of the brood gallery. Many species, particularly the polygamous ones, excavate a nuptial chamber for mating. Individual egg galleries for each female radiate from this chamber (Figure 2.3). The nuptial chamber is usually kept free of boring material and frass and may be visited from time to time by the females. By contrast, monogamous species generally mate on the bark surface near the entrance hole or in the tunnel in a "turning" niche. Although most monogamous species mate but once, some (e.g. most Dendroctonus species) can re-emerge and initiate a second gallery in the same or another host tree. It is not known whether re-emerging beetles will mate again but this phenomenon has important implications for population dynamics (Chap. 8).

Bark beetle tunnels or galleries are of many types and often so characteristic of a species that they serve for species identification (Figure 2.3d-g). The principal gallery is excavated for the deposition of eggs but many species also excavate ventilation and food tunnels. Ventilation tunnels are bored by the adult from the egg gallery through the bark to the outside. They are most common in species which have long egg galleries. It is assumed that the principal purpose is for ventilation but they serve other purposes such as outlets for frass and boring dust. They are also utilized by parasites and predators to gain entry into the bark beetle gallery. Food tunnels are usually extensions of the main egg gallery or of the pupal chambers that are excavated by those species which mature in the fall but which overwinter under the bark and do not emerge until the following spring. Some species feed in portions of the host other than where they establish their egg galleries, i.e. in buds, small twigs or outer bark.

The egg gallery of all bark beetles is almost invariably in the phloem; the larval galleries and pupal cells of most are also in the phloem but in Dendroctonus brevicomus, the western pine beetle, third and fourth instar larval mines and pupal cells are excavated in the outer bark. Some species keep the egg gallery free of frass and borings while others pack this material behind them as they excavate. This chore is usually done by the male, and in polygamous species with up to 8 or 10 females the task becomes comparable to that of Hercules in the Stygean stables! It has been suggested that clean galleries facilitate repeated copulation in polygamous forms. Special morphological adaptations have evolved to accomplish frass removal. For example, the elytral declivity

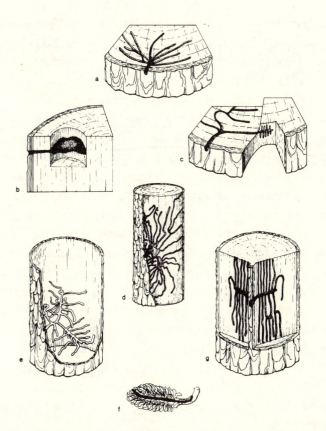

Figure 2.3.   Ambrosia (a-c) and bark (d-g) beetle galleries.
(a) branched type e.g. Xyleborus spp.   (b) cave type e.g.
Xyleborus spp.   (c) compound type e.g. Trypendron,
Gnathortricuss spp. (d) cave type, Renocis heterodoxus. (e)
radiate   or   star-shaped,   Pityokteines   elegans   and
Taenioglyptes, Ips spp.   (f) cone beetle type Conophthorus
spp.   (g) forked tunnels, Leperisinus californicus and
Scolytus, Alniphagus, Pseudohylesinus spp.   Redrawn from
Chamberlin (1958).

in _Ips_ and _Scolytus_ at the posterior is modified somewhat
like a specialized bulldozer or shovel (Figure 3.4). Other
modifications include rake-like tibiae, flattened or concave
head capsules, and dense hairiness (a mobile broom!).

The number of eggs laid by bark beetles varies widely from
as few as 6-8 in _Hylastinus_ _obscurus_ to as high as 300 in
Dendroctonus _ponderosae_. The more destructive species are
the more fecund. Eggs are laid in various patterns, along
each side of the egg gallery (Figure 3.4). These may be
spaced regularly, close together or distinctly separated, and
they may be opposite one another, or alternated. A single
egg is placed in each niche and covered with frass. Egg
grooves are made by the Douglas-fir and Engelmann spruce
beetles in which 20 or more eggs are placed. In cave type
galleries, eggs may be deposited loosely without any apparent
order, in niches or at the end of short tunnels, in small
clumps around the margin, or in masses encompassing all parts
of the margin (Figure 2.3d). The period during which egg
deposition occurs varies considerably among species but
invariably is restricted to spring, summer or early autumn.
Only a few species over-winter in the egg stage and then only
in warmer climes.

Larval mines may be bored in the same plane as the egg,
e.g. entirely in the phloem (Figure 2.3), extending directly
to the outer bark or sapwood, or show a combination of both
characteristics. The mines increase in size as larvae grow,
and are usually packed with frass and borings. Typical mines
vary in length from a few mm to up to 5 cms. When the larva
has reached maturity it excavates a slightly larger cavity, a
pupal cell, in which it transforms to the pupal stage (Figure
2.1).

Newly transformed adults of bark beetles pass through a stage as callow or teneral adults during which the exoskeleton hardens and sexual maturation is reached (Figure 2.1). Soon after they mature, they excavate an exit hole from their pupal cell to the outside. Some species feed for a few days or even weeks before exiting. Species maturing in the fall may pass the winter as young adults, emerging when external temperatures are suitable. The adult of other species however, may leave the host tree and over-winter in the duff and litter of the forest floor, in deep cracks or crevices of the bark, or may even excavate hibernation tunnels in the bark near ground level. Various cone-inhabiting species may hibernate one or more winters in cones on the forest floor, or on lichen-covered limbs. One bark beetle species constructs special hibernating tunnels in which 12-20 other species may be found!

As might be expected from the latitudinal and altitudinal distribution of bark beetles, a great diversity in seasonal development occurs. This variation occurs within species as well as between. For example, in Canada the western pine beetle has only one or one and a partial second generation per year. In its most southerly location in California, three and a partial fourth generations may develop (Miller and Keen 1960). The mountain pine beetle (D. ponderosae) normally has one generation per year, but at higher altitudes and latitudes may require two years to complete a generation (Amman 1973). For the southern pine beetle, the number of generations per year may range from 2-3 in the northern part of its range to 7 or 8 in southeastern Texas, depending on the warmth and length of the season (Thatcher 1960, Coulson et al. 1979a). Hibernating adults usually emerge first, followed by those that passed the winter as newly formed adults, then those that passed the winter as pupae, then those that passed the winter as larvae.   Only a few generalizations can be made about the timing of emergence. One is that scolytids are remarkably adapted to prevailing weather and climate and can adjust their emergence accordingly, although all species seem to have "peak" periods of emergence and attack. Another is that, in colder portions of their range, few, if any, over-winter in the two most vulnerable stages, eggs and pupae.   The cone beetles (Conophthorus spp.) and perhaps others, exhibit an unusual survival mechanism in that not all individuals of a brood emerge the first year.   This is probably an adaptation to cope with the normal periodicity of cone crops. Cone beetles may also, on occasion, become twig borers or they may feed on the bark and phloem at the juncture of two small twigs.

SCOLYTIDS IN FOREST ECOSYSTEMS

Since ecology is a relatively young science, few fundamental principles go unchallenged.   Ecology is really a debate. This complex subject must be approached with some temerity, particularly since the ecological role of insects in forest ecosystems has been poorly studied.   Further, scolytids are but one small component of the abundant and diversified insect fauna inhabiting forests.   Although we are concerned here only with conifers, many broad-leaved trees, plants and shrubs also serve as hosts for bark beetles.   Thus, the total

influence of scolytids on forests cannot be fully described
in this restricted treatment. We can, however, suggest some
probable significant influences of tree-killing scolytids on
three fundamental ecological processes: natural selection,
primary production, and forest succession.

Natural Selection

A forest ecosystem consists of a community of organisms plus
its physical environment. The community, in turn, consists
of populations of several species whose activities are
interdependent. The genetic structures of the populations
strongly influence the ways in which the populations
interact, and the levels of genetic diversity that we see in
the populations are, in turn, strongly influenced by spatial
and temporal variation in the environment. Changes in the
environment affect different genotypes differently and the
differential reproduction of genotypes is what we call
natural selection.

The tree-killing Scolytids, e.g. most Dendroctonus, act as
one of the forces of natural selection exerting substantial
pressures on their host trees. Of course, so too have the
trees exerted selective pressures on the beetles (Chap. 10).
Over the long history of their coexistence, the beetles have
eliminated highly susceptible trees from the forest. By
their preference for physiologically weakened trees, bark
beetles help to keep the forest free of diseases. But other
forces continue to generate trees which vary in their ability
to resist attack. Older trees are less resistant, as are
trees that have been weakened by stresses such as disease,

fire, or smog. Other trees vary in their ability to produce large quantities or different types of toxins. As long as there is a genetic component to the differences in susceptibility among the trees, bark beetles will continue to influence the evolution of their hosts.

Primary Production

Primary production is viewed as a product of the amount of photosynthetic biomass present in the system and the average rate of net photosynthesis. Scolytids influence primary production directly by reducing biomass and net photosynthesis by killing whole trees (e.g. Dendroctonus), or by killing the tops of trees (e.g. by Ips and Pityophthorus). However, often these trees are already weakened or senescent; their photosynthesis is not in excess of their respiration, so they are not growing. Tremendous quantities of organic molecules (and energy) are stored in these trees and, by killing them, scolytids accelerate the recycling of these nutrients. Elimination of these trees allows growth to occur in the more vigorous trees in the stand. Therefore, cropping by beetles may actually increase primary production in the forest. In comparison with vertebrate and other invertebrate herbivores, insects consume proportionately more energy and, therefore, they make more energy available to other consumers in the ecosystem (Price 1975). Mattson and Addy conclude that

Insect grazers function much like cybernetic regulators of primary production in natural ecosystems. That is, they tend to ensure consistent and optimal output of plant production over the long term for a particular site. Their actions or activities seem to vary inversely with the vigor and productivity of the system. This inverse relation is probably a consequence of the long history of coevolution between plant systems and their usual consumers.

(Mattson and Addy 1975)

Forest Succession

Plant species may occupy a particular area if the physical and biotic factors meet the requirements of the plant. If the environment stayed constant the species could theoretically remain there forever. Environmental conditions do change, however, and many of the changes are induced by the species naturally inhabiting the site. As conditions change so do the plants occupying any particular site. For example, the first species of tree established at a site may be replaced by more shade-tolerant species. The process of change in plant communities over time is called succession. Scolytids, particularly the tree-killing Dendroctonus species, play an integral role in natural forest succession. For example, extensive outbreaks of the spruce beetle D. rufipennis Kby. in Alaska (Baker and Kemperman 1974) and Colorado (Schmid and Hinds 1974, Schmid and Frye 1977) killed most of the spruce, which dominated the stands, and returned the forests to earlier stages of succession dominated by shade intolerant species.

The effects of scolytids on forest succession is best explained by examining a single example in some detail.

Lodgepole pine ecosystems and the role of mountain pine beetle in them have been studied extensively (Amman, 1977, Pfister and Daubenmire 1975, Roe and Amman 1970, Wellner 1978). Lodgepole pine (LP) plays four basic successional roles in various plant series (see Chapter 9 for management implications). These are:

Minor seral. LP is a minor component of young even-aged mixed stands. Due to its rapid growth habit it assumes dominance early in succession but it does not regenerate well. It is gradually eliminated from the forest as mortality increases from age 50-200 years. The mountain pine beetle hastens this process by attacking the larger (60-80 year old) LP.

Dominant seral. LP is the dominant cover type of even-aged stands with a vigorous understory of shade-tolerant species that will normally replace LP in 100-200 years. The beetle could hasten this process as above, but, because LP is the dominant cover type, extensive outbreaks lead to a build-up of dead trees that constitutes an extreme fire hazard. The beetle kill, followed by an intense wild fire, returns the forest to the earliest stage of succession where LP once again dominates.

Persistent. Even-aged stands of LP are the dominant cover type and there is little evidence of replacement by other species. Again the beetle and fire have acted in concert over long periods to favor LP. Elimination of LP by the mountain pine beetle returns the forest to an earlier stage of succession, e.g. brushfields, which will be taken over by LP once again.

Climax. LP is self-regenerating. Such climax forests of LP are infrequent, occurring only in habitats unsuitable for other conifers. In these, destruction of the LP results in conversion of the habitat to other plant species.

Lodgepole pine and the mountain pine beetle have probably coexisted since the trees' earliest existence (Amman 1977, Roe and Amman 1970) so the potential for their coevolution has been great. Peterman (1978) suggested that mountain pine beetles may decrease the probability that stands of LP with predominately serotinous cones would stagnate. The age at which LP becomes susceptible to the mountain pine beetle (70-80 years) coincides with maximum cone production. Continued cone production long after this time results ultimately in so much germination that competition among seedlings reduces survival. Thus, beetle kill at age 70-80, followed by fire, releases the seeds from the serotinous cones at a time that ensures maximum survival of the lodgepole pines and therefore of the food supply of the beetle.

This brief overview of the major roles of scolytids in forest ecosystems does not do justice to the multitude of intricate relationships of the many scolytids present in forests. Nor can one point to a single reference, or even several, where such information is available. Long term ecological research necessary to elucidate involvement of scolytids in ecological processes is generally lacking. There is little question, however, that tree-killing scolytids and their coniferous hosts have co-evolved.

3. Taxonomy and Geographic Variation

D. E. BRIGHT and M. W. STOCK

The amount of diversity in even a relatively small family of insects such as the Scolytidae is staggering. Worldwide, over 6000 species are now described and many undoubtedly await discovery. Clearly, it would be impossible to deal with this enormous diversity were it not ordered and classified. Systematic zoologists endeavor to order diversity and to develop methods and principles to make this task possible. Simpson (1961) defines systematics as "the scientific study of the kinds and diversity of organisms and of any and all relationships among them"; or, as Mayr (1969) states more simply, "Systematics is the science of the diversity of organisms." Because "relationships" include all biological interactions among organisms, a broad area of common interest has developed among systematics, evolutionary biologists, ecologists, and behavioral biologists. This chapter considers the development of bark beetle systematics and illustrates the evolution of a holistic approach to problem-solving in this area. Problems remaining in bark beetle systematics and directions for future research are outlined.

ORIGIN OF SCOLYTIDAE

The great diversification of insect orders began about 300 million years ago in the Carboniferous Period when swamp forests containing seed ferns, sphenophylls, calamites, lycopods, cordaites, mosses, and liverworts dominated the landscape. Vegetative parts, pollen grains, and spores of these early plants provided food for the expanding insect fauna and decaying litter provided them with shelter. No one disputes a common Carboniferous origin for all insects that have complete metamorphosis (the Endopterygota). However, origins of the orders within the Endopterygota, and the relationships of beetles to other orders within this group, are not as clear. Early theorists (Handlirsch 1906-1908, Tillyard 1924, Zeuner 1933) believed Coleoptera arose independently of other endopterygote orders from a cockroach-like stock. Others (e.g., Ponomarenko 1969) believe that the structural similarities between exopterygote and endopterygote beetles are a result of convergence of adaptations for living in enclosed spaces -- in plant debris, dead wood, and litter -- rather than from close common ancestry. Most contemporary workers favor a derivation of Coleoptera from Neuroptera-like ancestors (Crowson 1960, Mickoleit 1973).

During the transition from the Carboniferous to the Permian Period (over 200 million years ago), conifers superseded the ancestral cordaites and became more conspicuous in forests of the northern hemisphere. During the Permian Period, Europe was uplifted and extensive mountain ranges were formed in North America. Swamps were drained and extensive glaciation occurred. The rate of diversification of insect orders, which reached its peak at this time, was correlated with an increased diversity of habitats and vegetative food sources. A number of insect orders, including the Coleoptera, originated in the Permian.

Because of the uniqueness of their elytra (wing covers) and their durability in the fossil record, great numbers of undisputed beetle fossils have been described from this period (Ponomarenko 1969).

The pines, which originated during the Permian, split into two major subgroups, the Haploxylon and Diploxylon pines, during the Jurrassic Period (100 million years ago) (Mirov 1953), and although angiosperms rose to dominance in the Cretaceous Period, conifers continued to remain important, thus providing an even greater diversity of habitats for adaptive radiation of insect groups.

The Scolytidae are believed to have arisen from primitive curculionoid stock late in the Triassic or early in the Jurassic Period, although no undisputed fossils as yet support this idea. Scolytid generic radiation probably occurred in the Cretaceous Period. Brongniart (1877) and Walker (1938) described bark beetle galleries in fossilized gymnosperm wood from this period. The earliest actual fossils ascribed to the Scolytidae date from the Eocene Epoch (early Tertiary Period) and were found in Colorado and Wyoming. Somewhat more recent fossil records are particularly revealing. During several million years of the Oligocene Epoch in Europe, numerous insects became trapped in resin extruded by the extinct pine tree, Pinites succinifera. The hardened resin, now called Baltic amber, is found in alluvial soil or along the shores of the Baltic Sea and provides some of the best examples of scolytid fossils yet identified. Bark beetle fauna of Baltic amber have close affinities with modern genera and genera that, although structurally distinct from modern genera, can easily be related to a modern classification. All bark beetles found in Baltic amber belong to the subfamily Hylesininae, generally considered to be the more primitive of the two presently recognized subfamilies of Scolytidae.

Insect associates of the present-day Araucariaceae (the family that includes the monkey puzzle tree) provide additional clues to early evolution of Scolytidae. The geologically old plant genus Araucaria occurs today in South America, New Zealand, Australia, and New Guinea. This conifer genus serves as host for several primitive scolytid genera in the subfamily Hylesininae and several curculionid genera in the tribe Araucariini. It is from this scolytid-Araucaria relationship that the modern-day genus Dendroctonus is thought to have originated.

Early evolution of certain Scolytidae is thus closely correlated with the evolution of conifers, with the adaptive radiation of the beetles occurring as floral elements gained in complexity. Not all Scolytidae originated in conifers, however. Evidence suggest that the genus Ips originated in tropical angiosperms, secondarily invaded pines in North America, then later invaded Europe and Asia from North America (Wood 1982). Amber from the Dominican Republic, probably not of coniferous origin, is as old as Baltic amber and contains at least four genera of Scolytidae, including two of the most advanced genera in the subfamily.

SCOLYTIDAE TODAY

The superfamily Curculionoidea, containing weevils, bark beetles, and several related families, is generally considered the most highly specialized beetle group. This superfamily is differentiated from other Coleoptera primarily by the presence of an anterior prolongation of the head (the snout or rostum) and by several other features. The Curculionoidea is a large, economically important group of beetles with more than 3100 species in North America alone, practically all of which are phytophagous (Borrer et al. 1976). Crowson (1968) regards the Scolytidae as a subfamily within the Curculionidae (weevils), but most authorities (e.g., Morimoto 1976; Wood 1963, 1982) recognize the Scolytidae as a separate family. The Scolytidae is most closely related to the family Platypodidae, the pin-hole borers. These two families are distinct from other families of Curculionoidea in having pregular sutures defined internally by an inflection of the cuticle and by the scarcely developed snout (Borrer et al. 1976, Wood 1978).

The Scolytidae include the bark or engraver beetles and the ambrosia or timber beetles. Bark beetle adults are small, compact, cylindrical insects ranging in size from 1-18 mm long. Their antennae are elbowed and the outer segments are enlarged and clublike. Most species are brown or black. Some have variegated markings comprised of grayish or brownish scales. The head, which is partially to completely hidden from above by the thorax, has strong jaws (mandibles) for boring. Morphologically, the bark and ambrosia beetles are very similar. The major difference between the two groups is biological. Bark beetles live beneath the bark of trees, mining on the surface of the sapwood but not entering it. These scolytids are often called engraver beetles because of the elaborate patterns they excavate beneath the bark. In contrast, the ambrosia or timber beetles bore into the sapwood of trees and feed on fungi ("ambrosia") that they cultivate in their galleries. They do not eat the wood but damage it with their tunnelling and with the stain produced by the fungi.

The relationship between Scolytidae and fungi is an ancient one. Bark beetles introduce pathogenic fungi into their hosts and feed on the host plant tissues. Ambrosia beetles live within their host trees but they feed on the fungi, not on the plant tissues. The ambrosial habit has arisen independently at least eight times within the Scolytidae.

Scolytidae are found in all forested regions of the world. Of the approximately 185 scolytid genera (Wood 1978), a vast majority are found in the tropics where there is a great diversity of potential hosts and habitats (Schedl 1978). A breakdown of scolytid fauna by tribe reveals that the wood-boring ambrosia beetle habit is dominant in the tropics, whereas the bark-boring habit prevails in the northern hemisphere. Only about 10 percent of all scolytid species in Asia, Europe, and North America are wood borers.

CHARACTERS USEFUL IN BARK BEETLE IDENTIFICATION

Basic bark beetle structure is illustrated in Figure 3.1. More complete treatments are given by Hopkins (1909), Schedl (1931), Kaston (1936), and Wood (1982).

On the scolytid head, the frons (Figure 3.2), between the eyes and above the mandibles, displays a vast range of structural modifications and is one of the most important areas for taxonomic discrimination. Its characters are useful for distinguishing major and minor taxonomic divisions, species, and sexes.

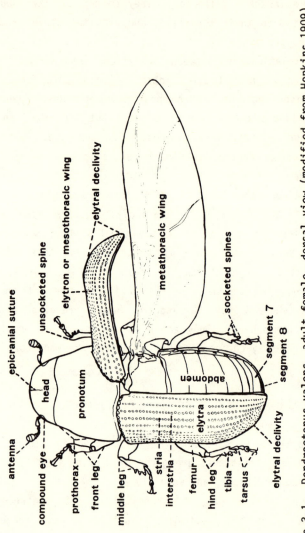

Figure 3.1.   Dendroctonus valens, adult female, dorsal view (modified from Hopkins 1909).

Another head region, the epistoma (Figure 3.2), represented by a thickened, heavily sclerotized region below the frons and above the mandibles, serves as a rigid support for articulation of the mandibles. It varies in shape and in structure and is often used to distinguish Dendroctonus species, although it is useful for taxonomic purposes in other scolytid genera as well.

Antennal structures vary greatly (Figure 3.3). The scape is generally club-shaped; the funicle, which varies from 1-7 segments, is frequently used for generic and tribal characterization. The club is the most variable feature of

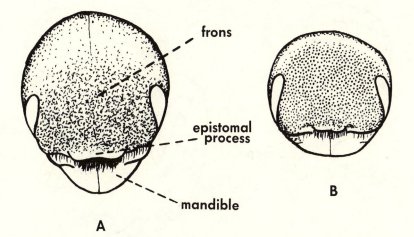

Figure 3.2. Anterior view of heads of male Dendroctonus valens (A) and D. ponderosae (B). In D. valens, the epistoma is broad (about half the width of the base of the mandibles) and the frons is granulate. In D. ponderosae, the epistoma is small (less than a third the width of the base of the mandibles) and the frons is punctate.

the antennae and is one of the most important features used in scolytid classification.

The pronotum, the dorsal portion of the prothorax (first thoracic segment), varies considerably in shape and sculpture and often has taxonomic significance. The pronotum may be smooth or punctured and may bear various types of setae and scales. A lateral and/or basal line may be present and may serve as a convenient character to define tribes and, in some cases, genera. Three factors of broad evolutionary significance can be noted on the pronotum (Wood 1978). First, the opening in which the head fits, in primitive genera, is rather large and its axis is nearly vertical

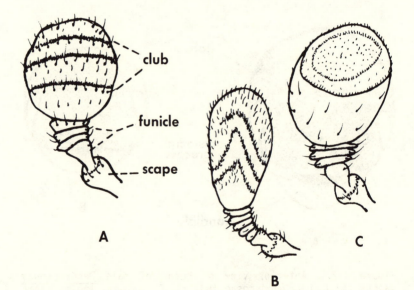

Figure 3.3. Examples of diversity in antennal structure. In Dendroctonus valens (A), the club is not flattened as it is in Scolytus ventralis (B). In Xyleborus saxeseni (C), the club is truncate. (From Bright and Stark 1973).

(Figure 3.4A); in specialized genera, it is smaller and the axis is more oblique (Figure 3.4B). The restructuring of the pronotum to accomodate this shift has shortened the sternal or ventral area and caused the dorsal area to become more rounded. Second, as a consequence of the boring habit, the sternal area had to be modified. In most primitive genera, the forecoxae (basal segments of the front legs) are widely separated, whereas they are contiguous in most specialized groups. Various other modifications accomodate this change. Third, the lateral raised line serves as a strengthening device in those groups having concave pronotal sides and evidently does not serve the same purpose in those groups having convex sides.

Legs are used extensively in scolytid classification and appear to offer a great deal of modification. The number and position of spines on the tibia (distal leg segment) are useful for some tribal distinctions. The spines may be socketed or unsocketed (Figure 3.1). Socketed spines are presumably of setal origin (Wood 1978). A major character separating the Scolytini from most other tribes is the presence of only one large, curved spine on the outer apical angle of the tibia. The shape of the foretibia itself is important in separating the Xyleborini and the Micracini from other tribes.

The elytra (Figure 3.1) are extremely variable in structure and sculpture. The striae may be punctured in even rows or the elytra may be randomly punctured with no indication of striae. The interstriae, the region between two striae, may be punctured or smooth or ridged or variously sculptured and pubescent. The sloping posterior area of the elytra (the elytral declivity) is one of the most important

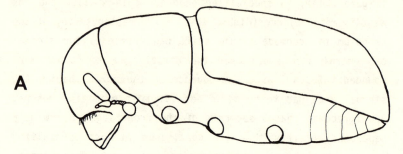

A

Dendroctonus valens

B

Ips pini

1 mm

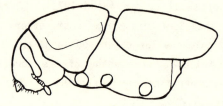

C

Scolytus ventralis

Figure 3.4.   Side views of three common scolytids showing variation in size and general morphology.

areas used in classification. The elytral declivity is usually steep (as in Figures 3.4A, B), convex to deeply concave, and may be smooth or punctured as in the remainder of the elytra or may bear various combinations of spines, teeth, granules, setae, or other special modifications. The elytral declivity is greatly shortened in the genus Scolytus (Figure 3.4C), but members of this genus have an abdominal or ventral declivity which may bear spines in one or both sexes.

Some interesting evolutionary developments are associated with the structure of the elytral declivity. In most of the primitive groups, the parental gallery is left filled with frass (fecal pellets and boring dust), as in Dendroctonus. Only the end of the gallery where the female is working is kept free of frass. In these groups, the elytral declivity is generally simple, convex, and usually without any conspicuous teeth or spines (Figure 3.4A). Some of the interstriae may be slightly elevated, or small to large granules may be present. In the more specialized groups, as in Ips, the adult galleries are kept clean and free of frass. In these groups, the elytral declivity is more concave with the lateral margins elevated and often bearing conspicuous teeth or spines (Figure 3.4B). The scooplike elytral declivity of Ips is particularly suited for pushing frass out of the gallery.

Because very little modification occurs on the abdomen, this region is not generally used in classification, except for species in the genus Scolytus. However, the abdomen is often useful in sex determination. As in most curculionoids, the male scolytid has 8 visible abdominal terga (dorsal segments) (Figure 3.5); the female has 7, with segment 8 reduced and concealed beneath segment 7 (Figure 3.1).

Exceptions occur in several tribes (e.g., Ipini, Xyleborini), but this basic structure holds true in the vast majority of North American species.

Larvae of Scolytidae are apodous (legless) and resemble those of other curculionoids. Little has been done with larvae, and this area offers considerable opportunity for further research. The most recent comprehensive works on scolytid larvae are by Thomas (1957, 1960, 1965) and Lekander (1968).

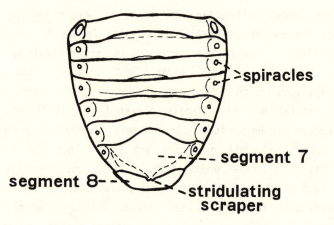

Figure 3.5. Dendroctonus valens, adult male, dorsal view of abdomen. (Modified from Hopkins, 1909).

In his taxonomic revision of the genus Dendroctonus, Wood (1963) found that characteristics of the egg gallery, position and arrangement of egg niches and grooves, and the character and position of the larval mines provided features for field recognition of species that were equal, if not superior, to anatomical characteristics. For example, several species ordinarily confine their attacks to the basal portion of the tree, seldom attacking more than a meter above ground level (e.g., Dendroctonus valens). In other species (e.g., D. ponderosae), attacks commonly begin on the lower third of the tree and progress upward. In still others (e.g., D. frontalis), the attack begins in the upper midtrunk area and progresses upward and downward from that point. Some groups (e.g., Ips) keep the adult egg galleries clean of frass and others pack their galleries with it (e.g., Dendroctonus). Parental galleries of Dendroctonus spp. are usually more or less straight and follow the grain of the wood (Figure 3.6A); however, a few species, such as the southern pine beetle and the western pine beetle, produce strongly sinuate galleries (Figures 3.6B, C), the whole complex forming an intertwining network of winding, branching galleries. Larval mines are as distinct as those of their parents.

BARK BEETLE SYSTEMATICS

Pre-1900

The dominant theory of classification for many centuries and up until the late 1800s was based on Aristotelian logic. This method attempted to assign the variability of nature to a fixed number of basic types at various levels. It postulated that all members of a taxon reflect the same essential nature; in other words, they conform to the same

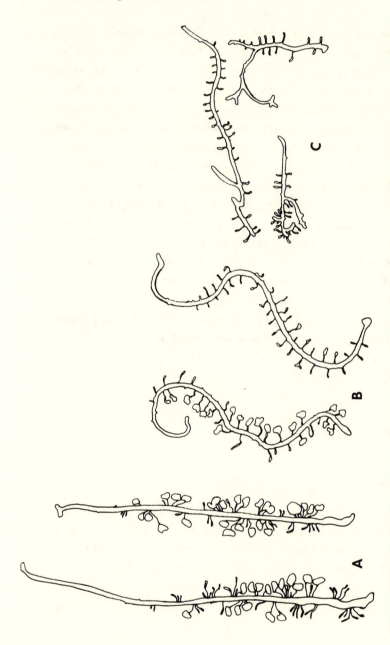

Figure 3.6. Galleries of the mountain pine beetle, Dendroctonus ponderosae (A), the southern pine beetle, D. frontalis (B), and the western pine beetle, D. brevicomis (C). (From Wood 1963).

hypothetical type.   This method is also referred to as typology.   Variation was considered by the typologists to be trivial and irrelevant; the constancy of taxa and the sharpness of gaps between them tended to be exaggerated. Significant variations, regardless of their cause, were considered to represent different taxa.   Thus males and females from one population could be assigned to different species.

Rapid advances in descriptive insect taxonomy occurred in the early 19th century when travel became more common.   Large outbreaks of forest insects were witnessed in different parts of the world and it became necessary to describe and name the insects.   For these early taxonomists, working in isolation with little material, typology was a convenient way to organize observed variation.

The development of concepts that influenced early bark beetle systematics are discussed succinctly by Schwerdtfeger (1973).   The first North American bark beetles were described by Johann Fabricius and G. W. F. Panzer in the late 1700s and by Thomas Say, W. F. Erichson, and C. G. Mannerheim in the early 1800s, followed by W. J. Eichoff later in the 19th century.   J. L. LeConte, a medical doctor and the leading coleopterist of this time, also published numerous works covering the entire field of North American Coleoptera.   In 1876 he published his "Rhynchophora of North America," in which he divided Scolytidae into two subfamilies, Platypodidae and Scolytidae.   At this time, the North American scolytid fauna contained 123 described species. During the late 1800s, we also begin to see the development and refinement of a higher classification, following the initial efforts of P. A. Latreille, Erichson, and C. G.

Thompson. Subsequent authors of the 19th century added a few more species to the fauna but did not significantly alter the LeConte classification.   At the close of the century, Blandford (1895-1905) published the Scolytidae portion of the Biologia Centrali-American series.   This was the first compilation of the neotropical fauna and it had an important influence on future studies.

Because these and other early scolytid taxonomists used a purely morphological (typological) species concept, variation in appearance was assumed to be indicative of species status. Moreover, workers were collecting and describing new species without the benefit of examining material upon which contemporaries based their new species. LeConte and Eichhoff described many species almost simultaneously. Consequently, synonyms were numerous, and geographic races and sibling species with no apparent morphological differences were not recognized. Clinal variation was overlooked, misunderstood, or misinterpreted.

Development of a Modern Classification

In The Origin of Species, Darwin (1859) established the theoretical basis for a natural classification. Darwin's theory was based on several main ideas. These included overproduction of offspring, variation among organisms, even those of the same kind, a struggle for survival on limited resources, and natural selection of variations most beneficial to survival. This process leads to gradual change in populations, or evolution (Chap. 1). For divergence to occur, groups of organisms must be isolated from other groups long enough to become reproductively different. Isolation may be geographical, morphological, ecological, physiological, or ethological. This irreversible splitting of evolutionary lines is called speciation because it results in a new species, a group of individuals that can interbreed to produce fertile offspring but which are reproductively isolated from other such groups. The biological species concept, a modern outgrowth of Darwinian theory, stresses that species form (1) a reproductive community in which individuals recognize and seek each other as potential mates, (2) an ecological unit in which the species interacts as a unit with other species sharing the same environment, and (3) a genetic unit of populations sharing a common gene pool and reproductively isolated from other such groups (Mayr 1969). Species can, therefore, include host or geographic races -- populations showing some degree of ecological, behavioral, or morphological differentiation but not yet exhibiting the total reproductive isolation of species. Conversely, organisms which are genetically divergent and reproductively isolated, but which have not acquired conspicuous morphological differences, are also recognized as species.

Unfortunately, the principles detailed by Darwin were only very slowly incorporated into the practices of working taxonomists, even though the acceptance of evolution as a theoretical basis for classification caused little disruption of the systems developed by typologists. The grouping of species based upon their morphological similarities coincided fairly well with phylogenetic lines. However, the biological species concept led to the recognition that traditional morphological studies would have to be augmented with other approaches in order to delineate species. If current concepts and methods in genetic and molecular biology had been available to the scientists of Darwin's day, his ideas would undoubtedly have been assimilated much more rapidly.

To varying degrees, and with varying degrees of success, taxonomists of the late 19th and early 20th centuries began to incorporate additional criteria into their work. A. D. Hopkins placed considerable importance upon host preferences and other non-morphological characters. This resulted in almost as much taxonomic confusion as the early typologists' ideas had. Hopkins began his work in agricultural entomology but switched to forest entomology during the Dendroctonus frontalis epidemic in 1891 in West Virginia. He became head of the Division of Forest Insect Investigations established in 1902 within the U.S. Department of Agriculture and embarked on a vigorous program of bark beetle studies. From 1893-1921, Hopkins described 6 new genera and 133 new species from North America, plus a large number of genera and species from exotic areas of the world. Many of his species differed only in the locality or the host tree from which they were collected. These practices resulted in a proliferation of species synonyms and did not represent a significant advance from the morphological or typological approach. Of the 106 species named by Hopkins in the genera Hypothenemus and Stephanoderes, only 21 are now considered valid (Wood 1972). Likewise, in Conophthorus, Hopkins named 14 species (all after their host plants) when, it is now believed, only 5 or 6 actually exist. In spite of these problems, Hopkins contributed a great deal to scolytid systematics. He focused the attention of research workers on this economically important group of forest insects; his research into the systematics and biology of Dendroctonus species is a classic work and a number of his concepts are being revived today.

J. M. Swaine was working in Canada at about the same time that Hopkins was working in the United States. Swaine's first paper, a brief review of scolytid biology, was published in 1907. This work was followed in 1909 by a catalog of the described species of North American Scolytidae which included 191 species. His major work, published in 1918 and titled Canadian Bark Beetles, was a classic of practical and scientific forest entomology, enhanced by A. E. Kellett's superb drawings of various species of bark beetles. Hopkins and Swaine were the first of the modern-day taxonomists to study Scolytidae. Both men stand as giants in scolytid taxonomy, and their work formed the basis for all subsequent work on North American Scolytidae.

Hopkins' and Swaine's work was continued and augmented by M. W. Blackman at the New York State College of Forestry at Syracuse and later at the U.S. National Museum in Washington, D.C. Blackman continued to describe species and revise classifications. He was a prolific and meticulous worker; many of his papers are still useful today. A number of large and unknown genera were revised by Blackman. His keys and descriptions were well done, his type material usually clearly labelled, and his species concepts clearly outlined. As more information and specimens became available, some of Blackman's names fell into synonymy, but this was usually the result of expanded knowledge, not taxonomic errors.

In Europe, the two most prolific authors were Hans Eggers, who described more than 1100 species in about 120 papers from 1899-1949 on forest entomology and scolytid systematics, and Karl E. Schedl, who published nearly 340 papers on scolytid taxonomy and authored almost 2000 species names. These two authors are best known for the sheer number of species they described; little effort was expended in developing a workable classification. However, in the late 1800s, another European, Eichhoff, was an important scolytid taxonomist whose careful work is considered basically sound today.

CURRENT STATUS OF SCOLYTID TAXONOMY

In 1939, W. J. Chamberlin at Oregon State University gathered scattered descriptions, keys, biological notes, etc., and included them in his book The Bark and Timber Beetles of North America, North of Mexico. This book provided a compendium of current Scolytidae knowledge and was a major influence in stimulating additional taxonomic study of this group. During the subsequent 40 years, the systematics of North American bark beetles has greatly stabilized; almost all of the important genera of North American Scolytidae have been revised to reflect more clearly the biological species concept. The first revised group was Cryphalini (Wood 1954a), followed by the genera Carphoborus (Wood 1954b), followed by Trypodendron (Wood 1957), Pityoborus (Wood 1958), Dendroctonus (Wood 1963), Dryocoetes (Bright 1963), Ips (Hopping 1963-65), Xyleborus (Bright 1968), Pseudohylesinus (Bright 1969), and Pityophthorus (Bright 1981). In addition to these generic revisions, over 90 other papers clarifying taxonomic problems have been published by S. L. Wood and D. E. Bright. Wood examined the type specimens of almost all North American scolytid species, resulting in the placement of each valid species on a firm foundation with its synonyms clearly defined. These two authors have collected extensively in North America, in Mexico, and in Central and South America so that the number of specimens available for study has greatly increased, much biological information has been obtained, and distributional data have been greatly enhanced. Culminating all of this activity is the monograph on North and Central American Scolytidae by S. L. Wood (1982). Included are keys, descriptions, biological data, illustrations, etc., for 1,432 species of North and Central American bark beetles.

S. L. Wood (pers. comm.) estimates that 90-95% of the scolytid fauna of North America (north of Mexico) has been named, and that described species are, to a large extent, firmly established and recognizable.   Most of the real problems remaining the taxonomy of this group involve sibling (or cryptic) species or forms represented by too few specimens to evaluate their status. Taxonomic questions, for example, still occur in the Hylurgops pinifex - rufipennis group, in Pityogenes plagiatus - knechteli, in some Pseudohylesinus spp. and some Pityophthorus spp., in Dendroctonus frontalis, Polygraphus rufipennis, Scolytus tsugae - monticolae - reflexus, Conophthorus ponderosae, and in several other species. Further advances will therefore come as undescribed species are found in isolated or obscure hosts or localities and as species complexes, sibling species, and conspecific variants are more clearly defined, through augmentation of traditional morphological studies with behavioral, genetic, cytological, and biochemical studies.

Sibling Species

In sibling species, genetic divergence and reproductive isolation are not accompanied by acquisition of conspicuous

morphological differences. Lanier (1970) showed that the
bark beetle species known as Ips confusus LeConte, common in
pines throughout the western U.S., actually contains three
sibling species: 1) the true I. confusus in pinyon pines in
the Southwest, 2) I. paraconfusus in ponderosa pine and other
pines along the Pacific-facing slopes of mountains in
California and Oregon, and 3) I. hoppingi in Pinus cembroides
in Mexico and adjacent parts of Arizona, New Mexico, and
Texas.   These species were suspected based on their
ecological differences, but proof of their specific integrity
was obtained only by cytogenetic as well as morphological
studies.   Similarly, Ips plastographus seemed to contain
three distinct ecological races. One race infested Bishop,
Monterey, and shore pines along the California coast, one
attached lodgepole pine in the subalpine zone of the western
mountain ranges, and the third occurred in ponderosa pine
from British Columbia south through the Rocky Mountains into
Mexico. No external morphological differences could be found
between the coastal and subalpine populations.   However,
Lindquist (1969) believes that the parasitic mites associated
with each race of the insect are distinct subspecies of
Ipomenus plastographus, suggesting that the three ecological
races of Ips plastographus might be distinct species or
subspecies. In fact, further studies by Lanier (1970) showed
that the Rocky Mountain race was the species I. integer
(Eichhoff) and the subalpine and coastal races were
recognized as subspecies of I. plastographus.

Recent studies of the genus Pityophthorus indicated that
P. deletus may be a complex of 6 sibling species (Bright
1981). More material from the southwestern states is needed
before the status of the variants can be ascertained. The

species groups nitidus, nitidulus, cariniceps, confertus, lautus, and several others also contain species in which sibling species or well-marked geographic variants might also occur. Each of these offers ample opportunities for research. Sibling species will probably be found in most genera of Scolytidae when detailed studies of host races and of geographical variants are conducted.

Biochemical techniques are also proving useful in discriminating between apparent sibling species (Berlocher 1979). Allelic forms of enzymes, detected by gel electrophoresis, can be used as diagnostic markers for species identification (e.g., Ayala and Powell 1972, Ayala 1975, Huettel and Bush 1973, Higby and Stock 1982). The fact that two groups are fixed for different enzyme types produced at a gene locus tells us that the two groups are not interbreeding at a given location and, in conjunction with other types of biological and ecological information, can confirm the status of the two groups as valid species. The practicality of constructing biochemical keys to sibling species using diagnostic loci has been demonstrated by several authors (e.g., Mahon et al. 1976, Miles 1979, Berlocher 1980). Of particular value are biochemical keys for identification of immature forms, such as insect larvae, where different species may be difficult to distinguish by morphology.

Species Complexes

Other examples of integrated approaches in scolytid taxonomy are seen in recent studies of species complexes in Dendroctonus species. The mountain pine beetle, Dendroctonus ponderosae, exhibits a considerable range of morphological, behavioral, and physiological variation throughout its range. Morphological variation is seen in size differences in beetles from different localities and from different host trees, in the surface sculpturing of the elytral declivity, the pattern and curvature of interstriae and striae, and the size and spacing of tubercles and setae. Also, the size and spacing of punctures on the frons vary, as do the size and density of granules between or on the margins of the punctures (Sturgeon 1980). In addition, considerable variation occurs in host specificity. While preferred hosts of this insect include several tree species, local endemic beetle populations in mixed pine stands may concentrate on one pine species (Craighead 1921, Hopkins 1916, Wood 1963). For example, along the north slope of the Uinta Mountains in Utah, some outbreaks occur in ponderosa pine but not in lodgepole pine. However, mountain pine beetles commonly switch to other host species during outbreaks, possibly as a result of crowding and diminished availability of preferred host trees. Under such conditions, any local tendency to select one of a number of host trees owing to primary (tree-produced) attractants is overridden by more powerful, secondary (beetle-produced) attractants. For example, a mountain pine beetle emerging from lodgepole pine will be more attracted to a ponderosa pine under attack than to an unattacked lodgepole pine. The apparently conservative speciation record of many bark beetle groups may be attributed to similar flexibility in host preferences and resulting gene flow from one population to another.

Within a single host species, mountain pine beetles may show differences in tree diameter preference (Amman 1977, Cole and Amman 1969, McCambridge 1967, Sartwell and Stevens

1975, Stark et al. 1968) and, in stands of different host species, mountain pine beetle populations appear to respond differently to stereoisomers of host volatiles and beetle-produced pheromones (McKnight 1979).

This enormous diversity within one species is reflected in its taxonomic history. Prior to 1963, the mountain pine beetle (= D. monticolae) was believed to attack all pines, except Jeffrey pine, throughout its distribution, which extended from the northern range of coastal ponderosa pine in Canada southward along the Pacific coast to Baja California. In the interior mountainous regions, the Black Hills beetle (= D. ponderosae) was also considered a polyphagous species, attacking primarily inland varieties of ponderosa and lodgepole pine in the Black Hills of South Dakota and throughout the eastern and southern Rocky Mountains. D. jeffreyi was limited to southern Oregon, California, western Nevada, and northern Baja California by the distribution of its only host, Jeffrey pine. D. ponderosae was distinguished from monticolae by the larger size and larger, more strongly impressed pronotal punctures. D. jeffreyi was larger than ponderosae but the pronotal punctures were smaller and more lightly impressed than those of monticolae (Hopkins 1909).

In 1963, S. L. Wood synonymized D. monticolae and D. jeffreyi with D. ponderosae, forming the species complex now referred to as D. ponderosae, the mountain pine beetle. Wood's (1963) synonymy of monticolae with ponderosae had been indicated earlier by morphological (Blackman 1931) and hybridization (Hay 1956) investigations. Thomas (1965) concurred with both synonymies, based on his studies of larval and pupal morphology. However, observations of attack patterns (Eaton 1956) and the generally larger size (Hopkins

1909, Wood 1963) of specimens from Jeffrey pine in California raised doubts about the jeffreyi-ponderosae synonymy. In addition, Smith (1963, 1965) found that jeffreyi and monticolae responded differently to resin vapors from their own and the other's host tree and he recommended that jeffreyi and monticolae be recognized as distinct physiological races.

A comprehensive review of these three groups was undertaken by Lanier and D. L. Wood (1968). The complete failure of jeffreyi to produce fertile offspring with ponderosae and monticolae, together with differences in survival of laboratory broods, egg incubation time, karyotypes, and width and scupturing of the pronotum, confirmed and re-established the species status of D. jeffreyi. A genetic comparison of mountain pine beetle and Jeffrey pine beetle from northern California support their separate species designations (Higby and Stock 1982). The two beetle groups are fixed for different alleles at two loci, strongly suggesting that interbreeding does not occur.

While Lanier and Wood's investigations generally confirmed the synonymy of monticolae and ponderosae, it is obvious that many problems remain in defining the infraspecific groups. Recent and ongoing genetic studies of the mountain pine beetle (Stock and Guenther 1979, Stock and Amman 1980, Sturgeon 1980) are helping to elucidate the population structure of this insect. Genetic differences among populations are generally associated with geography; however, genetic differentiation among mountain pine beetle subgroups in local areas is at least partly related to host tree species.

The southern pine beetle, Dendroctonus frontalis, like the mountain pine beetle, has a wide geographic distribution over which it mainfests much behavioral variation. Based on findings of several reproductive, genetic, behavioral, and physiological differences among groups (Vite et al. 1974, Namkoong et al. 1979, Anderson et al. 1979), scientists have recommended that the current taxonomy of this species complex be reviewed.

A similar pattern of differentiation has been found in the Douglas-fir beetle, Dendroctonus pseudotsugae. Stock et al. (1979) found highly significant genetic differences between populations attacking the coastal form of Douglas-fir, Pseudotsuga menziesii var. menziesii, and those attacking the inland form, var. glauca. A similarity coefficient of 0.63 (on a scale where 1.0 equals genetic identity) indicated that these geographically separated populations represent well differentiated races, at least. Further taxonomic work on this species, using additional taxonomic criteria, may reveal more clearly the status of these two groups.

Cytological and genetic approaches to the study of insect evolution and systematics have the advantage of utilizing characters that may not be evident by external morphology. Because of the relative ease of rearing and sexing bark beetles, controlled breeding studies coupled with electrophoretic analyses and examination of karyological characters promise to add much to the understanding of species variability within the Scolytidae.

4. Aggregation Pheromones

J. H. BORDEN

An intricate complex of chemical messengers, or semiochemicals, are utilized by scolytid beetles in communication with members of their own species. The same or different chemicals may also be involved in communication with organisms in other species (Figure 4.1).

Pheromones are semiochemicals which induce a behavioral or physiological response in members of the same species (Shorey 1977). In the Scolytidae, the major pheromones are aggregation pheromones which cause both sexes to aggregate on a host tree or log (Borden 1974, 1977). In addition, there are epideictic or spacing pheromones (Prokopy 1980), which are also called antiattractants (Wright et al. 1961) or antiaggregation pheromones (Borden 1977). These serve to regulate attack and population density. In many species, there are even more complex systems involving both chemical and sonic communication.

Numerous scolytid-produced semiochemicals (many of them pheromones) and several host plant volatiles also serve as interspecific messengers. In this case they are called allomones (Brown 1968) or kairomones (Brown et al. 1970), depending on whether their action results in adaptive benefit to the emitting or perceiving organism, respectively.

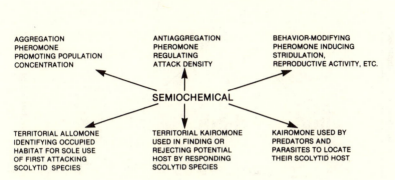

INTRA-SPECIFIC CHEMICAL MESSAGE

AGGREGATION
PHEROMONE
PROMOTING POPULATION
CONCENTRATION

ANTIAGGREGATION
PHEROMONE
REGULATING
ATTACK DENSITY

BEHAVIOR-MODIFYING
PHEROMONE INDUCING
STRIDULATION,
REPRODUCTIVE ACTIVITY, ETC.

SEMIOCHEMICAL

TERRITORIAL ALLOMONE
IDENTIFYING OCCUPIED
HABITAT FOR SOLE USE
OF FIRST ATTACKING
SCOLYTID SPECIES

TERRITORIAL KAIROMONE
USED IN FINDING OR
REJECTING POTENTIAL
HOST BY RESPONDING
SCOLYTID SPECIES

KAIROMONE USED BY
PREDATORS AND
PARASITES TO LOCATE
THEIR SCOLYTID HOST

INTER-SPECIFIC CHEMICAL MESSAGE

Figure 4.1.   Possible multiple functions of one or more semiochemicals produced by a scolytid beetle.

As in an earlier discourse (Borden 1974), this chapter will explore the biology of semiochemicals in a host selection context (Figure 4.2). It will attempt to describe how aggregation and antiaggregation pheromones fit into each step of this critical phase of scolytid biology, and how semiochemicals are utilized as allomones or kairomones in interspecific communication. It will not stress chemistry or the utilization of semiochemicals in pest management. Some license has been taken. Research on species which inhabit deciduous trees as well as conifers in both North America and Eurasia is cited. In addition, examples gained from research on individual species are applied to the whole family. These generalizations are, of course, open to the challenge of further research, as are many subjects identified in the chapter about which we know too little.

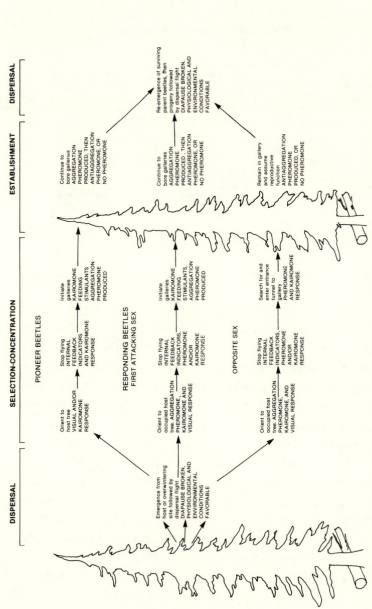

Figure 4.2. Generalized behavioral sequence in host selection and mass attack by scolytid beetles. Semiochemicals and other factors influencing behavior are identified.

THE HOST SELECTION SEQUENCE

Scolytids surmount monumental problems in host selection (Figure 4.2). They must emerge from their old host or overwintering site with adequate metabolic resources, often after a bracing winter, assess the ambient conditions, and launch themselves into a dispersal flight in the most favorable environmental conditions possible. They must then negotiate an often formidable air space on the way to their new hosts, toward which they orient, and which they must distinguish from non-hosts. The beetles need to aggregate in sufficient numbers to overcome host resistance (Anderson 1948, Chap. 7), or to utilize a host optimally (Atkins 1966a), and establish themselves successfully in the new host so that they can rear their broods. If the host is unsuitable, or if the parent beetles are still vigorous after raising their first brood, the whole sequence may be repeated.

It is generally accepted that the different behavioral phases of host selection (Figure 4.2) are genetically determined. However, the insect is far from passive. Rather, as it encounters a "series of take-it-or-leave-it situations" (Thorsteinson 1960), it must assess each situation externally and internally and make a series of "go or no go decisions". The beetle will proceed from one behavioral phase to another only if the check list is complete, i.e. if the external environment and internal, physiological condition of the insect permit a behavioral transition.

Wood (1972) grouped the host selection sequence for the western pine beetle, Dendroctonus brevicomis, into three phases: dispersal and selection, concentration (aggregation), and establishment. My concept of the sequence also includes three phases, but unites selection and concentration together in the second phase (Figure 4.2).

DISPERSAL

Emergence

The first task facing a beetle prior to dispersal is to emerge from its host or overwintering site. Usually, it will leave a host in which it has matured, but some may emerge from an over-wintering host or a non-host substrate (such as litter and duff on the forest floor) to which they had earlier dispersed.

Before emergence, the beetle must be in a state of physiological readiness for dispersal flight. Diapause must be broken and the flight muscles mature. Although it appears normal, the striped ambrosia beetle, Trypodendron lineatum will not mate while still in diapause (Fockler and Borden 1972). The flight muscles of the Douglas-fir beetle, Dendroctonus pseudotsugae, will not mature while the insect is in diapause underneath its host's bark (Ryan 1959). During its maturation feeding underneath the bark, the flight muscle volume of the California five-spined ips, Ips paraconfusus, increases 10-fold (Borden and Slater 1969). And adult D. pseudotsugae must build up their lipid reserve to 20% of total dry weight to be optimally flight positive (Atkins 1966b).

Emergence in a particular species is probably temperature induced (Cameron and Borden 1967) and follows a distinct diel periodicity (Gara and Vite 1962, Daterman et al. 1965, Cameron and Borden 1967). The basis for periodicity may lie in response to environmental conditions or in endogenous circadian rhythms, as demonstrated for Xyleborus ferrugineus (Saunders 1967) and the mountain pine beetle, Dendroctonus ponderosae (Billings and Gara 1975). In any case, emergence periodicities are probably reflected in flight patterns (Gara 1963, Vite et al. 1964, Daterman et al. 1965), which govern the timing of attack of new hosts or the response to pheromones.

Typically, the first emerging sex will not be the one which initiates the attack on the new host. Thus in the early stages of emergence, females predominate in I. paraconfusus (Cameron and Borden 1967) and males in D. ponderosae (Billings and Gara 1975). This phenomenon is probably an adaptation to ensure outbreeding. The first emergents will leave the immediate area without ever having a chance to perceive and respond to aggregation pheromones produced by their later-emerging siblings.

Flight Behavior

The emergent beetle is a miniature meterologist. It is capable of assessing a variable environment in which light, temperature, relative humidity, wind, and atmospheric pressure constantly fluctuate. Flight initiation by scolytids is known to be governed by temperature, light and humidity (Atkins 1959, Henson 1962, Perttunen and Bowman 1965, Borden 1967, McCambridge 1971). In an enclosed laboratory, Bennett and Borden (1971) found that D. pseudotsugae and T. lineatum either did not fly or flew erratically on cloudy or unsettled days. They hypothesized that air swallowing prior to and during flight produced a ventricular air bubble (Graham 1961) which served as an internal barometer. In support of this hypothesis, Lanier and Burns (1978) experimentally demonstrated that pheromone response by the smaller European elm bark beetle, Scolytus multistriatus, is significantly reduced following exposure to fluctuating barometric conditions.

Wind speed and rain are critical factors during flight. In wind over 8.3 K/h, catch of the southern pine beetle, Dendroctonus frontalis, was decreased by 51% over that in stiller air (Coster et al. 1978). Prolonged rain resulted in a marked reduction in flight. However, during light rain showers of short duration, catches of beetles in traps on attacked trees were 1.4 fold greater than in no rain (Coster et al. 1978), probably because beetles already in flight were induced to land at the onset of rain.

Assuming that all external and internal conditions are met, the beetle will take flight. Initially, the flight is not host directed. Only after a requisite flight exercise do most beetles become host and/or pheromone positive. This phenomenon was first noted for T. lineatum (Graham 1959, 1962). On the laboratory flight mills (Borden and Bennett 1969), T. lineatum and D. pseudotsugae required 30 and 90 min. of accumulated flight, respectively, before they would cease flight when exposed to the odor of attractive, female-produced frass (Bennett and Borden 1971). More recently, Choudhury and Kennedy (1980) demonstrated that initial flight by male S. multistriatus in a wind tunnel is positively phototactic rather than negatively geotactic, and that the propensity for upwind flight in response to pheromone odor increased as the time in flight increased.

The time in flight prior to the behavioral transition from a dispersal positive to a host positive condition is apparently related to lipid reserves on emergence and lipid metabolism in flight. Atkins (1966b) found that Douglas-fir beetles with less than 20% total lipid content on emergence were immediately able to search for and infest new hosts.

Those with higher lipid contents were flight positive and refractory to their hosts. Atkins (1969) hypothesized that behavioral changes during flight may be stimulated by the accumulation of certain metabolites. In support of this hypothesis, Thompson and Bennett (1971) disclosed that male D. pseudotsugae in flight selectively oxidize particular fatty acids, providing a mechanism for internal feedback detection of the amount of flight experience a beetle has had.

These insects are thus marvelously adapted to optimize their opportunities. Only vigorous beetles with an abundance of lipid reserves can afford the luxury of extended dispersal, and are by their nature commited to it for a reguisite period. This commitment to disperse ensures that the most vigorous individuals have a maximal chance of outbreeding and of utilizing all available hosts. On the other hand, lipid-poor beetles are able to disperse, but if they encounter suitable hosts or pheromone sources they can respond immediately, thus ensuring that nearby as well as distant hosts are exploited.

SELECTION-CONCENTRATION

Primary Attraction

Since Person's (1931) hypothesis that D. brevicomis is attracted to and attacks physiologically weakened trees, arguments have raged as to whether or not primary (i.e. host induced) attraction for scolytids actually exists. Several years of research have made the waters somewhat less muddy, and it now appears that interspecific differences preclude any sweeping generalizations.

For some species there is clear primary attraction. Ambrosia beetles attack logs in which anaerobic metabolism (Graham 1968) results in the production of ethanol, which can act as an attractant alone or in combination with monoterpenes such as alpha-pinene (Cade et al. 1970, Moeck 1970, 1971, Bauer and Vite 1975, Nijholt and Schonherr 1976). Bark beetles may also respond to primary attractants. D. pseudotsugae is attracted to host monoterpenes (Rudinsky 1966), and the fir engraver, Scolytus ventralis, to weakened trees (Ferrell 1971). However, Wood (1976) trapped D. brevicomis and D. ponderosae in equal numbers on diseased, physiologically weakened and dead ponderosa pines. And extensive, meticulous experiments in which trees were stressed artificially (e.g. by freezing the root collar) failed to induce primary attraction for any California bark or ambrosia beetle (Moeck et al. 1981).

While primary attraction may enhance attack efficiency, it may not be necessary to assure attack success. A single pioneer beetle may find a potential host tree, e.g. by visual response followed by detection of the correct feeding stimulants. It alone may produce sufficient aggregation pheromone to attract first one and then another beetle, until several beetles in concert induce a mass attack on the tree.

Secondary Attraction

Since the first successful characterization of a scolytid pheromone system for I. paraconfusus (Silverstein et al. 1966), many aggregation pheromones have been isolated and/or identified from scolytids, the majority from North American species (Table 4.1). Usually they form part of an attractive chemical complex, which also includes host-produced

Table 4.1. Pheromones and kairomones isolated and/or identified in the Scolytidae: their source and biological activity.

Species	Pheromone (P) or Kairomone (K)[a]	Source and Biological Activity
Blastophagus destruens	trans-verbenol (P) verbenone (P)	Identified in hindguts of both sexes (Carle et al. 1978). Both compounds tested seperately attractive to both sexes in laboratory (Carle 1978).
Blastophagus piniperda	trans-verbenol (P) verbenone (P)	Identified in hindguts of both sexes (Francke and Heeman 1976, Carle et al. 1978). Trans-verbenol attractive in laboratory alone (Kangas et al. 1970) or in combination with low concentration of verbenone (Carle 1978). Verbenone inhibitory at high concentration (Kangas et al. 1979, Carle 1978)

Species	Pheromones	Comments
Dendroctonus adjunctus	frontalin (P) exo-brevicomin (P) terpene mixture (K)	Frontalin identified from hindguts of 48 h fed females, and exo-brevicomin from emergent males. Both attractive with alpha-pinene, beta-pinene and myrcene (mixed), exo-brevicomin primarily to females and frontalin to males (Hughes et al. 1976).
Dendroctonus brevicomis	frontalin (P) exo-brevicomin (P) trans-verbenol (P) verbenone (P) myrcene (K)	Exo-brevicomin (Silverstein et al. 1968) and myrcene (Silverstein 1970) isolated and identified from frass, frontalin from hindguts (Kinzer et al. 1969). All 3 attractive in field (Bedard et al. 1969), exo-brevicomin mainly to males and frontalin to females (Vite and Pitman 1969). Enantiomeric composition primarily (-)-frontalin and (+)-exo-brevicomin (Stewart et al. 1977). 1S, 5R-(-)-frontalin and 1R, 5S, 7R-(+)-exo-brevicomin are the active isomers in field tests (Wood et al. 1976). Verbenone and trans-verbenol identified from hindguts of males and females, respectively (Renwick 1967).

Verbenone proposed as antiaggregation pheromone (Renwick and Vite 1970). Verbenone and trans-verbenol inhibit response (Wood 1972, Bedard et al. 1980a,b).

Frontalin identified in female hindguts (Kinzer et al. 1969), primarily (-) enantiomer (Stewart et al. 1977), trans-verbenol and verbenone from hindguts of males and females, respectively (Renwick 1967), endo-brevicomin in male hindguts (Vite and Renwick 1971a), myrtenol from female volatiles (Rudinsky et al. 1974a). Frontalin with alpha-pinene attractive to both sexes in field (Renwick and Vite 1969). Trans-verbenol can substitute for alpha-pinene as synergist (Renwick and Vite 1969, 1970, et al. 1978). Verbenone and myrtenol multifunctional pheromones, attractive with other components at low concentrations, inhibit response at high concentrations (Rudinsky 1973a, Rudinsky et al.

Dendroctonus
frontalis

frontalin (P)
endo-brevicomin (P)
myrtenol (P)
trans-verbenol (P)
verbenone (P)
alpha-pinene (K)

1974). _Endo_-brevicomin acts as antiaggregation pheromone (Vite and Renwick 1971a, Payne et al. 1978), and induces rivalry stridulation by males (Rudinsky and Michael 1974).

Dendroctonus _jeffreyi_	1-heptanol (P) 2-heptanol (P)	Identified from hindguts of attacking females. 1-Heptanol attractive to both sexes in field tests, 2-heptanol apparently inhibitory (Renwick and Pitman 1978).
Dendroctonus _ponderosae_	exo-brevicomin (P) _endo_-brevicomin (P) trans-verbenol (P) alpha-pinene (K) myrcene (K) terpinoline (K)	Trans-verbenol isolated and identified from female hindguts (Pitman et al. 1968), attractive to both sexes in field with alpha-pinene (Pitman 1971). Brevicomins identified from hindguts or beetle volatiles (Pitman et al. 1969, Rudinsky et al. 1974a). Exo-brevicomin attractive with other components at low concentrations, inhibitory at high concentrations (Rudinsky et al. 1974a) in white pine forests (Pitman et al. 1978). _Endo_-brevicomin inhibits response

Dendroctonus
pseudotsugae

frontalin (P)
trans-verbenol (P)
seudenol (P)
verbenone (P)
trans-pentenol (P)
3,2-MCH (P)
3,3-MCH (P)
methylheptenone (P)
alpha-pinene (K)
camphene (K)
ethanol (K)

(Rudinsky et al. 1974a). Myrcene and terpinoline better synergists for trans-verbenol than alpha-pinene (Billings et al. 1976).

Frontalin (Pitman and Vite 1970), trans-verbenol (Rudinsky et al. 1972a), seudenol (Vite et al. 1972a), verbenone (Rudinsky et al. 1974b) and trans-pentenol (Ryker et al. 1979) isolated and/or identified from female hindguts or volatiles; 3,3-MCH (Libby et al. 1976) and methylheptenone (Ryker et al. 1979) from male volatiles; 3,2-MCH from both sexes (kinzer et al. 1971, Pitman and Vite 1975). Frontalin, seudenol, trans-pentenol, alpha-pinene, camphene and ethanol all components of attractive mixtures (Pitman and Vite 1970, Furniss and Schmitz 1971, Pitman et al. 1975, Ryker et al. 1979). Trans-verbenol synergistic with frontalin (Rudinsky et al. 1972a), but role unclear in other tests (Rudinsky et al. 1972b,

Furniss et al. 1972). Verbenone (Rudinsky et al. 1974b) and 3,2-MCH (Rudinsky et al. 1972b, Rudinsky 1973b, Rudinsky and Ryker 1980) attractive with other components at low concentrations, inhibitory at high concentrations. 3,3-MCH (Rudinsky and Ryker 1979) and methylheptenone (Ryker et al. 1979) inhibit response.

Dendroctonus
rufipennis

frontalin (P)
seudenol (P)
alpha-pinene (K)

Frontalin (Pitman and Vite 1970) and seudenol (Vite et al. 1972) detected in female hindguts. Frontalin induced attack of baited trees (Dyer 1973, 1975). Seudenol with alpha-pinene more attractive than frontalin in traps (Furniss et al. 1976, Dyer and Lawko 1978).

Dendroctonus
vitei

frontalin (P)
1-phenylethanol (P)
1-heptanol (P)
2-heptanol (P)

Frontalin identified from hindguts of attacking females, 1-phenylethanol from emergent males, and the heptanols from attacking beetles of both sexes. Frontalin attractive in field tests,

Gnathotrichus retusus	sulcatol (P) alpha-pinene (K) ethanol (K)	1-phenylethanol inhibitory, especially to males, and the heptanols apparently inhibitory to males (Renwick et al. 1975). S-(+)-sulcatol isolated and identified from volatiles of boring males (Borden et al. 1980a), attractive alone in laboratory and field, antipode inhibitory (Borden et al. 1980a). Synergized in the field by ethanol and alpha-pinene (Borden et al. 1980b).
Gnathotrichus sulcatus	sulcatol (P) alpha-pinene (K) ethanol (K)	A 65:35 mixture of S-(+) and R-(-) sulcatol isolated and identified from male boring dust, hindguts and volatiles (Byrne et al. 1974). Both enantiomers synergistic and essential for response (Borden et al. 1976). Ethanol and alpha-pinene synergize response in the field (Borden et al. 1980b, 1982).

Species	Compounds	Description
Ips acuminatus	ipsenol (P) ipsdienol (P) cis-verbenol (P)	Identified from hindguts of boring males (Vite et al. 1972b, Bakke 1978). Synthetic ipsenol, S-(+)-ipsdienol and S-cis-verbenol with a Scots pine log competetive with log infested by 15 males in field tests; R-(-)-ipsdienol and R-cis-verbenol suppressed attraction (Bakke 1978).
Ips avulsus	ipsdienol (P)	Detected in hindguts of boring males (Vite et al. 1972, Hughes 1974). Both sexes attracted in the field (Renwick and Vite 1972). R-(-)-ipsdienol attractive, response inhibited by S-(+) enantiomer (Vite et al. 1978).
Ips calligraphus	ipsdienol (P) cis-verbenol (P) trans-verbenol (P)	Detected in hindguts of boring males (Renwick and Vite 1972, Hughes 1974). Ipsdienol attractive in field with trans- or (primarily) cis-verbenol. Addition of host volatiles produced comparable attraction to male-infested log (Renwick and Vite 1972). R-(-)-ipsdienol

Ips		
calligraphus	ipsdienol (P)	Detected in hindguts of boring males (Renwick and Vite 1972, Hughes 1974). Ipsdienol attractive in field with trans- or (primarily) cis-verbenol. Addition of host volatiles produced comparable attraction to male-infested log (Renwick and Vite 1972). R-(-)-ipsdienol (Vite et al. 1978) and S-cis-verbenol (Vite et al. 1976) attractive, S-(+)-ipsdienol inhibits response (Vite et al. 1978).
	cis-verbenol (P)	
	trans-verbenol (P)	

Ips		
cembrae	ipsenol (P)	Detected in hindguts of boring males. Ipsdienol and ipsenol attractive in field together, but not alone; addition of methylbutenol enhances attraction (Stoakley et al. 1978).
	ipsdienol (P)	
	3-methyl-3-buten -1-ol (P)	

(Vite et al. 1978) and S-cis-verbenol (Vite et al. 1976) attractive, S-(+)-ipsdienol inhibits response (Vite et al. 1978).

*Ips
confusus*

ispenol (P)
ipsdienol (P)
cis-verbenol (P)

Isolated and identified from male frass or volatiles. Ipsenol attractive alone in laboratory to both sexes, response greatest to ternary mixture (Young et al. 1973). Only ternary mixture attractive in field, but not competitive with pinion pine log containing 25 boring males (Birch et al. 1977a).

*Ips
duplicatus*

ipsdienol (P)

Detected in hindguts of boring males or after exposure to myrcene vapor. Beetles responded in field to traps baited with ipsdienol, but not to other known *Ips* pheromones (Bakke 1975).

*Ips
grandicollis*

ipsenol (P)

Detected in male hindguts (Vite and Renwick 1971b, Vite et al. 1972b, Hughes 1974). Both sexes respond in field to S-(-)-ipsenol (Vite et al. 1976b).

Ips
paraconfusus

ipsenol (P)
ipsdienol (P)
cis-verbenol (P)
2-phenylethanol (P)

S-(-)-Ipsenol, S-cis-verbenol and S-(+)
ipsdienol isolated and identified from male
frass (Silverstein et al. 1966), attractive in
laboratory and field at level comparable to
male-infested logs (Wood et al. 1968).
2-Phenylethanol identified from hindguts of
boring males; increased attraction to
male-infested logs in field (Renwick et al.
1976a). Response inhibited by R-(-)-ipsdienol
(Birch et al. 1980a).

Ips
pini

ipsdienol (P)

Isolated and identified from boring male
hindgust, abdomens or volatiles; California
(Birch et al. 1980a) and Idaho (Plummer et al.
1976) populations produce only R-(-)-ipsdienol,
New York populations a 65:35 mixture of S-(+)
and R-(-) enantiomers (Lanier et al. 1980).
R-(-)-ipsdienol attractive to California
beetles, S-(+)-ipsdienol inhibits response
(Birch et al. 1980a). Both enantiomers

		synergistic for attraction of New York populations (Lanier et al. 1980).
Ips sexdentatus	ipsenol (P) ipsdienol (P)	Identified from hindguts of males after 48 h feeding or 20 h exposure to myrcene vapor. Both sexes respond to ipsdienol in field, but response reduced when both compounds present (Vite et al. 1974).
Ips typographus	ipsdienol (P) 3-methyl-3-buten-2-ol(P) cis-verbenol (P)	Cis-verbenol and methylbutanol identified from hindguts of males initiating boring (Bakke et al. 1977), ipsdienol from males after completion of nuptial chamber (Bakke 1976). Ipsdienol attractive alone in field (Bakke 1976), but response greatest to ternary mixture (Bakke et al. 1977). S-cis-verbenol is active enantiomer (Vite et al. 1976a).
Pityogenes chalcographus	chalcogran (P)	Isolated and identified from volatiles produced by 100,000 juvenile hormone-treated males.

Pityokteines
curvidens

ipsenol (P)

Scolytus
multistriatus

methylheptanol (P)
alpha-multistriatin (P)
delta-multistriatin (P)
alpha-cubebene (K)

Racemic mixture of two pairs of diasteriomers comparable in attraction in field to spruce log containing 30 boring males (Francke et al. 1977).

Ipsenol identified as principal volatile produced by males boring in spruce logs or exposed to myrcene vapors. Attractive to both sexes in field (Harring and Vite 1975). $S-(-)$ enantiomer attractive, antipode inactive (Harring and Mori 1977).

Isolated and identified from frass and volatiles from female-infested elm logs (Pearce et al. 1975, Peacock et al. 1975). Absolute configurations determined from synthetic compounds as 3S, 4S-(-)-methylheptanol and 1S, 2R, 4S, 5R-(-)-alpha-multistriatin (Pearce et al. 1976, Mori 1976c, 1977, Elliott et al. 1979). Alpha-cubebene present as (-) enantiomer

		(Pearce et al. 1975). Ternary mixture necessary for optimal activity (Pearce et al. 1975, Lanier et al. 1977). In Germany (Gerken et al. 1978), but not England (Blight et al. 1980), delta-multistriatin is the active isomer.
Scolytus scolytus	methylheptanol (P) alpha-multistriatin (P) alpha-cubebene (K)	Isolated and identified from volatiles of boring beetles (Blight et al. 1977, 1978a). Males produce both 3S, 4S-(-)- and 3R, 4S-(-) isomers of methylheptanol, females only 3R, 4S-(-)-methylheptanol (Blight et al. 1979a). Only 3S, 4S-(-)-methylheptanol attractive in field (Blight et al. 1979b), and is synergized by alpha-cubebene (Blight et al. 1978b). Alpha-multistriatin produced by females, acts as antiaggregation pheromone (Blight et al. 1978b, 1979c).
Trypodendron domesticum	methylbutanone (P)	Isolated and identified from both sexes (Francke et al. 1974). Attractive in laboratory at

concentration equivalent to 5 beetles (Francke and Heeman (1974). Apparently inhibits response in field tests (Nijholt and Schonherr 1976).

| Trypodendron lineatum | lineatin (P) alpha-pinene (K) ethanol (K) | Lineatin isolated and identified from female frass (MacConnell et al. 1977). Absolute configuration of active enantiomer (Borden et al. 1980c) determined from synthetic enantiomers as $1R$, $4S$, $5R$, $7R-(+)$-lineatin (Slessor et al. 1980). Highly attractive to both sexes in field (Borden et al. 1979, Vite and Bakke 1979, Klimetzek et al. 1980. Ethanol and alpha-pinene synergistic with lineatin for European populations (Vite and Bakke 1979; Klimetzek et al. 1980), but not North American populations (Borden and McLean 1980). |

[a]Only compounds which have been tested in laboratory and/or field bioassays to verify their status as pheromones are listed. Chemical names for pheromones listed by common or trivial name are as follows:

exo-brevicomin: exo-7-ethyl-5-methyl-6,8-dioxabicyclo [3.2.1] octane

endo-brevicomin: endo-7-ethyl-5-methyl-6,8-dioxabicyclo [3.2.1] octane

Chalcogran: 2-ethyl-1,6-dioxaspiro [4.4] nonane

frontalin: 1,5-dimethyl-6,8-dioxabicycle [3.2.1] octane

ipsenol: 2-methyl-6-methylene-7-octen-4-ol

ipsdienol: 2-methyl-6-methylene-2, 7-octadien-4-ol

lineatin: 3,3,7-trimethyl-2,9-dioxatricyclo [3.3.1.0^{4,7}] nonane

3,2-MCH: 3-methylcyclohex-2-en-1-one

3,3-MCH: 3-methylcyclohex-3-en-1-one

methylbutanone: 3-hydroxy-3-methylbutan-2-one

methylheptanol: 4-methylheptan-3-ol

methylheptenone: 6-methylhept-5-en-2-one

alpha-multistriatin: 2,4-dimethyl-5-ethyl-6,8-dioxabicyclo [3.2.1] octane

myrtenol: 4,6,6-trimethylbicyclo [3.1.1] hept-3-en-10-ol

trans-pentenol: trans-pent-3-en-2-ol

seudenol: 3-methyl-2-cyclohexen-1-ol

sulcatol: 6-methylhept-5-en-2-ol

cis-verbenol: cis-4,6,6-trimethylbicyclo [3.1.1] hept-3-en-2-ol

trans-verbenol: trans-4,6,6-trimethylbicyclo [3.1.1] hept-3-en-2-ol

verbenone: 4,6,6-trimethylbicyclo [3.1.1] hept-3-en-2-one

kairomones. Secondary attraction is the term applied to this complex (Borden et al. 1975). It serves to induce and regulate the mass aggregation and attack phases of host selection (Figure 4.2).

Many pheromones have been discovered through painstaking and systematic isolations (Silverstein et al. 1967) in which an attractive starting material (e.g. frass or captured volatiles from infested logs) has been repeatedly fractionated, each fraction and the recombined mixture being tested for activity in a laboratory bioassay. In this way, no major component of the chemical message is overlooked. Usually more than one pheromone is discovered (Silverstein and Young 1976), but some scolytids, e.g. the ambrosia beetles T. lineatum (MacConnell et al. 1977) and Gnathotrichus retusus (Borden et al. 1980a), appear to use only one. Only two kairomones have been systematically isolated and identified, myrcene for D. brevicomis (Silverstein 1970) and alpha-cubebene for S. multistriatus (Pearce et al. 1975).

Once certain pheromones or kairomones are known, investigators can search for them or related compounds in other beetle or host species. This procedure has resulted in the identification of attractants for many scolytids (Table 4.1), but for many species, elucidation of the precise nature of secondary attraction must await more rigorous chemical isolations.

Several pheromones are shared by more than one species, e.g. frontalin in Dendroctonus spp. and ipsenol and ipsdienol in Ips spp. (Table 4.1). This phenomenon suggests that pheromones might be used in the future to disclose or confirm taxonomic affinities and differences.

Precise chemical identifications to the structural and optical isomer level have been essential to an understanding of pheromone biology. For example, cis-verbenol is an aggregation pheromone for I. paraconfusus (Silverstein et al. 1966), whereas trans-verbenol is an aggregation pheromone for D. ponderosae (Pitman et al. 1968), and an antiaggregation pheromone for D. brevicomis (Wood 1972, Bedard et al. 1980). Similarly, G. retusus utilizes only the S-(+) enantiomer of sulcatol (Borden et al. 1980a), while the pheromone message of the closely related species, Gnathotrichus sulcatus, comprises both enantiomers (Byrne et al. 1974), which are synergistic with each other (Borden et al. 1976). And the "ubiquitous" ipsenol and/or ipsdienol in Ips spp. may be less ubiquitous than it appears. With a choice of one or two compounds, each of which can be present as separate or combined enantiomers, there are 15 possible combinations. Add to this other pheromones such as S-cis-verbenol in I. paraconfusus (Silverstein et al. 1966) and methylbutanol in I. typographus (Bakke et al. 1977) and there is ample provision for species specificity in the chemical message.

The importance of host kairomones in secondary attraction appears to vary with the nature of the host being attacked. Aggressive species in the genus Dendroctonus appear to rely on monoterpenes as synergists in the chemical message, bringing them quickly to a host that must be subdued by a mass attack before it can resist by a copious resin flow (Chap. 7). Less aggressive species in the genus Ips attack weak, broken or cut trees, or attack in the upper crown or branches where resin pressure is reduced. Attack can proceed more slowly on these less resistant hosts. While host kairomones may be involved in secondary attraction for Ips spp., they are obviously much less important, and have yet to be discovered. Ambrosia beetles, however, utilize ethanol and alpha-pinene (and probably other monoterpenes) as synergists for their pheromones (Vite and Bakke 1979, Klimetzek et al. 1980, Borden et al. 1980b), assisting them in orienting to properly aged and conditioned hosts.

Pioneer Beetles. "Pioneer beetles" are the first few to attack a new host, and thus to produce aggregation pheromones (Figure 4.2). They are always of the same sex within a species, and usually within a genus. For example, male Ips spp. are invariably the pioneers as are female Dendroctonus spp. However, males of the large European elm bark beetle, Scolytus scolytus apparently initiate the attack and definitely produce aggregation pheromone(s) (Borden and King 1977, Gerken et al. 1978), but in the same genus, female S. multistriatus are the pheromone-producing pioneers (Peacock et al. 1971, 1973).

With or without the aid of primary attractants, the pioneer beetles must find a potential new host and assess its species and suitability for infestation. Host selection by responding beetles is considerably easier, for they can respond to secondary attraction resulting from the action of pioneer beetles in a proven host.

In some species, e.g. the red turpentine beetle, Dendroctonus valens, there are no evident aggregation pheromones. It may solely utilize host kairomones, or further research may disclose the occurrence of pheromones. Both laboratory (Oksanen et al. 1970) and field tests (Perttunen et al. 1970) failed to indicate the presence of pheromones in Blastophagus piniperda, but later studies disclosed positive evidence for their occurrence (Carle 1974, Carle et al. 1978). An interesting question is whether the parthenogenetic or spanandrous ambrosia beetles, e.g. Xyleborus spp., utilize aggregation pheromones, for this would demand kin selection (Beaver 1977).

Pheromone Production Modes. Attacking beetles in aggressive scolytid species may release aggregation pheromones immediately upon encountering a host tree. The habit of swallowing

air before and during flight produces a large ventricular air
bubble within the flying beetle (Graham 1961, Bennett and
Borden 1971). These "flatulent" beetles thus carry with them
the means for rapidly expelling pheromones from their
hindguts. Some Dendroctonus spp. apparently have this
capability. Crushed, emergent D. frontalis females, D.
brevicomis males and D. pseudotsugae females, all of which
produce frontalin (Table 4.1) were attractive to flying D.
frontalis in field tests (Renwick and Vite 1968). Ryker et
al. (1979) showed that frontalin is present in attacking
female D. pseudotsugae and can be released at the moment of
attack. In the absence of any immediate exogenous
precursors, D. brevicomis females treated with juvenile
hormone (JH) synthesized large quantities of exo- brevicomin
(Hughes and Renwick 1977a). Thus, it is possible that
pheromones like frontalin and exo-brevicomin are synthesized
de novo by the beetle, or that they are synthesized over time
from host-derived precursors ingested during maturation
feeding in the previous host.

In addition to pheromones, female D. pseudotsugae harbor
host terpenes (Ryker et al. 1979), one of which, alpha-
pinene, acts as a synergist with the pheromones frontalin and
seudenol (Furniss and Schmitz 1971, Pitman et al. 1975, Dyer
and Lawko 1978). Whether alpha-pinene should be classed as a
pheromone or a kairomone in this case is a moot point.
Regardless, it is doubtless soon overwhelmed by the release
of additional terpenes from the ruptured bark tissues as the
beetles begin to bore into a tree.

Attacking beetles exposed to the vapors of toxic monoterpenes (Smith 1965, 1966) have apparently acquired the ability
to detoxify them to less toxic terpene alcohols (Hughes
1973a, 1975). This phenomenon was unveiled by Vite
et al's. (1972b) perceptive experiments which demonstrated

that simply exposing males of 12 Ips spp. to the vapors of pine oleoresin resulted in the production of cis- and trans-verbenol in their hindguts. The detection of terpene alcohols in the hemolymph of D. valens and D. ponderosae following exposure to monterpene vapors led to the hypotheses that the vapors are inhaled through the tracheal system, the detoxification process occurs (or at least begins) in the hemolymph, and the resultant terpene alcohols are removed from the hemolymph and secreted into the hindgut by the Malpighian tubules.

It would have been a natural evolutionary step for the beetles to utilize some of the terpene alcohols and ketones as pheromones. Thus, when exposed to the vapors of alpha-pinene, D. ponderosae females produce the aggregation pheromone trans-verbenol (Hughes 1973b), and D. frontalis males produce the antiaggregation pheromone verbenone (Hughes 1975). Male Ips spp. produce ipsdienol and/or ipsenol when exposed to the vapors of myrcene (Hughes 1974).

The conversion of vaporized monoterpenes to terpene alcohols via the hemolymph and Malpighian tubules is in opposition to early research which noted that feeding was a prerequisite to pheromone production in I. paraconfusus, and implicated the alimentary system as a site of pheromone production (Wood and Bushing 1963, Pitman et al. 1965). This research led to the hypothesis that "either a precursor is ingested and metabolized to the attractant or the metabolism of food material causes secretory activity in specialized cells" (Wood et al. 1966). Therefore, it is quite possible that there are two modes of production for some pheromones.

The Possible Role of Microorganisms in Pheromone Synthesis. The hypothesis that microorganisms produce compounds that are involved in secondary attraction has shifted in and out of favor. Originally, Person (1931) hypothesized that secondary attraction for D. brevicomis is created by fermentation products from yeasts introduced into the inner bark by the first attacking beetles. This hypothesis is supported by the fact that many yeasts are associated with bark beetles, and some of them render innoculated phloem attractive to bark beetles (Callaham and Shifrine 1960). However, Vite and Gara (1962) found that impregnation of ponderosa pine logs with chemicals toxic to yeasts did not reduce their attractiveness after infestation by male I. paraconfusus or female D. brevicomis. Graham (1967) contended that the role of yeasts in secondary attraction is questionable, and the early explosion of knowledge on pheromone production in scolytids produced no evidence implicating microorganisms in pheromone production (Borden 1974).

More recently, Brand et al. (1975) demonstrated that Bacillus cereus isolated from the guts of both sexes of I. paraconfusus produced both cis- and trans-verbenol when alpha-pinene was added to the growth medium. They contended that bacterial conversion of ingested or inhaled monoterpenes in the gut is a likely method of pheromone production.

The role of microorganisms in pheromone production is further supported by evidence that: mycangial fungi associated with D. frontalis can transform cis- and trans-verbenol to the anti aggregation pheromone verbenone (Brand et al. 1976), metabolites of yeasts associated with D. frontalis can greatly enhance the attraction of frontalin with trans-verbenol and turpentine in a laboratory bioassay (Brand et al. 1977) and mycangial fungi associated with D. frontalis can produce sulcatol and "sulcatone" (Brand and Barras 1977), the aggregation pheromone and its possible precursor, respectively, in the ambrosia beetles, G. sulcatus (Bryne et al. 1974) and Platypus flavicornis (Renwick et al. 1977).

The fact that juvenile hormone (JH) can stimulate pheromone synthesis in scolytids (Borden et al. 1969, Hughes and Renwick 1977a, b, Blight et al. 1979c) suggests that pheromone-synthesizing microorganisms are responsive to their host's JH. Alternatively, there may be two systems, a hormonally-controlled beetle system, and an independent microorganism system. The latter possibility is supported by the fact that exposure of I. paraconfusus males to precocene [which causes degeneration of the corpora allata (Unnithan et al. 1977)] for 24 h prior to their entering a new host results in only partial inhibition of pheromone production (Kiehlmann et al. 1982). Moreover, in Ips cembrae JH stimulates production of the pheromone, methybutenol, but not ipsenol and ipsdienol, which are produced when males attack a new host or are exposed to myrcene vapors (Renwick and Dickens 1979), again indicating two production systems.

Ultimately, the relationship between scolytids and symbiotic microorganisms will be resolved by separating them. Studies on isolated microorganisms have yielded promising evidence. It is now possible to rear microbe-free beetles (Barras 1973, Chap. 6). Definitive experiments should disclose to what degree these microorganism-free beetles are capable of producing pheromones.

Ips paraconfusus: a Model System. In I. paraconfusus the complex control systems and biogenetic pathways for pheromone synthesis have been more researched than for any other scolytid (Figure 4.3).

On boring into a new host, males are exposed to the released vapors of monoterpenes, and also begin to feed. At some point after the initiation of attack JH is synthesized and released from the corpora allata. The stimulus for JH synthesis is probably gut distension following feeding. This

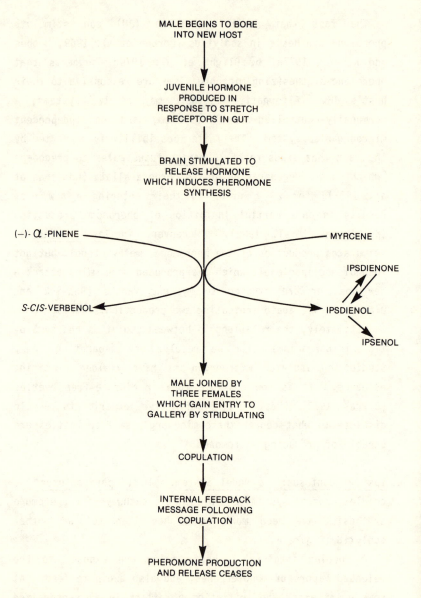

Figure 4.3.   Pheromone biogenesis and control systems in *Ips paraconfusus*.

event can be mimicked by distending the gut with air, which is then followed by the conversion of myrcene to ipsenol (Hughes and Renwick 1977b). However, there may also be a chemical stimulus as found for D. brevicomis females in methanolic extracts of ponderosa pine phloem (Hughes and Renwick 1977a).

The act of feeding can be bypassed by applying JH externally, inducing the male guts to become attractive (Borden et al. 1969). Later experiments demonstrated that JH treatment induced the production of ipsdienol, ipsenol and 2-phenylethanol in beetles that were denied an exogenous source of pheromone precursors (Hughes and Renwick 1977b). It also enhanced the conversions of alpha-pinene to cis-verbenol and of myrcene to ipsdienol and then ipsenol. The endogenous sources of pheromone precursors are unknown. They may be ingested during maturation feeding, or be derived from conjugated terpenes sequestered by the pupae. Pupae of the black terpentine beetle, Dendroctonus terebrans, and D. frontalis can apparently convert alpha-pinene to a conjugated form which is later deconjugated and converted to trans-verbenol by the adults (Hughes 1975). The implication of the corpora allata as a control center has been further substantiated by the inhibition of pheromone synthesis by I. paraconfusus males (Kiehlmann et al. 1982) and D. ponderosae females (Conn 1981) following their exposure to precocene.

Hughes and Renwick (1977b) obtained evidence that JH does not act directly on pheromone-producing cells, but induces the brain to release a pheromone synthesis factor. Decapitated beetles failed to produce the pheromones even when treated with JH or implanted corpora allata, but did so when corpora cardiaca, neurohemal organs in which brain hormone is stored, were implanted.

Lest one generalize too quickly, it should be noted that for female D. brevicomis, gut distension has no apparent effect on pheromone production while a host-derived chemical stimulus does. Treatment of decapitated beetles with synthetic JH induced the production of brevicomin, ruling out any effect of a brain hormone, and implantation of corpora allata-corpora cardiaca from newly emerged or pheromone producing females into newly emerged females had no effect on pheromone production (Hughes and Renwick 1977a).

The conversion of monoterpenes to the pheromones cis-verbenol, ipsdienol and ipsenol (Figure 4.3) has been evident in numerous experiments (Vite et al. 1972b, Hughes 1974, Renwick et al. 1976b, Byers et al. 1979). The conversion of myrcene to ipsdienol and then ipsenol has been confirmed by the use of deuterium-labelled myrcene (Hendry et al. 1980). Demonstration of the reversible conversion of ipsdienol to ipsdienone has provided evidence of an alternative biosynthetic pathway to the pheromones (Fish et al. 1979).

Myrcene is achiral, yet I. paraconfusus males produce S-(+)-ipsdienol and S-(-)-ipsenol (Silverstein et al. 1966; Mori 1976a,b). Alternatively, the beetles convert (+)-alpha-pinene into trans-verbenol, but convert (-)-alpha-pinene into the pheromone (+)-cis-verbenol (Renwick et al. 1976b) [=S-cis-verbenol (Mori et al. 1976)]. In the latter case, the chirality of the pheromone precursor determines the nature of the pheromone produced, and thus limits the host range of the beetle to pines which produce (-)-alpha-pinene. In both cases, the chiral nature of the pheromones must be determined by specific enzymes which can recognize the appropriate precursors and produce chirally correct pheromones.

The three aggregation pheromones attract I. paraconfusus of both sexes in the field (Wood et al. 1968), females to join the males and males to initiate other galleries on a proven host. The males deny entrance to the nuptial chamber by placing themselves in the entrance tunnel. Only after the females have announced their presence by stridulating (Barr 1969) does the male admit up to three of them into the nuptial chamber to mate and construct egg galleries. When a fourth female was imprisoned above entrance tunnels into which three females had already been admitted, the males denied half of them entry, forcing 60% of the rejected females to bore separate entrance tunnels into the nuptial chamber (Borden 1967). As with Ips calligraphus (Vite et al. 1972b) and S. multistriatus (Elliott et al. 1975), mated beetles cease to produce pheromone. With each successive female the male I. paraconfusus frass becomes less and less attractive, with attraction being lost after three or four females have joined the male (Borden 1967). This result suggests a mating-induced, negative feedback mechanism that causes the cessation of pheromone production and terminates the attraction of natural populations to a saturated host.

Response Mechanisms

In the selection-concentration phase (Figure 4.2) the majority of beetles of both sexes respond to the complex of beetle-produced pheromones plus host-produced kairomones that comprise the various components of secondary attraction. The intensity and nature of the stimulus will depend on the numbers of pheromone-producing beetles boring in the host and whether or not they have been joined by beetles of the opposite sex. The olfactory stimulus may be supplemented by other stimuli, primarily visual. The ultimate result is a concentration of beetles in a mass attack culminating in the establishment of a population in the new host.

The responding beetle is a remarkable organism. It is a
finely tuned machine, capable of perceiving and discriminating between specific odors at great distances and responding
to them with precise, efficient, yet flexible behavior.

Perception. Experiments on attennectomy and ablation of
sensilla (Borden and Wood 1966) and recordings of electroan-
tennograms (EAG's) (the summed depolarization of many receptor cells) (Payne 1970, 1971, 1974a) unequivocally established scolytid antennae as sites of pheromone perception.

More extensive electrophysiological research has elucida-
ted some of the intricacies of pheromone perception. EAG's
of the pine engraver, Ips pini, to (+)- and (-)-ipsenol
disclosed extremely low receptor thresholds, suggesting
perception and response capability at great distances from a
pheromone source (Angst and Lanier 1979). Greater antennal
response in both sexes of S. scolytus to (-)-threo and (-)-
erythro isomers of methylheptanol than to their antipodes
provides evidence for chiral receptors in scolytid antennae
(Blight et al. 1979). Antennal muscle potentials recorded
simultaneously with EAG's from D. frontalis antennae stimulated by frontalin suggest that antennal movement and other
orientation behavior is induced by the pheromone stimulus
(Payne 1974b, Dickens and Payne 1977, 1978). The fact that
D. frontalis and D. brevicomis antennae of both sexes respond
equally to frontalin even though there are marked behavioral
differences between the sexes to the same stimulus indicates
that electrophysiological data should not be taken as
strictly valid indicators of response behavior (Payne 1975).

Single cell recordings have also been made from bark
beetle antennal sensilla. Mustaparta et al. (1979)
discovered that in recordings from 95 I. pini cells, 45 were
specific receptors for ipsdienol and 15 for ipsenol; 23
responded to trans-verbenol, cis-verbenol or verbenone, 6 to

myrcene and a few to camphor and linalool. Simultaneous exposure of ipsdienol-sensitive cells to the aggregation pheromone ipsdienol and the antiaggregation pheromone ipsenol failed to inhibit the response to ipsdienol, suggesting that response inhibition in I. pini is not at the primary receptor level, but resides in central nervous system integration of stimuli (Mustaparta et al. 1977, 1979). EAG data on chiral response capability are supported by single cell recordings from D. frontalis sensilla basiconica, which are 40% more responsive to (-)- than (+)-exo-brevicomin (Dickens and Payne 1977).

A technique called differential adaptation has been used to determine whether there are one or two acceptor sites on the same sensory cell (Payne and Dickens 1976). An antennal preparation is adapted to one compound and then exposed within milliseconds to another. No response to the second compound indicates specificity for the first; response indicates two acceptor sites on the same cell or a combination of specific and nonspecific acceptor sites. Data obtained using this procedure indicate that cells associated with the long sensilla basiconica in D. frontalis are specific to frontalin, whereas more generalist cells can also respond to other compounds such as exo- and endo-brevicomin, trans-verbenol, verbenone, alpha-pinene, and 3-carene (Payne and Dickens 1976, Dickens and Payne 1977).

The existence of generalist cells suggests that an appropriate mixture of compounds might block their perception of frontalin, thus at least partially inhibiting response behavior (Payne 1979). This hypothesis is supported by the fact that decreased antennal muscle potentials are recorded from D. frontalis exposed to verbenone, trans-verbenol or host terpenes, suggesting a possible arrestment response (Dickens and Payne 1977).

Electrophysiological studies have yielded definitive data on the response capabilities of receptor cells of various species and both sexes under a variety of conditions. They have also yielded tantalizing suggestions on the nature of behavioral thresholds, changes and inhibitions. However, to be most useful to behaviorists (and ultimately to pest managers) electrophysiological recordings should be obtained from both the receptors and the motor nerves associated with the leg or wing muscles, and possibly the genitalia. Only then will realistic interpretations be possible on how sensory impulses are processed through the central nervous system to promote or inhibit various types of behavioral responses.

Response Behavior. Once it has perceived the chemical stimulus, a beetle must orient toward the source of secondary attraction. The orientation flight and orientation by walking beetles in the laboratory can be characterized as klinotaxes (Wood and Bushing 1963, Borden and Wood 1966; Borden 1977). For T. lineatum in flight it has been described as steady, with frequent turns not exceeding two meters to either side of the straight line of approach (Rudinsky and Daterman 1964).

When they have been observed, orientation flights are upwind. In fact, the dispersal flight of bark beetles is predominantly downwind (Vite and Gara 1962). For I. paraconfusus, it reverses rapidly when the beetle encounters an attractive stimulus on the wind (Vite et al. 1963). The hypothesis that the upwind flight is anemotactic (Borden 1974), rather than to an increasing concentration of pheromone, has recently been confirmed for S. multistriatus. Male beetles in a wind tunnel, in which the pheromone concentration was uniform throughout, oriented upwind in a positive anemotaxis (Choudhury and Kennedy 1980). As discussed above, the intrinsic nature of the flying beetle appears to govern its ability first to disperse, and then to shift to host- and pheromone-positive behavior.

Visual stimuli may influence the olfactory orientation of scolytids, but few definitive data exist. Pitman and Vite (1969) reported that horizontal, tree-trunk-simulating olfactometers were better than vertical ones for D. ponderosae, but Billings et al. (1976) found the opposite. Vite and Bakke (1979) concluded that for T. lineatum there is a three-way interaction between lineatin, the synergists alpha-pinene and ethanol, and the dark silhouette of a tree-trunk-simulating olfactometer. However, other results indicate that glass barrier or wire mesh traps are either better or no different from silhouette traps (Klimetzek et al. 1980, Borden and McLean 1980).

Characteristically, the responding sex is more frequently caught in pheromone-baited traps than the sex that produces the pheromone. The difference may lie in the fact that beetles of the responding sex orient directly to a pheromone source, as they would to a prospective mate. Beetles of the first attacking sex would exhibit a less precise, area response (Wood and Bushing 1963), which would be more likely to bring them to an uninfested area of a host tree or a nearby unattacked tree. This hypothesis is supported by field experiments on D. brevicomis and D. frontalis. Many D. brevicomis can be caught on unbaited traps surrounding a pheromone source, with males (the responding sex) penetrating nearest to the source before being caught, indicating a more direct flight path than females (Tilden et al. 1979). When the dose of frontalin was lowered and trap size increased many D. frontalis were caught (presumably on the outer area of the trap), correcting a previously unbalanced sex ratio (Hughes 1976).

Other factors may influence the sex ratio of responding beetles. When host kairomones are offered as stimuli with aggregation pheromones, the response often shifts in favor of the first attacking sex (Pitman and Vite 1970, Billings et al. 1976, Borden et al. 1980b). This shift reflects the fact that the first attacking sex must initiate new attacks on a host, whereas the responding sex need only find a mate. Trap design may produce sex ratio artifacts. A glass barrier (Nijholt and Chapman 1968) or sticky trap (Browne 1977) may catch roughly equal numbers of both sexes. However, the clyindrical Scandinavian drainpipe traps (Bakke and Saether 1978) in which the responding beetles must find and enter holes may select for the responding sex which exhibit hole-finding behavior in nature (Klimetzek et al. 1980).

At close range to an attractive host or pheromone-producing beetle gallery, the orienting beetle must be arrested. In laboratory experiments, flight may cease in response to attractive odors (Bennett and Borden 1971), and walking beetles may be arrested over a pheromone source (Jantz and Rudinsky 1965, Borden et al. 1968, Peacock et al. 1973). In extremely high concentrations of pheromone, $\underline{I}$. paraconfusus is arrested, and will not orient upwind (Wood et al. 1966), and in the odor of attractive frass, flight initiation by both sexes is inhibited (Borden 1967). This behavior will hold the beetles in the presence of a potential host or mate. However, it must be reversible should the prospective host or mate prove unsuitable.

Interaction Between Attacking Beetles, and Attack Density Regulation. Bark beetle galleries are nonrandom in their distribution on the bark (Shepherd 1965, Berryman 1968). Because their habitat, the host phloem tissue, is essentially two dimensional and thus limited, some sort of spacing mechanism is of major adaptive advantage. Overcrowding could result in limited food resources for the next generation, cannibalism and possible danger from other factors, such as excessive drying and deterioration of the bark, or rapid spread of an epizootic through a captive and dense larval population. On the other hand, ambrosia beetles infest a three-dimensional habitat, the sapwood, and could populate a host at much higher densities than bark beetles. There would be far less adaptive advantage in a spacing mechanism, particularly if their potential hosts were limited.

Both chemical and acoustic spacing mechanisms exist and interact in the bark beetles. This interaction has been most extensively studied in the Douglas-fir beetle (Figure 4.4).

Rudinsky (1968, 1969) reported the startling discovery that female D. pseudotsugae could "mask" their attractive pheromone if stridulating males were placed outside their galleries. If the females underneath the bark were killed with a hammer, no masking occurred, and attraction persisted due to residual aggregation pheromones in the frass. Later research also disclosed that males also contribute to the mask (Pitman and Vité 1975, Rudinsky et al. 1976), and that the mask is due to two compounds, 3,2-MCH (Vité et al. 1972a) and 3,3-MCH (Libbey et al. 1976), which act as antiaggregation pheromones (Furniss et al. 1972, Rudinsky et al. 1972b, Rudinsky and Ryker 1979). The females also stridulate, apparently as an attack density regulating mechanism (Figure 4.4) (Rudinsky and Michael 1973). To verify the function of sonic stimuli, recorded stridulation sounds were used experimentally to induce the same effects as stridulating beetles (Rudinsky et al. 1973, Rudinsky and Ryker 1977).

The interaction of sonic and chemical communication has not been extensively studied in species other than D. pseudotsugae. However, beetles of the first attacking sex in

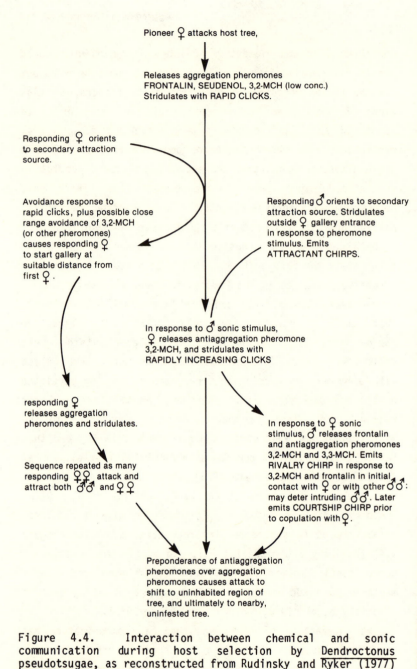

Pioneer ♀ attacks host tree,

Releases aggregation pheromones
FRONTALIN, SEUDENOL, 3,2-MCH (low conc.)
Stridulates with RAPID CLICKS.

Responding ♀ orients
to secondary attraction
source.

Avoidance response to
rapid clicks, plus possible close
range avoidance of 3,2-MCH
(or other pheromones)
causes responding ♀
to start gallery at
suitable distance from
first ♀ .

Responding ♂ orients to secondary
attraction source. Stridulates
outside ♀ gallery entrance
in response to pheromone
stimulus. Emits
ATTRACTANT CHIRPS.

In response to ♂ sonic stimulus,
♀ releases antiaggregation pheromone
3,2-MCH, and stridulates with
RAPIDLY INCREASING CLICKS

responding ♀
releases aggregation
pheromones and stridulates.

In response to ♀ sonic
stimulus, ♂ releases frontalin
and antiaggregation pheromones
3,2-MCH and 3,3-MCH. Emits
RIVALRY CHIRP in response to
3,2-MCH and frontalin in initial
contact with ♀ or with other ♂♂:
may deter intruding ♂♂. Later
emits COURTSHIP CHIRP prior
to copulation with ♀.

Sequence repeated as many
responding ♀♀ attack and
attract both ♂♂ and ♀♀

Preponderance of antiaggregation
pheromones over aggregation
pheromones causes attack to
shift to uninhabited region of
tree, and ultimately to nearby,
uninfested tree.

Figure 4.4. Interaction between chemical and sonic
communication during host selection by Dendroctonus
pseudotsugae, as reconstructed from Rudinsky and Ryker (1977)
and references therein.

numerous species stridulate, providing evidence for potential spacing mechanisms, and in many species there is sonic communication prior to mating, rivalry behavior or admission of a responding beetle to a gallery (Barr 1969, Rudinsky and Michael 1973, Rudinsky and Ryker 1976, 1977, Ryker and Rudinsky 1976, Oester et al. 1978, Rudinsky et al. 1978, Rudinsky 1979, Oester and Rudinsky 1979).

Antiaggregation pheromones have been discovered in many scolytid species (Table 4.1), e.g. verbenone and trans-verbenol for D. brevicomis (Wood 1972, Bedard et al. 1980a,b) endo-brevicomin for D. frontalis (Vite and Renwick 1971a, Payne et al. 1978) and exo-brevicomin for D. ponderosae (Pitman et al. 1978). These Dendroctonus spp. are aggressive and monogamous. Once a host tree is colonized to an appropriate density and mating pairs are established, antiaggregation pheromone promotes the switching of attack to nearby trees (Gara and Coster 1968, Geiszler and Gara 1978).

In the polygamous Ips spp. there is no evidence for antiaggregation pheromones. While there must be some (as yet undisclosed) spacing mechanism, it would be counter productive to utilize an antiaggregation pheromone when one female joins a male, thereby inhibiting the response of subsequent females. Rather, Ips spp. are better served by an internal feedback mechanism resulting from mating, that results in a progressively weaker pheromone signal as a male is joined by successive females (Borden 1967).

S. multistriatus, which is more like Dendroctonus than Ips spp. in its habits, is unusual in that its attack termination is not caused by an antiaggregation pheromone. Apparently, the promiscuous males rapidly inseminate all available females, which soon cease to produce or release pheromone (Elliott et al. 1975). Of all the scolytids, S. multistriatus is the only one in which a pheromone-producing gland (for alpha-multistriatin) has been implicated (Gore et al. 1977), a structure amenable to immediate cessation of pheromone release. It is not yet known whether antiaggregation pheromones act as repellents or disrupt orientation behavior in some manner. If they are repellent, they might be used practically to exclude beetles from a particular area, e.g. via an olfactory barrier (Borden 1977), a more attractive outcome to a forest manager than disrupting orientation within a target area.

Interspecific Communication with Other Scolytids

Birch and Wood (1975) published a landmark paper on interspecific communication between $\underline{I}$. paraconfusus and $\underline{I}$. pini which broadened the conception of chemical communication in insects. It is now apparent that interspecific chemical communication occurs among numerous sympatric scolytid species, particularly those that infest the same host (Table 4.2). The same compound may have numerous functional roles and thus definitions (Figure 4.1) following Blum's (1970) rule of "pheromonal parsimony."

Examples of communication between species infesting different host species have not been extensively studied and the following interpretations are definitely hypothetical. Interspecific inhibition of response could be kairomonal in action, assisting one species in rejecting a non-host, e.g. the inhibitory effect of methylbutenol produced by the spruce-inhibiting Ips typographus on response to aggregation pheromones by the pine-infesting I. acuminatus (Bakke 1978). Cross attraction, e.g. between Trypodendron domesticum and T. lineatum (Francke and Heeman 1974, Klimetzek et al. 1980) or D. pseudotsugae and the spruce beetle, Dendroctonus

Table 4.2. Summary of interspecific chemical communication between sympatric species in the Scolytidae.

Species producing semiochemical(s)	Compounds produced with demonstrated or implicated interspecific activity, and/or source of natural interspecific semiochemical	Interspecific action
Dendroctonus pseudotsugae	frontalin, seudenol female-infested logs	D. pseudotsugae females introduced into either Picea glauca or Pseudotsuga menziesii attract beetles of both species in field tests; D. rufipennis in P. glauca attract both species but won't attack P. menziesii (Chapman and Dyer 1969). Although D. rufipennis attracted to frontalin-baited trees (Dyer 1973, 1975),
Dendroctonus rufipennis	frontalin, seudenol female-infested logs	

responds only to seudenol in traps, while D. pseudotsugae responds to seudenol and frontalin alone or combined (Dyer and Lawko 1978).

Dendroctonus
  brevicomis

verbenone
naturally-infested
Pinus ponderosa log

Logs infested by male I. paraconfusus inhibit response by D. brevicomis to naturally-infested log or frontalin, exo-brevicomin and myrcene (Byers and Wood 1980). I. paraconfusus pheromones attractive to D. brevicomis at 10^{-9} grams per microliter in the laboratory, but inhibitory at higher concentrations (Byers and Wood 1981). Logs infested by D. brevicomis inhibit response by I. paraconfusus to male-infested log (Byers and Wood 1980). Inhibition caused by verbenone released by male D. brevicomis early in attack phase (Byers and Wood 1980, 1981). Mutual inhibition of response probably segregates each species in distinct area of same host tree (Byers and Wood 1980).

Ips paraconfusus

S-(-)-ipsenol
S-(+)-ipsdienol
S-cis-verbenol
Pinus ponderosa log
infested with males

Dendroctonus frontalis	frontalin female-infested Pinus taeda log	D. frontalis not attracted to odor of any male Ips. sp., male response inhibited by male I. grandicollis, females' odor inhibitory to I. grandicollis (Birch et al. 1980b), probably due to frontalin inhibiting response of I. grandicollis to ipsenol and host terpenes (Werner 1972, Birch and Svihra 1979). Response of I. avulsus enhanced by presence of boring male I. grandicollis (Hedden et al. 1976, Birch et al. 1980b), response greater to combined ipsdienol and (-)-ipsenol than to either compound alone (Hedden et al. 1976). I. calligraphus inhibit response of I. grandicollis and attract I. avulsus, response inhibited by presence of I. avulsus (Birch et al. 1980b). Interaction related to sequence of arrival at and partitioning of host tree; see Birch et al. (1980b) for additional details.
Ips avulsus	R-(-)-ipsdienol male-infested Pinus taeda log	
Ips calligraphus	R-(-)-ipsdienol male-infested Pinus taeda log	
Ips grandicollis	S-(-)-ipsenol	

Species	Semiochemical	Response
Gnathotrichus sulcatus	65:35 mixture of S-(+)- and R-(-)-sulcatol	G. sulcatus unresponsive or very slightly responsive to S-(+)-sulcatol. G. retusus response inhibited by high concentration of R-(-)-sulcatol (Borden et al. 1980a). Both Gnathotrichus spp. tolerate lineatin; T. lineatum either tolerates sulcatol (Borden and McLean 1980, Borden et al. 1981), or is partially inhibited by it (T. L. Shore and J. A. McLean, pers. comm.).
Gnathotrichus retusus	S-(+)-sulcatol	
Trypodendron lineatum	lineatum	
Ips paraconfusus	S-(-)-ipsenol S-(+)-ipsdienol male-infested Pinus ponderosa log	Neither species attracted to logs infested by males of both species indicating mutual inhibition which reserves host for first arriving species (Birch and Wood 1975). S-(-)-Ipsenol (Birch and Light, Birch et al. 1977b) and (+)-ipsdienol (Birch et al. 1980a) inhibitory to I. pini. (-)-Ipsdienol inhibitory to I. paraconfusus (Light and Birch 1979). Report that linalool is deterrent to I.
Ips pini (California population)	R-(-)-ipsdienol linalool male-infested Pinus ponderosa log	

Species	Compound	Notes
Ips typographus	methylbutenol	paraconfusus (Birch and Wood 1975) not supported by further research (Birch and Light 1977). Response of I. acuminatus to ipsenol, ipsdienol and cis-verbenol suppressed by addition of methylbutenol (Bakke 1978).
Scolytus multistriatus	alpha-multistriatin	Response of S. scolytus to methylheptenol and alpha-cubebene inhibited by presence of alpha-multistriatin (Blight et al. 1978b, 1979c).
Trypodendron domesticum	methylbutanone	Methylbutanone slightly attractive to T. lineatum in the laboratory; 50 mg treatment enhanced attraction of log moderately infested by T. lineatum (Francke and Heeman 1974), but inhibited response to ethanol and alpha-pinene (Nijholt and Schonherr 1976). Lineatin slightly attractive to T. domesticum in field (Klimetzek et al. 1980).

rufipennis (Chapman and Dyer 1969, Dyer and Lawko 1978), suggests shared pheromones or pheromone analogues. Reproductive isolation between these species will be better understood when the entire complex of pheromones and host kairomones mediating behavior are known. It may reside in a number of factors, e.g. a scenario in which cross attraction is kairomonal, aiding both species in finding a suitable habitat, such as a patch of windthrown trees, which are then selected on the basis of species-specific boring stimulants and/or deterrents.

For species infesting the same host tree, interspecific communication ranges from mutual inhibition of response to a host, to tolerance of the pheromone of another species, and finally to cross attraction.

Inhibition occurs when a compound (often a pheromone) produced by one species inhibits or interrupts the response of another. Such a compound is clearly an allomone (Brown 1968) of adaptive advantage to the producing species. It functions to reserve a host for the first arriving species, ensuring that those beetles have a maximal chance of reproducing. However, the same compound may also function as a kairomone (Brown et al. 1970) of benefit to the perceiving organism. Birch (1978, 1980) has shown that brood production of both I. pini and I. paraconfusus inhabiting the same log is over 65% lower than in separate logs. The basis for this effect is unknown, but it is of obvious advantage to individuals in the perceiving species to avoid a log already infested by beetles of another species, and thus to avoid the risk of greatly decreased brood production.

A clear example of mutual inhibition of response occurs between I. paraconfusus and I. pini, each species being inhibited from responding to its own aggregation pheromone odor by pheromones of the other species (Table 4.2). Between

G. retusus and G. sulcatus the communication mechanisms are less well defined (Table 4.2). G. sulcatus is generally unresponsive to the G. retusus aggregation pheromone, while G. retusus response is inhibited by relatively large amounts of R-(-)-sulcatol, such as the 35% present in the natural pheromone complex of G. sulcatus (Byrne et al. 1974).

The above two examples illustrate the critical need to know the precise isomeric nature of semiochemicals, and the need for superior chemical syntheses, which yield structurally and optically pure products. When R-(-)-ipsdienol was contaminated with 2.2% of the S-(+) enantiomer the response of Californian I. pini was normal, but at 5% of the S-(+) enantiomer, response to R-(-)-ipsdienol was completely inhibited (Birch et al. 1980a). If even 1% of R-(-)-sulcatol is in mixture with 99% S-(+)-sucatol, the response of G. sulcatus rises markedly (Borden et al. 1980a) due to synergism between the two enantiomers (Borden et al. 1976).

Tolerance to the pheromone of a sympatric species infesting the same host occurs between T. lineatum and either G. retusus or G. sulcatus (Table 4.2). There is no apparent need for chemical communication in maintaining reproductive isolation of beetles in these distinct genera, and there is relatively little adaptive advantage in reserving the spacious three-dimensional sapwood habitat for the first arriving species.

The complex of beetles which co-inhabit southern pines display several different types of chemical communication (Table 4.2), ranging from mutual inhibition of response between Ips grandicolis and D. frontalis to strong kairomonal cross attraction of Ips avulsus to I. calligraphus (Birch et al. 1980b). By such a complex system, these beetles are able to cooperate in mass attacking a tree, while still ensuring that each species occupies a distinct portion of that host (Birch 1978, Birch et al. 1980b).

Lanier and Burkholder (1974) see no evidence that differences in pheromone communication precipitated speciation in the genus Ips. Indeed, the most closely related Ips spp. are seldom sympatric, but tend to be cross attractive. This observation suggests that some other mechanism of speciation occurred, such as an increased tolerance to climatic extremes, or the ability to utilize an alternate host species. The most distantly related the species, the less cross attraction is evident, and the more likely the species are to be sympatric. I. paraconfusus and I. pini, which are in separate species groups (Hopping 1963), exemplify the latter situation. However, it is possible that enantiomeric specificity, as between I. paraconfusus and I. pini, reflects at least a partial basis for separation between species groups, rather than individual species.

Knowledge of interspecific chemical communication among scolytids has added a new dimension to host selection. In orienting to and landing on a host, both pioneer and responding beetles may need to assess the host for the presence of a competing species (Birch 1980). If an offending infestation is present, the beetles must resume dispersal (Figure 4.2) until they are able to find an acceptable host. On the other hand, the orientation of both pioneer and responding individuals may be enhanced by the presence of a cooperating species.

Communication With Insects Other Than Scolytids

Many species of insects are associated with scolytids in their host trees. Among them are other phytophagous insects, which may compete for the host resource. There are commensals, which may feed on microorganisms or dead and moribund insects inside the galleries. And there are a

Table 4.3. Kairomone response to natural or synthetic scolytid aggregation pheromones by entomophagous insects.

Associated Insect	Kairomone or natural source of kairomone	Response Information
COLEOPTERA		
Enoclerus lecontei (Cleridae)	ipsenol ipsdienol cis-verbenol Pinus ponderosa logs infested with Ips pini	Trapped in California to mixture of the 3 pheromones (Wood et al. 1968). In California and Idaho field tests, more attracted to odor of boring I. pini males from New York than from California or Idaho (Lanier et al. 1972). Attraction to logs infested by male I. pini enhanced when ipsenol added as a stimulus (Furniss and Livingston 1979). Response to logs infested by I. paraconfusus and Dendroctonus brevicomis greater than to logs infested by I. paraconfusus alone (Byers and Wood 1980a).

Stigmatium nakanei (Cleridae)	Pinus densiflora logs infested with 0,10,50 or 150 females Taenioglyptes fulvus	Response to traps baited with logs infested by female T. fulvus greater than to control log. Number of predators caught increased markedly to logs infested with 50 or 150 females (Sasakawa et al. 1976).
Thanasimus dubius (Cleridae)	frontalin trans-verbenol verbenone	Responds in same diel rhythm as its prey, Dendroctonus frontalis, to frontalin alone or with oleoresin or trans-verbenol; verbenone raised male:female response ratio, but lowered overall response (Vite and Williamson 1970, Dixon and Payne 1980). Greater aggregation in low density host populations attributed to kairomone response (Reeve et al. 1980). Males predominate during first 3 days of attack, suggesting that they attract females (Dixon and Payne 1979).
Thanasimus formicarius (Cleridae)	Norway spruce log baited with methylbutenol	Response to baited logs in field tests significantly greater than to unbaited control logs (Bakke and Kvamme 1978).

Species	Chemical	Response
	cis-verbenol, & ipsdienol	
Thanasimus rufipes (Cleridae)	Norway spruce log baited with methylbutenol cis-verbenol, & ipsdienol	Response to baited logs in field tests significantly greater than to unbaited control logs (Bakke and Kvamme 1978).
Thanasimus undatulus (Cleridae)	frontalin	Attracted in large numbers to frontalin-baited spruce trees (Dyer 1973, 1975) or traps (Kline et al. 1974).
Lasconotus pusillus (Colydiidae)	frontalin loblolly pine turpentine	Response to baited traps in field tests (Dixon and Payne 1980).
Abraeus sp. (Histeridae)	frontalin	Response to baited traps in field tests (Dixon and Payne 1980).

Cylistix attenuata (Histeridae)	frontalin	Response to baited traps in field tests (Dixon and Payne 1980).
Plegaderus sp. (Histeridae)	endo-brevicomin	Response to baited traps in field tests (Dixon and Payne 1980).
Leptacinus paurumpunctatus (Staphylinidae)	frontalin exo-brevicomin loblolly pine turpentine	Response to baited traps in field tests (Dixon and Payne 1980).
Corticeus glaber (Tenebrionidae)	frontalin exo-brevicomin endo-brevicomin loblolly pine turpentine	Response to baited traps in field tests (Dixon and Payne 1980).
Temnochila chlorodia (Trogositidae)	exo-brevicomin frontalin myrcene	Responds to traps baited with exo-brevicomin alone (Bedard et al. 1969, Pitman and Vite 1971) or with frontalin and myrcene (Bedard and

| | trans-verbenol
verbenone | Wood 1974). Trans-verbenol with verbenone apparently interrupt response to exo-brevicomin (Bedard et al. 1980b). |

HYMENOPTERA

Dendrosoter protuberans (Braconidae)	alpha-multistriatin methylheptanol alpha-cubebene	Responds to traps baited with binary or ternary combinations of multilure components (Kennedy 1979).
Coeloides pissodes (Braconidae)	frontalin loblolly pine turpentine	Response to baited traps in field tests (Dixon and Payne 1980).
Spathius benefactor (Braconidae)	alpha-multistriatin methylheptanol alpha-cubebene	More individuals captured on multilure-baited than unbaited traps (Kennedy 1979).
Spathius pallidus	trans-verbenol exo-brevicomin	Response to baited traps in field tests (Dixon and Payne 1980).

Species	Semiochemicals	Response
	endo-brevicomin, loblolly pine, turpentine	
Entedon leucogramma (Eulophidae)	alpha-multistriatin, methylheptanol, alpha-cubebene	Responds to traps baited with multilure components alone (except methylheptanol) or in binary or ternary combinations (Kennedy 1979).
Cerocephala rufa (Pteromalidae)	alpha-multistriatin, methylheptanol, alpha-cubebene	Responds to traps baited with multilure components alone or in binary and ternary combinations (Kennedy 1979).
Cheiropachus colon (Pteromalidae)	alpha-multistriatin, methylheptanol, alpha-cubebene	Responds to traps baited with multilure components alone or in binary and ternary combinations (Kennedy 1979).
Heydenia unica (Pteromalidae)	frontalin, endo-brevicomin, loblolly pine, turpentine	Response to baited traps in field tests (Dixon and Payne 1980).

Tomicobia tibialis (Pteromalidae)	Pinus ponderosa logs infested with male Ips paraconfusus or male Ips pini	Attracted to odor of boring male I. paraconfusus (Bedard 1965, Rice 1969). I. pini from California or Idaho more attractive than those from New York (Lanier et al. 1972).

DIPTERA

Medetera aldrichii (Dolicopodidae)	3,2-MCH	Reduced attack density on Dendroctonus pseudotsugae when host tree treated with high concentrations of 3,2-MCH (Furniss et al. 1974).
Medetera bistriata (Dolicopodidae)	frontalin trans-verbenol verbenone Pinus taeda logs infested with male Ips grandicollis or female Dendroctonus frontalis	Attracted to infested logs, frontalin plus alpha-pinene, trans-verbenol plus alpha-pinene, but not to frontalin or alpha-pinene alone; verbenone caused change in sex ratio in favor of females (Williamson 1971).

HEMIPTERA

| Scolopscelis mississippensis (Anthocoridae) | frontalin trans-verbenol exo-brevicomin endo-brevicomin loblolly pine turpentine | Response to baited traps in field tests (Dixon and Payne 1980). |

multitude of entomophagous insects, which parasitize or prey on the scolytids; these must follow a host selection sequence which involves dispersal, host habitat finding and host finding phases (Borden 1974).

Remarkably, some of the entomophagous insects have evolved the ability to utilize the scolytid aggregation pheromones as kairomones, which lead them to their prey (Table 4.3). These are generally species with life cycles closely synchronized with those of their hosts, and which are adapted to host finding during the early phases of the scolytid attack.

Clerid and trogositid beetles are often found preying on aggregating bark beetles, even as they land on a new host, or laying eggs at the mouth of newly formed galleries. Thanasimus dubius is so closely adapted to its host, D. frontalis, that its respone to frontalin occurs with the same diel rhythm (Vite and Williamson 1970), and its peak arrival on trees attacked by D. frontalis is on day four of the attack, one day after the peak arrival of colonizing bark beetles (Dixon and Payne 1979).

The pteromalid parasite, Tomicobia tibialis, must find adult ips so that it can mount and oviposit through their pronota or elytra before they begin to bore into a new host (Bedard 1965). The dipteran, Medetera bistriata, is adapted to oviposit on the bark around freshly formed D. frontalis or I. avulsus galleries, and the newly hatched larvae enter the gallery to prey on the scolytid larvae (Thatcher 1960). However, the strong response of Cheiropachus colon to multilure is a mystery, because it parasitizes late stage S. multistriatus larvae, long after pheromones cease to be produced in nature (Kennedy 1979). One enterprising pteromalid hyperparasite, Cerocephala rufa, uses alpha-multistriatin, an aggregation pheromone of its host's host, as a kairomone to locate the habitat of its host, Dendrosoter protuberans, a primary parasite of S. multistriatus (Kennedy 1979).

If entomophagous insects can utilize their hosts' aggrega- tion pheromones as kairomones, what is their response to antiaggregation pheromones?   Medetera aldrichii, which arrives at host trees too late to respond to the aggregation pheromones of its host, D. pseudotsugae (Fitzgerald and Nagel 1972), were found in reduced densities on felled Douglas-fir trees treated continuously with 3,2-MCH (Furniss et al. 1974).   Enoclerus sphegus, Thanasimus undatulus, Temnochila chlorodia and Coeloides brunneri were not affected by 3,2-MCH.   Quite possibly, M. aldrichii is able to use the antiaggregation pheromone (and aggregation pheromones?) as signals not to respond, thus delaying its response until the development of more mature D. pseudotsugae larvae.   The clerids and trogositids would be expected to ignore antiaggregation pheromone and continue to respond to infested trees as long as aggregation pheromones are released, and there are still many susceptible scolytid prey in the early stage of attack.   This hypothesis is supported by the observation that 3,2-MCH did not deter T. undatulus from responding to frontalin-baited traps (Kline et al. 1974) and T. dubius was not inhibited from landing on trees treated with the antiaggregation pheromones verbenone, exo- and endo-brevicomin mixed, or exo- and endo-brevicomin with verbenone (Richerson and Payne 1979).   However, T. dubius response is inhibited somewhat in the presence of verbenol (Vite and Williamson 1970), and the response of T. chlorodia appears to be interrupted by trans-verbenol with verbenone (Bedard et al. 1980b).   C. brunneri would not be expected to respond to either aggregation or antiaggregation pheromones of its hosts, since it arrives late in the attack and oviposits through the bark onto relatively mature scolytid larvae which it locates by heat perception (Richerson and Borden 1972).

The capture of large numbers of predators on pheromone-baited traps or trap trees is somewhat alarming. In a large trapping program for D. brevicomis, 86,000 T. chlorodia were trapped at a ratio of one predator for every seven bark beetles (Bedard and Wood 1974). Similar figures have been obtained for other predator-prey systems, e.g. one T. undatulus to every two or seven D. rufipennis in two tests (Dyer 1973) or one for every four D. pseudotsugae (Pitman 1973), and one Thanasimus sp. (formicarius or rufipes) to every four I. typographus (Bakke and Kvamme 1978). Thus, there is a great probability of removing many predators in a mass trapping program aimed at reducing bark beetle numbers. Whether or not this removal of predators with their prey would have any effect on the success of the mass trapping operation or on subsequent bark beetle generations has not been investigated.

Host selection sequences by scolytids and by some of their entomophagous enemies are clearly parallel and interlocking systems. They lead not only to the establishment of scolytid beetles in a new host, but also to the establishment of a whole community of species.

ESTABLISHMENT

The selection and concentration phases of host selection serve to aggregate a scolytid population on a host rapidly and at an attack density which will permit successful brood production. The next critical event is the establishment of the parent beetles in their new galleries.

The beetles have only a short time (probably hours) to test the quality of the host, e.g. for adequate moisture, feeding stimulants, and lack of toxic materials. Then within 24 hours (Bhakthan et al. 1970), their flight muscles begin to degenerate, locking them inexorably into association with their host (Chapman 1957, Reid 1958, Henson 1961, Atkins and Farris 1962, Borden and Slater 1969). In I. paraconfusus, the muscle degeneration occurs in both sexes (Borden and Slater 1969, Bhakthan et al. 1970) and is induced by JH (Borden and Slater 1968, Unnithan and Nair 1977), the same hormone that promotes the production of aggregation pheromone. At the same time, as demonstrated for female T. lineatum, JH stimulates oogenesis and maturation of the ovarioles (Fockler and Borden 1973).

A lasting bond usually forms between parent beetles. It is common to find both sexes inhabiting the galleries and engaging in tunnel construction, protection against predators, brood tending, and as for I. pini, frequent copulation (Schmitz 1972). Pheromone-based communication may occur between the parents. However, reproducing I. paraconfusus adults become refractory to aggregation pheromones. Moreover, the atmosphere within the galleries would be quickly saturated with any volatile chemical, soon resulting in sensory adaptation. Therefore, any chemical communication would probably involve low volatility compounds perceived by contact chemoreception.

As brood production proceeds, the flight muscles in I. paraconfusus begin to regenerate through two means, the differentiation of new myoblasts, and the regeneration of the old flight muscle itself (Bhakthan et al. 1971). In most species, the most vigorous parents "re-emerge" to disperse, select new hosts and reproduce (Figure 4.2). Little research has been done on re-emerged beetles, but they have larger flight muscles than the original parents in $\underline{I}$. paraconfusus (Borden and Slater 1969), re-emerged $\underline{D}$. frontalis can produce (Coster 1970) and perceive pheromones (Coulson et al. 1978), and many re-emerged parent $\underline{T}$. lineatum are caught in pheromone-baited traps late in the summer (Lindgren 1980). Therefore, re-emerged parents probably make a significant contribution to future scolytid populations.

The brood beetles, which are much more numerous than re-emerged parents, reinitiate the entire host selection sequence.

5. Relationships between Bark Beetles and Their Natural Enemies

D. L. DAHLSTEN

The coevolutionary relationships between bark beetles and their hosts have been studied from a diversity of perspectives. The purpose of this chapter is to look at the complex of organisms that has co-evolved with the bark beetles and that affects their mortality and natality through parasitoidism, predation or competition. I will describe the relationships between bark beetles and nematodes, mites, parasitoids, predators, and pathogens associated with them; the role of pathogenic microorganisms is described in Chapter 6. While presenting the relationships among species, I will focus upon the potential importance of associated species for the regulation of bark beetle populations, and the development of integrated control strategies. The work done to date on a number of different bark beetle species suggests that the complex of organisms associated with any one of the species are specialized in their habits and have evolved with the beetles over many years.

A salient feature of bark beetle communities is the staggering number of organisms associated with them. Over 70 species of insect natural enemies and associates have been recorded for the western pine beetle, Dendroctonus brevicomis (Dahlsten 1970), and 60 species were associated with the mountain pine beetle, D. ponderosae, attacking sugar pine (Dahlsten and Stephen 1974). Several investigators have studied the associates of the southern pine beetle, D. frontalis, at several locations in the southern United States. Overgaard (1968) found 84 species of insects, representing 42 families, associated with this beetle. Moore (1972) found an additional 21 genera of mites, including 13 species that were found on and around the eggs and larvae of the southern pine beetle. Moser et al. (1971b), working with loblolly pine infested with both D. frontalis and Ips spp., found 96 species of organisms associated with them. Of the 96 species, 29 were identified as potential predators, but the habits of the remaining species were unknown.

European and Asian studies reveal patterns similar to those found in North America. Kozak (1976) found 44 predators and 10 parasitoids of bark beetles in the Volynsk region of the Ukraine (USSR). Nuorteva (1971), working in North Finland with the natural enemies of 34 species of scolytids, described 34 species of Coleoptera (beetles), 16 species of Hymenoptera (wasps), and 14 species of Diptera (flies) associated with these bark beetles. Other workers have described the organisms associated with particular species of bark beetles or with species found in specific regions (Polozhentsev and Kozlov 1975, Carle 1975, Egger 1974). These studies show a remarkable taxonomic similarity among the complexes of bark beetle associates seen in different parts of the world. Although the numbers of insect species in bark beetle communities at first appear to be bewildering, there are patterns to the timing of arrival and in the abundance of the species. Several studies in the United States have focused attention on documenting which species arrive at a tree infested with bark beetles rather than simply observing those which emerge from infested bolts (a log or a section of the tree stem). The arrival studies show that there is a succession of insects arriving on a tree from the time the tree is initially under attack until the brood emerges and that this succession continues until the log is decomposed (Shelford 1913, Savely 1939). Both qualitative and quantitative differences were found in over 100 associates of the early (spring-summer) and late (summer-fall) generations of the western pine beetle (Stephen and Dahlsten 1976). There were definite patterns of arrival in this study that were related to the developmental stage of the western pine beetle. For example, a group of predators known to feed on adult bark beetles arrived while the tree was first under attack, but parasitoids of the bark beetle larvae arrived much later in the successional sequence. The arrival sequence of the more common associates of the western pine beetle is shown for the spring-summer generation (Figure 5.1). Similar arrival patterns were seen in the associates of the southern pine beetle (Camors and Payne 1973, Dixon and Payne 1979) and mountain pine beetle (Edson 1978).

The community beneath the bark has been well described in a taxonomic sense, but the difficulties of conducting studies in the cryptic habitat beneath the bark have limited our knowledge of interactions among species or the role that the organisms play in the dynamics of bark beetle populations. The forest entomologist is dealing with vast acreages of trees that are in mixed age and/or mixed species complexes and, because collecting normally involves taking bolts from trees, trees must be cut or climbed. Once the bolts are

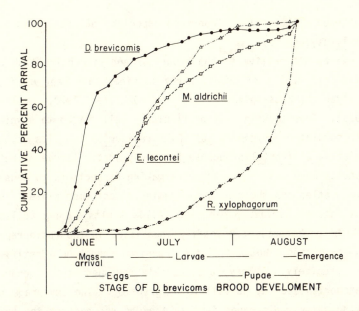

Figure 5.1.   Arrival pattern of the western pine beetle (D. brevicomis) and three of its natural enemies, E. lecontei, a beetle predator, M. aldrichii, a fly predator, and R. xylophagorum, a parasitoid.   The approximate stages of bark beetle development are shown.

obtained they are awkward to handle and much space is necessary for rearing. Furthermore, there are many problems in maintaining the proper conditions for rearing. Dissection or removal of bark may create conditions that are too dry for insects to tolerate.   Dissections of bolts is also a laborious process. And finally, laboratory rearing of bolts from selected trees may create unusual conditions under the bark which may limit the natural occurrence of associates and may produce observations not particularly relevant to natural environments.   The cost and labor involved in ecological studies of bark beetle communities have limited these studies to the economically important species of Dendroctonus, Scolytus, and Ips.

As an alternative to the laborious and disruptive task of dissecting bolts or trees, x-ray machines have been used to analyze bark samples (Stark and Dahlsten 1970, Coulson 1979). This technique is particularly well suited to studies of population dynamics of D. brevicomis and its close relative D. frontalis because the early instar larvae of both species move into the outer bark making dissections to census the beetles and their natural enemies almost impossible. The late instar larvae, pupae, and callow adults (young adults) of the bark beetle are readily identified in the radiographs (x-ray plates), but associated organisms cannot be identified as precisely. They are usually grouped into general categories such as parasitoid or predator. The advantage of using x-ray radiographs is that they do not destroy the bark samples which can then be placed in rearing cartons.

Describing and identifying the associates of bark beetles is a difficult task, but it is even more difficult to understand the interactions among the species in the bark beetle community. Coulson (1979) believes that natural enemies are of great importance but he does not believe that a quantitative assessment of the role of these organisms in the dynamics of bark beetle populations is logistically possible. As will be seen below in the discussion of the different types of natural enemies, some attempts have been made to quantify the relationships, but the logistical problems and the belief of many of the investigators that natural enemies constitute little more than "background noise" explain why these potentially important organisms have not been studied in greater detail.

PARASITOIDS

The term "parasitoid", though not commonly used in the entomological literature, is used here to distinguish insect parasitoids from parasites in the classical zoological sense. Parasitoids kill their host and normally complete development on or within a single organism while parasites usually do not kill their hosts. In addition, the larva and the adult of parasitoids have different feeding habits, the larva feeding on the host insect and the adult feeding on honeydew or floral nectaries. Adult parasitoids may also feed on fluids of potential hosts (host feeding) and often sting many more hosts than they parasitize (Clausen, 1940). Since it is cumbersome to use "parasitoidize", I will use "parasitize", "hyperparasitism", etc. from here on.

All bark beetle parasitoids are wasps (Figure 5.2), members of the insect order Hymenoptera. Most of the parasitoids of bark beetles belong to two families, the Braconidae and Pteromalidae, although two other families, the Eurytomidae and the Torymidae, are also important. Bushing (1965) lists 15 families of parasitoids associated with bark beetles, but because many of the records were established from mass rearings, these records may not be reliable. For example, the encyrtid, Microterys montinus, is listed as a parasitoid of D. rufipennis, but has also been reared from a scale insect as well as a satyrid moth. Proctotrupid wasps that are listed as parasitoids of bark beetles are no doubt parasitoids of cecidomyid or sciarid midges (flies) that are associated with bark beetles, and ichneumonid wasps are parasitoids of weevils, wood borers, and various lepidopterans and probably do not parasitize bark beetles. The braconid, Coeloides pissodes, is a parasitoid of the southern pine beetle, several Ips. spp., and the white pine

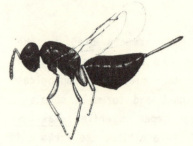

Roptrocerus xylophagorum ♀

Eurytoma conica ♂

Dinotiscus burkei ♀

Eurytoma conica ♀

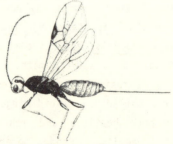

Coeloides sympitys ♀

Figure 5.2.   Parasitoids of the western pine beetle.   (From Stark and Dahlsten 1970).

weevil, Pissodes strobi, a weevil attacking the terminal
leader.   Several   pteromalids   (Dinotischus   dendroctoni,
Cheiropachus quadrum, Rhopalicus pulchripennis, and Heydemia
unica) that are common parasitoids of several scolytids have
also been recorded from various weevils.   That bark beetle
parasitoids also parasitize weevils is not too surprising,
for weevils and bark beetles are very closely related;
however, many weevils occupy very different habitats than do
bark beetles so host switching in some cases would be
considered unusual.   It appears that there are only one or
two host-specific bark beetle parasitoids. Most of them are
polyphagous (attack many species) or oligophagous (attack a
few species).   The host trees are often attacked by two or
more beetle species, so that alternate hosts for the
parasitoids are readily available.   There is, however, some
evidence that two parasitoids, Dinotiscus acuta and
Macromesus americanum, show some degree of host beetle
preference (Ball and Dahlsten 1973).   In this study, the two
parasitoids were more abundant on Ips latidens and
Pityophthorus sp. larvae even though Ips paraconfusus larvae
were more abundant.

In comparison to defoliators or open feeding insects, bark
beetles, like other cryptic insects, have few parasitoids.
For   example,   the   Douglas-fir   tussock   moth,   Orgyia
pseudotsugata, has over 50 parasitoids and hyperparasitoids
associated with it (Torgersen 1978).   By contrast, the
western pine beetle has four, and only two of these,
Dinotiscus burkei and Roptrocerus eylophagorum, are really
common. Moser et al. (1971b) list 9 parasitoid species for
the southern pine beetle. Berisford et al. (1971) found a
total of 12 species of parasitoids associated with 4 species
of Ips. Only 1 parasitoid was found on all 4 species of Ips,
while 3 of them were found on only 1 species of Ips. Bushing

(1965) lists 13 species of parasitoids for D. ponderosae, 10 for D. ponderosae, 10 for D. pseudotsugae and 4 for D. rufipennis. The pattern of low parasitoid numbers is similar for other genera of bark beetles. For example, Berisford (1974) lists only 9 parasitoids for Phloeosinus dentatus. It is obvious that the parasitoid complexes are small. The cryptic habitat certainly confers some degree of protection to bark beetles and it is interesting to note that closely related scolytids that are concealed to an even greater degree than bark beetles have even fewer parasitoids. Bushing (1965) lists no more than one or two parasitoids for several species of Conophthorus, the cone beetles, and for Trypodendron lineatum, an ambrosia beetle. Eichhorn and Graf (1974) could find but two species of parasitoids on two species of Trypodendron and two species of Xyleborus; these ambrosia beetles mine deep into wood. The cryptic habit of bark beetles may also affect the incidence of hyperparasitism, superparasitism, and multiple parasitism. Hyperparasitism, a rare phenomenon (Bushing 1965), is the parasitization of one parasitoid by another. Most of the hyperparasitoids (secondary parasitoids) are facultative, but some are obligatory. For example, there are several records of species of Eurytoma acting as secondaries and parasitizing species of Coeloides, which are well-established primary parasitoids of a number of bark beetle species. Observations of parasitoid behavior on bark have been made at the time when larvae are large and parasitoid activity is at its peak (Dahlsten and Bushing 1970). One parasitoid, Eurytoma conica, would commonly sit near Coeloides sympitys while the Coeloides were ovipositing. Once the Coeloides completed oviposition, the Eurytoma female would move to the oviposition hole in the beetle made by Coeloides and insert her ovipositor. The eggs of Eurytoma were found near the eggs and larvae of Coeloides on the beetle larva; the Eurytoma are literally stealing the host from Coeloides. This is a special case of secondary or hyperparasitism known as cleptoparasitism.

There is very little in the literature regarding superparasitism and multiple parasitism of scolytids. Superparasitism refers to two or more parasitoids of the same species on a single host and multiple parasitism refers to two or more parasitoids of different species on the same host. According to Fiske (1908), superparasitism is less frequent in wood borers than in external feeders, but the lack of records may be due in part to problems with studying cryptic insects as mentioned above. Berisford et al. (1970), in a study of southern Ips in which host larvae were dissected and reared separately, found only one instance of superparasitism; two Roptrocerus xylophagorum adults were reared from one host larvae. One possible case of multiple parasitism is that in which R. eccoptogastri was found in association with Heydenia unica on a single beetle host (Hopkins 1899). These associations are especially difficult to establish since most of the parasitoid species easily become detached from the host larva upon removal for study (Berisford et al. 1970).

Host Finding

Evidence suggests that most parasitoids are attracted to tree odors rather than to the host beetle. By using sticky traps near Ips-infested bolts it was discovered that the most common Ips parasitoid, Rhopalicus pulchripennis, frequented sugar pine commonly, but was rarely caught in the traps around ponderosa pine, suggesting that tree odor is important. Thus, parasitoids attacking bark beetles in one species of tree may not attack the same species of bark beetle on another tree. In a study of the parasitoids of Ips paraconfusus on two tree hosts, the bark beetle was found to have ten parasitoids on sugar pine, but only five on ponderosa pine (Ball and Dahlsten 1973). Coeloides dendroctoni, an important parasitoid of the mountain pine beetle, attacks its host in three species of pine but not in ponderosa pine (DeLeon 1935). In Norway, Pettersen (1976) noted that there were several species of parasitoids more frequently associated with bark beetles in pine than those in spruce. As can be seen from the examples above, the host-finding behavior of bark beetle parasitoids may rely more upon the host tree than the host beetle species. This suggests that different bark beetles attacking the same species of pine are likely to share the same or similar complex of parasitoids.

Parasitoids use, among other cues, tree height and bark thickness to locate appropriate hosts. Bark gets thicker going from the top to the bottom of a tree. Since most bark beetle parasitoids oviposit through the bark, the thickness of the bark and the proximity of bark beetle larvae and pupae to the surface may be critical to oviposition success. It should be noted that on large diameter trees with deep fissures in the bark some parasitoids may go into these fissures for oviposition. Height per se does not appear to be a limiting factor. Bark thickness is also related to bark texture which may itself be a factor in successful host finding. Ball and Dahlsten (1973) found that Rhopalicus pulchripennis was only found parasitizing Ips where there was smooth bark and that these parasitoids were more efficient on sugar pine due to the structure of its bark. It was suggested, but not proven, that differences in the levels of parasitism in the tree for all parasitoids ovipositing through the bark were due to bark thickness. Berisford et al. (1971) found parasitism to be highest in the upper part of the bole, but could not correlate this with bark thickness. On the other hand, Moore (1972), working with southern pine beetle parasitoids, claims that bark thickness is not a factor. It has been noted by others that parasitoid activity on Ips paraconfusus (Ball and Dahlsten 1973), on D. brevicomis (Dahlsten, unpublished data) and D. ponderosa (Dahlsten and Stephen 1974), was greater in the upper portions of the tree where the bark is thinner.

In addition to factors associated with the tree, parasitoids also use bark beetle pheromones (Chap. 4) to locate the host beetle. The wasp, Tomicobia tibialis, is a parasitoid of adult Ips and has been shown to be attracted to their pheromones (Bedard 1965, Rice 1969). A pheromone, when used by another species for its benefit, is referred to as a kairomone. Parasitoids of bark beetle larvae are also attracted to bark beetle pheromones, as has been shown with Scolytus multistriatus (Kennedy 1979). The constituents of multilure (the S. multistriatus attractant), either alone or in combination, functioned as kairomones for three larval parasitoids (Cheiropachus quadrum, Dendrosoter protuberans, and Entedon leucogramma) and for a hyperparasite of D. protuberans, Cerocephala rufa. Since most larval and pupal parasitoids, as well as hyperparasitoids, parasitize later developmental stages of the bark beetle, the general lack of attraction during the mass attack of the beetle is not perplexing.

Parasitoids beneath the bark are in a protected environment, and so it is not surprising that the larval and pupal parasitoids are ectoparasitoids (they feed on the outside of the host insect). Almost all adult parasitoids oviposit through the bark, so it is likely that the parasitoid eggs, in many cases are not placed directly on a host beetle larva or pupa, but rather that the early instar larvae of the parasitoid seek out the host. Most bark beetles have clumped distributions so that parasitoid larvae in the general vicinity of the beetle larvae would have an excellent chance of finding a host.

Parasitoid Biologies

The biology of parasitoids in the genus Coeloides has been studied more than any other of the parasitoids of bark beetles and has been reviewed by Bushing (1965). Most species of Coeloides larvae remain attached to their bark beetle host. Berisford et al. (1970), in a study of Ips parasitoids, found C. pissodes usually remains attached to the host while many of the other Ips parasitoids were found unattached to a host larva or pupa. Some, like Roptrocerus xylophagorum, were found considerable distances from any potential host. Unattached larvae are probably capable of moving between hosts.

The only parasitoids known to enter the galleries of the beetles to oviposit are the torymid wasps Roptrocerus spp. (Dahlsten and Bushing 1970, Berisford et al. 1970, Reid 1957) and the eulophid wasp Entedon leucogramma (Beaver 1966). Reid observed a R. xylophagorum female ovipositing through an egg-niche plug into a larval gallery of the mountain pine beetle. When the bark is thin these parasitoids will also oviposit through the bark (Gyorfi 1952).

Coeloides may use either vibration or temperature to locate hosts. Ryan (1961) found that scratching the underside of the bark would cause female Coeloides brunneri (now C. vancouverensis) to oviposit through the bark. Richerson and Borden (1972) showed that heat perception was important with this parasitoid and that a 2-3°C temperature difference on the bark surface resulted from beetle metabolic heat. In this study, a small heating element in the bark of a tree elicited oviposition.

Bark provides adequate insulation from the cold for the overwintering stages of the beetle as well as for their parasitoids during most winters. Bark beetle parasitoids overwinter as late larva, prepupae, pupae or adults. Diapause is known to occur in each generation of several Coeloides species and this would provide additional protection from the cold. Although not known for other braconid and chalcidoid parasitoids of bark beetles, diapause is not doubt common in temperate regions for these organisms. A study in California on the effect of an abnormal cold spell in December 1972 on survival of various stages of the western pine beetle and its natural enemies showed that all stages of overwintering brood of the beetle were killed but the parasitoids were unaffected (Dahlsten et al. unpublished).

Different species of parasitoids are partitioned along a gradient of bark thickness and texture. On sugar pine, Rhopalicus pulchripennis occurs in the upper infested portions of the bole and in smaller diameter trees where the bark is smooth and thin (see above), while Coeloides brunneri (now C. sympitys) dominates in the lower bole and in larger diameter trees (Ball and Dahlsten 1973). Coeloides has a longer ovipositor and is capable of working in the bark fissures. Roptrocerus xylophagorum was evenly distributed over the diameter classes except it was absent in the smallest one (Figure 5.3). Berisford et al. (1970) also noted Roptrocerus to be abundant over the entire bole. This suggests that Roptrocerus and Coeloides do not compete well with Rhopalicus in thin, smooth bark.

All stages of bark beetles are attacked by parasitoids. Perhaps the most interesting, but the least studied, are those attacking the adults and the eggs.   The pteromalid

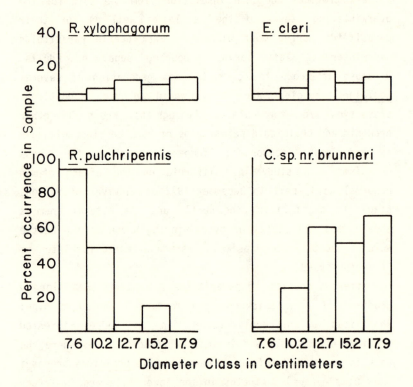

Figure 5.3.   Effect of tree diameter on the distribution of four parasitoid species attacking Ips paraconfusus in sugar pine.   (From Ball and Dahlsten 1973).

wasp, Tomicobia tibialis, and the braconid wasp, Cruptoxilos convergens (Bushing 1965) attack adults and Berisford et al. (1970) observed T. tibialis attacking Ips spp. before the adult beetles entered the tree. They found that parasitized females that were capable of making only short galleries and laying a few eggs. T. tibialis oviposits through the elytral suture and the adult bores out through the elytral declivity, leaving a round hole (Reid 1957). The braconid, C. convergens, consumes the contents of the adult, emerges from a small hole near the end of the elytra, and spins a cocoon in the beetle gallery (Muesebeck 1936). All adult parasitoids are internal.

The fact that most beetles lay their eggs individually in a niche that is covered with a frass plug probably helps to explain why egg parasitoids are not common. Egg parasitoids develop inside the host egg and so are internal parasitoids. Bushing (1965) lists two mymarid species as parasitoids of eggs, but the records are questionable since there is little biological information and no host associations. However, a species of Trichogramma has been reared from three bark beetle species occurring on ash in Poland (Michalski and Seniczak 1974). T. semblidis parasitized 10-15 percent of the bark beetles in one field study and some individual galleries suffered up to 98% egg parasitization. This is the only documented case of egg parasitism of bark beetles in the literature. Beaver (1966) found that Entedon leucogramma parasitized the eggs of Scolytus scolytus but that the parasitoid larvae did not develop until the beetle larvae were well grown.

Bark beetle larvae are the stage most commonly parasitized and species that parasitize the larvae will also parasitize pupae (Bushing 1965). In a study of Ips parasitoids, many parasitoids were found in the pupal chambers and the

parasitoid   larva   consumed   the   beetle   pupa   before
sclerotization or hardening had commenced (Berisford et al.
1970).

Impact on Bark Beetles

In most cases there is not much evidence that any of the
parasitoids respond to changes in beetle density.   Since
parasitism is normally very low, most workers feel that
parasitoids play little role in the dynamics of bark beetle
populations.   For example, Hetrick (1933) concluded that
parasitoids were relatively unimportant in control of the
southern pine beetle.   High rates of parasitism have been
observed on occasion but few workers have been able to
determine if change in parasitization rates were due to
changes in host density.   Moore (1972) found parasitism of
the southern pine beetle to average four percent, but found
rates as high as 25-30 percent on individual trees.   The
parasitoids of the western pine beetle cause only a small
fraction of the total mortality (Figure 5.4) and there is no
evidence that they respond to host density (Dahlsten and
Bushing 1970).   The braconid, Coeloides dendroctoni, a
parasitoid of the mountain pine beetle, parasitized an
average of 4 to 32 percent of the brood and some reached
rates as high as 98 percent, but no mention was made of
responses to host density (DeLeon 1935).   With several
species of Ips, parasitoid abundance generally was not
related to host density (Dahlsten and Bushing 1970, Ball and
Dahlsten 1973); however, one of the parasitoids, Roptrocerus
xylophagorum, was a potentially effective regulator since it
responded quickly to changes in host numbers (Berisford et

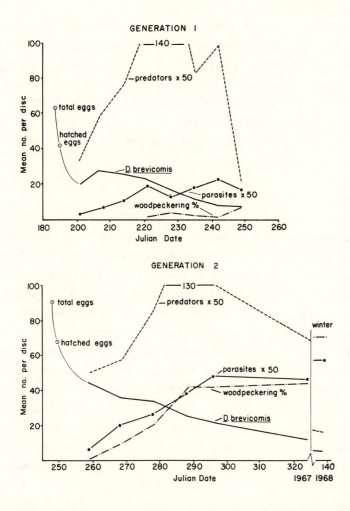

Figure 5.4.   Relative abundance of western pine beetle and its natural enemies for an early (spring generation) and late (overwintering generation) in California.

al. 1970). Beaver (1967) considered host density in his
studies in Great Britain and found a density-dependent
response of the ectoparasitoids to the host, Scolytus
scolytus. Studies of host density effects on parasitism are
made difficult because of the cryptic habitat of the bark
beetle, the thickness of bark, the height of trees, and the
presence of several species of bark beetles in one tree.

Parasitoids for Bark Beetle Control

Several types of pest management strategies using parasitoids
have been suggested or attempted. As early as the 1930's
there were recommendations to conserve parasitoids of bark
beetles. DeLeon (1935) found that Coeloides dendroctoni
parasitizing mountain pine beetle in lodgepole pine are
always in higher numbers in the areas of the oldest
infestations. The bark beetles gradually leave these areas
and leave the parasitoids behind. Thus, if direct control
procedures were to be implemented, it would be simplest and
most efficient to control beetles in newly infested areas but
not the old. Bedard (1933) found that many wind-thrown trees
have high populations of parasitoids and suggested not
disturbing these trees. Ryan and Rudinsky (1962) recommended
leaving small-diameter trees infested with Douglas-fir beetle
since Coeloides brunneri parasitization was higher in
thin-barked trees. Kozak (1976) found parasitoids to be more
common in mixed stands in the Ukraine, indicating that
silvicultural techniques might be feasible to encourage bark
beetle parasitoids. In this case, the forest manager would
manage for mixed species stands rather than single species
stands as a protective measure.

The classical approach to biological control of bark
beetles has hardly been attempted, but given the limited

efficacy of parasitoids in the studies cited above, this is not surprising. In Canada, Rhopalicus tutela was imported for release against the eastern spruce beetle, Dendroctonus piceaperda (Baird 1938). Most biological control attempts have focused on the smaller European elm bark beetle, Scolytus multistriatus, an introduced species that vectors the Dutch elm disease. One of its parasitoids, a European braconid, Dendrosoter protuberans, has been released at several locations in the United States (Kennedy 1970). The results of these releases have not been promising but releasing natural enemies for control of insects that vector diseases may be a questionable practice to begin with. The low parasitization rates observed in most studies are not encouraging since a high rate of vector control is needed to control a disease organism. For example, Schroeder (1974), in a study of parasitoids of Scolytus multistriatus and S. scolytus in Austria, found that 30 percent of the total mortality occurred during the third to fifth instars and the 7 species of ectoparasitoids were responsible for most of this mortality. Even though he found such a relatively high rate of parasitism, he concluded that the biological control of the elm bark beetle was unlikely to have much effect on the spread of Dutch elm disease. The potential for biological control of S. multistriatus has been reviewed by Peacock (1975). It has been suggested by Kennedy (1979) that since some of the parasitoids of S. multistriatus respond to the beetle pheromone, the attractants could be used to encourage parasitism in certain areas, but the use of parasitoids of bark beetles has not been studied further. The role of parasitoids has not been studied thoroughly, and characteristically low levels of parasitism do not negate the possibility of some response to increasing host density.

DISEASES

The role of pathogens in the dynamics of bark beetle populations is not well known and has not been studied to the extent that parasitoids and predators have. Bark beetles have a number of different relationships with microorganisms, several of which are discussed in Chapter 4. For those microorganisms that are pathogenic to bark beetles, little has been done beyond their isolation and identification. It is certain that the presence of these organisms is commonly overlooked. The discovery of diseased beetles would require careful dissection of infested bolts.

In North America, most of the work on bark beetle diseases has been done with the southern pine beetle. Moore (1971) studied the pathogenic fungi and bacteria of the southern pine beetle and found that these organisms caused about 20 percent mortality. This is high relative to many of the other mortality factors and is evidence that diseases may be more effective than previously thought. Sikorowski et al. (1979) made an intensive study of the diseases of the southern pine beetle in Mississippi and Alabama and found 22 percent mortality over a two-year period caused by a variety of organisms. The most common pathogen was a microsporidian, Unikaryon minutum.

Purrini (1978b) studied eight scolytid species in the Bavarian Alps and found four infected with protozoans; in addition, Purrini lists 16 species of Protozoa that are known as pathogens of scolytids. In the same area, three percent of the larvae and adults of a spruce bark beetle, Dryocoetes autographus, were found to be infected with a protozoan, Malameba locustae. This protozoan has ten other hosts which are mainly grasshoppers and locusts (Purrini 1978a).

In Yugoslavia, the midguts of 60% of the adults of S. scolytus were found to be infected with a microsporidian, Stempellia scolyti, but the larvae collected at the same time were not infected (Purrini 1975). This worker also reviewed the literature on Microsporidia infecting S. scolytus in other countries. Purrini (1975) concluded that diseased beetles appeared to be less resistant than healthy ones to other environmental stress factors.

The common insect fungus, Beauveria bassiana, was occasionally isolated from body surfaces of Scolytus multistriatus in Great Britain, and the fungus was shown to kill the larvae of S. scolytus in the laboratory (Barson 1977). Another fungus of second instar and older larvae of S. scolytus, Verticillium lecanii, was frequently isolated from field-collected material but pathogenic organisms were found to kill less than 7.5 percent of the beetle population (Barson 1976).

An attempt at biological control with a disease was made in Russia with a bark beetle, Dendroctonus micans (Kurashvili et al. 1974). In this test, Boverin, which is a preparation of the fungus, Beauveria bassiana, was dusted on the bark of infested trees. These workers recorded reductions of all stages of D. micans of up to 40 percent and concluded that this fungus could play a positive role in reducing numbers of D. micans if used in conjunction with other means of control. Other than this study, very little has been done with pathogens as a means of control. Since these organisms presumably have little in the way of side effects in the environment, there may be a great potential for their inclusion in control programs (Kurashvili et al. 1974).

PREDATORS

There are a number of invertebrates and some vertebrate predators of bark beetles. These will be discussed by their taxonomic groupings. The knowledge of predation of bark beetles is almost entirely from those studies of predators occurring beneath the bark or on the surface of the bark of bark beetle-infested trees. Many insects are associated with bark beetle galleries, but for many of these, we know little about their feeding habits or prey. In-flight mortality of bark beetle adults is greatly in need of study; we can only speculate about sources of mortality from the time the adults emerge from a tree until they attack another tree. The list of predators would probably be much larger if the predators of the exposed adults were known in their entirety.

Insects

There are a number of insects known to prey on bark beetles. Some predators feed on bark beetles both as adults and larvae, while others only feed on bark beetles either in the adult or larval stage. Many will feed on the surface of the bark as beetles attack or emerge while others feed on the larvae beneath the bark. The majority of the insect predators are in the order Coleoptera, but others are in the order Diptera and the order Hemiptera. Typical beetle predators are shown in Figure 5.5.

Moser et al. (1971b) speculated that as many as 29 species of insects were predators of the southern pine beetle and of several species of associated Ips in loblolly pine in southern United States. They showed that the seasonal abundance of most predators paralleled that of the southern

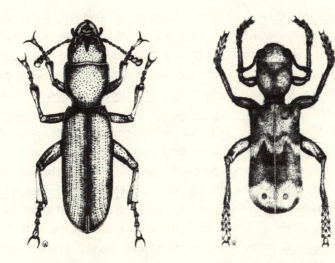

Temnochila chlorodia   Enoclerus leconti

Figure 5.5.   Common predators of the western pine beetle.
Temnochila chlorodia (blue-green trogositid) and Enoclerus
leconti (checkered beetles or clerids).

pine beetle.   In another study of the southern pine beetle,
only seven predators were listed but Aulonium spp. and
Lasconotus spp. (both colyiids or cylindrical bark beetles)
were listed as scavengers (Overgaard 1968).   Others have felt
that species in these two genera were predators of bark
beetles (Dahlsten 1970, Moser et al. 1971b).   In England,
Aulonium trisulcum was found to prey on Scolytus
multristriatus and S. scolytus (Allen 1976).   Larval
Lasconotus subcostulatus feed on fungi until the third instar
at which time they begin feeding on Ips larvae and on pupae
(Hackwell 1973).

Sequence of Arrival. The sequence of arrival of insect predators of western pine beetle has been studied in detail (Stephen and Dahlsten 1976). In this study, four predators arrived as soon as the tree was under attack. They were the clerid or checkered beetle, Enoclerus lecontei, the trogostid, Temnochila chlorodia, the colydiid, Aulonium longum (Figure 5.6), and the dolichopodid, or long-legged fly, Medetera aldrichii. These insects were undoubtedly responding to the bark beetle pheromone. Two beetles in the family Histeridae, or hister beetles, Platysoma punctigerum and Plegaderus nitidus, arrived after the initial four and fed on western pine beetle eggs and larvae. The colydiid Lasconotus subconstulatus, as mentioned above, arrived much later but fed on later instar bark beetle larvae. The cues used by the later arriving species are not known but many workers have established that early arriving predators rely on kairomones to find their hosts. Most of these species feed as adults on bark beetle adults as they are attacking trees, and then their larval progeny feed on the immature stages of bark beetles. Five predators of the southern pine beetle were caught during the initial stages of beetle attack and it was speculated that odors produced by the beetle attracted the predators (Camors and Payne 1973).

Host Finding. Some interesting studies have been made of the kairomone response of bark beetle predators and this seems to have evolved in many bark beetle communities. In Norway, logs baited with the pheromone of Ips typographus and then sprayed with Lindane (an insecticide applied to kill the bark beetles) attracted 17 times as many adults of both sexes of the clerids Thanismus formicarius and T. femoralis as did logs without pheromone. A similar response to an aggregation pheromone was shown by Stigmatium nakanei, a predator of the

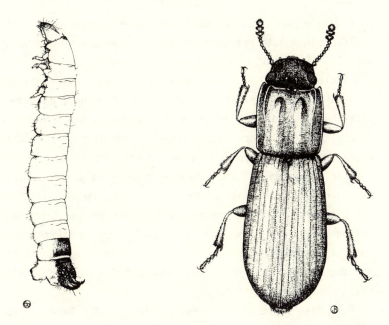

Figure 5.6.   Aulonium longum LeConte, a common predaceous associate of pine-infesting bark beetles.   Top: adult beetle. Bottom: last-instar larva. (From Stark and Dahlsten 1970).

minute pine bark beetle, Taenioglyptes fulvus, in Japan (Sasakawa et al. 1976).   In California, the trogostid, Temnochila chlorodia, a predator of the western pine beetle was attracted by the bark beetle pheromone (Bedard et al. 1969). Three common predators of Ips typographus, Thanisimus formicarius, Medetera signaticornis, and Epuraea pygmaea were attracted to terpenes and a male-produced pheromone in Europe (Rudinsky et al. 1971). Predator response to pheromones of bark beetle has been well established but it is not known how specific these responses are.

In comparison to the parasitoids of bark beetles, the predators feed on a wider range of prey.   Although the parasitoids apparently move about beneath the bark to some degree (Berisford et al. 1970), most of the predators are

more mobile and feed on many more prey during their life
cycle. Often both predator adults and larvae feed on one or
more life stages of bark beetles.

Little is known of the strategies used by these predators
to locate their prey. Since the bark beetles have a clumped
distribution and the predators are quite mobile, they may
locate prey by chance encounters. Since they are general
feeders, host synchrony does not appear to be a problem.
Some of the predator larvae such as Medetera spp. are capable
of moving between the xylem and phloem-cambial interface
while others, like the anthocorids (Hemiptera or true bugs
called minute pirate bugs), Scoloposcelis spp. and Lyctocoris
spp., are commonly found in the parent adult or egg galleries
and are thought to prey on eggs and early instar larvae with
their sucking mouth parts.

Some insect predators overwinter beneath the bark in late
larval to adult stages. DeMars et al. (1970) showed that
Enoclerus lecontei, an important predator of the western pine
beetle, moved down the bole of infested trees with the onset
of winter and that some of the adults overwintered in the
duff beneath infested trees.

Predator Impact. In almost all cases, predators are more
common than parasitoids. This has been shown with the
western pine beetle (Dahlsten 1970), southern pine beetle
(Moore 1972), and mountain pine beetle (DeLeon 1934). Moore
(1972) found clerids to be the most abundant of all natural
enemies of the southern pine beetle, taking 12.8 percent of
the brood. He also stated that the abundance of the clerid,
Thanasimus, appeared to be density-dependent, for clerid
abundance in the bole of the tree increased with southern
pine beetle density. Thatcher and Pickard (1966) estimated,
from clerid feeding studies, that 15 clerids per 100 southern

pine beetles would suppress a bark beetle population. Berryman (1966) estimated that each Enoclerus lecontei larva could consume seven prey during its larval development, and that the adult consumed 44 to 158 adult bark beetles during its life. Person (1940), in his studies with the same insect, stated that more than 25 beetles are eaten by each adult clerid and at least an equal number of beetles are eaten by clerid larvae. Person also found that western pine beetles were three times as abundant in logs from which clerids were excluded.

Insect Predators for Bark Beetle Control. Several suggestions have appeared in the literature regarding management practices to conserve the numbers of bark beetle predators. Williamson and Vite (1971) warned against the use of insecticides, for the chemicals kill predators of the southern pine beetle as well as the beetle. Berryman (1967) recommended that the bottom section of trees infested with western pine beetle not be treated chemically because clerids move to the lower portion of the tree. Moore (1972) found that clerids emerged later than southern pine beetles and recommended that infested trees not be removed until after the clerids had emerged. Since several of the predators are attracted to bark beetle pheromones it is conceivable that kairomones could be used to enhance predator populations. An attempt to use the beetles, Rhizophagus spp., for control of the bark beetle, Hylastes ater, was unsuccessful and the effectiveness of Thanasimus formicarious is as yet to be evaluated on the same bark beetle in New Zealand (Milligan 1978). The clerid, T. formicarious, has been reared and released in Germany to control two bark beetles and an ambrosia beetle (Reisch 1975), but it is not exactly clear how effective this program has been.   As with the parasitoids, relatively little has been done in the way of direct or indirect manipulation of bark beetle predators.

Spiders

Spiders are common on the bark of trees including those that are under attack by bark beetles. Spider predation on the surface of the bark has not been quantified, but Jennings and Pace (1975) observed both a hunting spider, Oxyopes scalaris, and a web-building spider, Theridion goodnighttorum, feeding on Ips pini adults. Spiders are extremely abundant in forest environments and their role in forest ecosystems has long been ignored; it is likely that they may be important in bark beetle population dynamics.

Mites

Organisms in the arthropod order Acarina are extremely common beneath the bark and many species are associated with bark beetles. These minute creatures have a diversity of habits and some are predators of the various life stages of bark beetles. Several scientists have made important contributions to understanding the biology of the mites associated with bark beetles (Lindquist 1969, Kinn 1970, 1971, Moser and Roton 1971).

Mites are phoretic and this is the means by which these animals get to their bark beetle hosts. Phoresy is a type of interrelationship between organisms in which a smaller organism is carried on the body of a larger animal, but the former does not feed on the latter. Phoresy is extremely common in many groups of animals and has been reviewed by Wilson (1980). In one study (Moser et al. 1971a), a mite, Pyemotes parviscolyti, was found to be phoretic only on the bark beetle Pityophthorus bisculcatus, indicating that this phoretic relationship may be quite specific. They found that the mite attacked all stages of this beetle except the adult.

Kinn (1970) found mites associated with the western pine beetle, and Moser and Roton (1971) found 96 mite species associated with the southern pine beetle and allied scolytids on Pinus taeda in Louisiana, and five of these 96 species were known predators of bark beetles (three Iponemus species, Pyemotes paraviscolyti, and Eugamasus lypriformis). Moser et al. (1974) found 57 species of mites in association with the southern pine beetle on pines in Honduras, Guatemala, and Mexico. Eleven of the mites were known to attack D. frontalis in Louisiana and at least 13 others were thought to be natural enemies.

Not all of the mites are predators of various bark beetle stages; mites have many different feeding habits (Lindquist 1969). The mites in the genus Iponemus attack the eggs of bark beetles. Lindquist (1969) found 14 of 16 species to have a specific relationship with their carrier. The female of these mites feeds on a single bark beetle egg and this one egg provides sufficient nourishment for both the adult and her progeny. Kinn (1971) studied a phoretic mite, Cercoleipus coelonotus, of two Ips species but this mite normally did not prey on any life stage of the bark beetles. These mites dismount as soon as the bark beetles reach the host tree, wander over the bark surface, and then enter the bark beetle galleries. The adults prefer to feed on the free-living instars of Contortylenchus elongatus, as endoparasitic nematode of Ips paraconfusus. All stages feed on nematodes (roundworms) in the family Diplogasteridae. In addition, the adult mite also feeds on a mite predator, Digamasellus quadrisetus, of the eggs and larvae of the bark beetle.

There is much to be learned of the biology of the many species of mites in association with bark beetles and some of the relationships are extremely complicated, as can be seen from the discussion above. None of the mites have been used in any biological control attempts, but Moser and Roton (1971) have been evaluating species of mites for this purpose. Moser et al. (1978) evaluated the potential of a mite, Pyemotes dryas, from Poland for its potential as a biological control agent. This mite is phoretic on and attacks a wide range of European bark beetles that infest conifers. The mite readily consumed the brood of the southern pine beetle, but it was found not to be phoretic on the southern pine beetle or six associated bark beetle species. The introduction of a phoretic mite poses an additional problem since a mite must be found that will both prey on the target pest and get from tree to tree. Therefore, a phoretic relationship must be established for there to be a successful introduction. Clearly, much work remains to be done on the ecology of the predaceous mites before any operational biological control programs can be implemented.

Nematodes

The nematodes, or roundworms, are potentially important natural enemies of bark beetles (Massey 1966, and Poiner 1975). Nickle (1978) discussed the genera of round worms, Parasitylenchus and Contortylenchus, which parasitize bark beetles and reduce or suppress beetle egg production. Gurando and Tsarichkova (1974) reviewed the importance of nematodes as parasites of bark beetles and then reported on their investigation of three parasitic nematodes of Tomicus minor. This beetle is a pest of pines in the Ukraine, and the authors discussed how Allantonematus sp., Parasitaphelenchus sp., and Parasitylenchus sp. enter the tissues of the ovaries of the female beetles and partly or completely destroy these organs. Ashraf and Berryman (1970) have described this phenomenon with Scolytus ventralis and in this case both males and females are sterilized by nematode activity. Parasitism of Conophthorus monophyllae, a scolytid cone beetle, by Neoparasitylenchus amvlocerus caused a reduction in fat body, ovaries, and length of life (Poinar and Caylor 1974). The biology of Contortylenchus pseudodiplogaster, a parasite of Ips sexdentatus, has been reviewed by Vosylyte (1978) and this author considers nematodes in the genus Contortylenchus to be highly effective parasites of bark beetles.

The activity of some nematodes does not inhibit the ability of certain bark beetles to produce offspring so, presumably, the effects are more subtle. In these cases, one might expect the parasitized bark beetles to be more susceptible to other biotic and abiotic stresses. In eastern Texas, three species of Ips were found to contain four specific internal nematodes (Hoffard and Coster 1976). In all three species of Ips the infestation was at its peak in July and August, when 50 to 58 percent of the adult beetles in naturally attacked pines contained nematodes. The nematode-infected beetles were delayed in their emergence in some cases, but nematodes infecting Ips grandicolis did not diminish either the beetle's ability to construct egg galleries or the number of offspring produced. Similar results were noted in England with Scolytus scolytus and S. multistriatus, where high levels of parasitism (44 to 61 percent) by Parasitaphelenchus oldhami were found but no pathological effects on the beetles were documented (Hunt and Hague 1974).

A strain of Neoaplectana (referred to as DD-136) was applied as a spray of 5000 nematodes per square meter to control Scolytus scolytus in England but was found to be insufficient to reduce the overwintering population of the bark beetle (Finney and Walker 1979). Although this particular attempt was a failure, some authors believe that nematodes possess great potential for bark beetle control; the situation is the same as with all natural enemies of bark beetles in that considerable biological and ecological background work must be done before their full potential as biological control agents can be realized.

Birds

The role of birds in the dynamics of forest insect populations has been considered by many, and it is generally accepted that birds may have an impact on sparse insect populations, but as insect numbers increase the impact of avian predation becomes less and less important. A good review of the role of birds in forest ecosystems has recently been published (Dickson et al. 1979).

Various birds feed on bark beetle stages occurring beneath the bark and most studies have concentrated on this phase of avian predation. Since it is difficult to document predation of beetles in flight or beetles that are attacking or emerging on the surface of the bark, researchers have only estimated the in-flight mortality due to avian predators. Stallcup (1963) claims that eight species of birds prey on the adult spruce beetle, Dendroctonus rufipennis, when it is flying and that this reduces the population by approximately ten percent. Otvos (1969) estimated in-flight mortality due to avian predators of western pine beetles ranges between 8 and 26 percent.

Woodpeckers have received the most attention from researchers studying the dynamics of bark beetle populations. This is undoubtedly because they have the greatest impact on bark beetle populations, but it is also because it is easy to document woodpecker activity. Woodpeckers prey on adults and larvae, leaving telltale signs of their activity in the bark of trees, while those birds that prey on flying beetles leave no trace of their feeding. Most of the work has been concentrated on the Dendroctonus species, presumably because of the pest status of several of these bark beetles. The most important woodpeckers on the spruce beetle are the hairy (Picoides villosus), downy (P. pubescens), and northern three-toed woodpecker (P. tridactylus). Koplin (1969) recorded a 50-fold increase in woodpecker density in fire-damaged trees attacked by beetles and an 85-fold increase in woodpecker density in epidemic populations versus endemic populations. The woodpeckers consume 20 to 30 percent of the beetles, and in epidemic populations they may consume up to 98 percent of the beetles in standing trees (Baldwin 1960, Koplin and Baldwin 1970). Stand density may also influence the extent of predation; open and semiopen areas had much higher rates of predation (71 to 83 percent of the beetles) than dense areas (52 percent) (Shook and Baldwin 1970).

The most important woodpeckers consuming southern pine beetles are the hairy, downy, and pileated woodpeckers (Kroll and Fleet 1979). Exclusion studies showed that woodpeckers had a significant impact on southern pine beetle pupae and brood adults, especially at midbole where the beetle populations were highest. Moore (1972) estimated that woodpeckers (primarily downy) caused an average reduction of 5 percent in southern pine beetle populations in North Carolina, but the reduction was as great as 50 percent on individual trees.

There are four woodpeckers that feed commonly on the western pine beetle. In addition to the hairy, downy, and pileated (Dryocopus pileatus) woodpeckers, there is the white-headed woodpecker (P. albolarvatus) (Otvos 1965). Otvos (1965) found these woodpeckers to cause 32 percent mortality during the early part of an outbreak. Woodpeckers at first increased gradually in western pine beetle-infested areas due to immigration, but later, the bark beetle-killed trees increased the availability of nesting and roosting sites (Otvos 1969). In one area where the old dead trees were removed, woodpecker populations declined considerably (Otvos 1969). With the woodpeckers and other cavity-nesters, food as well as nesting sites are limiting.

There are some interesting side effects due to woodpecker feeding since these birds puncture, flake, or drill the bark in their search for food. This can affect the survival of the bark beetles that remain in the tree or are flaked off in bark chips. With the western pine beetle, approximately 58 percent of the larvae are located in the portion of the bark removed by the woodpeckers, and the woodpeckers remove the bark to a more or less uniform thickness (Otvos 1970) (Figure 5.7). The insect inclusions in the bark flakes were few but included predators and parasitoids in addition to bark beetles. Hardly any of these insect inclusions emerged if they spent the winter on the ground; however, if the chips were flaked off during the first generation (early summer) of bark beetles, the insects in the chips on the ground had little difficulty surviving (Otvos 1965). Woodpecker activity has been shown to be greater on those trees in which bark beetles overwinter than those attacked in early summer (Figure 5.4). Kroll and Fleet (1979) found woodpeckers to dislodge large numbers of bark flakes containing southern pine beetle brood and recorded only a 20 percent survival of those insects dislodged.

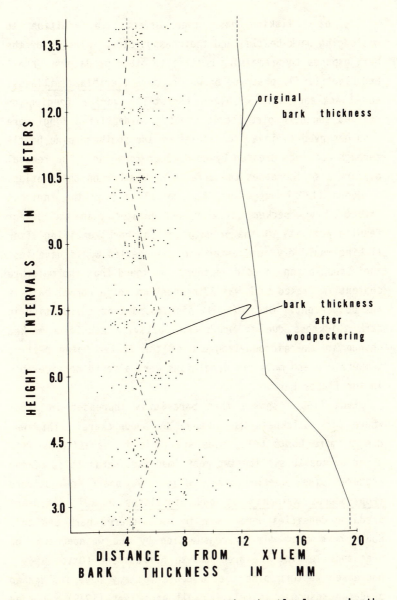

Figure 5.7.   Locations of western pine beetle larvae in the bark, showing effects of bark thinning by woodpeckers. (From Otvos 1970).

The bark flaking has other effects in addition to dislodging bark beetles and their associates. Flaking of the bark exposes the remaining beetles to other predators. Kroll and Fleet (1979) observed brown creepers (Certhia familiarus) and black-and-white warblers (Mniotilta varia) feeding on exposed southern pine beetle brood. Unidentified fungi were also observed to invade galleries of the southern pine beetle through openings created by woodpecker foraging. The removal of bark also increases the effects of weather on the beetles.

Moore (1972) suggested that mortality from the indirect effect of woodpecker activity was greater than the direct feeding activity of the predators. The bark remaining after flaking would dry out faster and the beetles would have lost their insulation in cold weather. He found that the moisture content in flaked bark was 33 percent while in normal bark it was 60 percent. Otvos (1969) also found that a reduction in bark thickness due to woodpecker activity caused a marked change in the microenvironment of the western pine beetle; temperature and moisture conditions were altered considerably in the flaked bark.

Otvos (1965) showed that parasitism increased in areas where bark thickness was reduced by woodpeckers. This was due to more hosts being made available to parasitoids with short ovipositors. The two most important parasitoids of the western pine beetle have relatively short ovipositors (Roptrocerus xylophagorum and Dinotiscus burkei). Insect predator densities were lower in woodpeckered bark and this could have been due to consumption by the woodpeckers, or predators leaving the area due to harrassment (Otvos 1969). Decreased western pine beetle densities may have influenced predator response as well. Kroll and Fleet (1979) recorded increases in both parasitoids and predators due to woodpecker activity on the southern pine beetle.

From the above it can be seen that birds, and in particular woodpeckers, are capable of causing considerable mortality of bark beetles both directly and indirectly. Management programs for increasing birds have not been proposed in the United States but they are common in Europe. Otvos (1969) had evidence to show that the removal of snags caused a reduction in woodpecker populations. It is conceivable that a woodpecker encouragement program could be successful either by leaving dead trees for roosting or nesting or by providing artificial nesting sites (boxes). The woodpeckers appear to be important enough that they should be at least considered when pest or forest management strategies are contemplated.

Lizards

Otvos (1977) recorded several species of lizards feeding on adult bark beetles on the surface of the bark. The proportion of beetles removed in all cases is surely very small. Since not much is known about mortality of bark beetles on the surface of the bark, it is certain that there are other invertebrate and vertebrate predators that have not been mentioned in the discussion above.

COMPETITION

Since bark beetles attack their hosts in large numbers and since there are many other phloem-feeding insects attracted to bark beetle-infested trees, intra- and interspecific competition are potentially important mortality factors.

Individual trees are commonly infested by several bark beetle species and often these beetles partition the resource. However, the species often overlap and therefore there is competition for food and space. For example, the western pine beetle commonly attacks ponderosa pine at mid-bole and shortly after, Ips spp. attack the top of the tree and the red turpentine beetles, Dendroctonus valens, fill in at the base. The western pine beetle galleries overlap with galleries made by Ips at the top and galleries made by the red turpentine beetle at the bottom of the tree. Oftentimes a tree will have what the entomologists called mixed broods. In this instance, a ponderosa pine will have both the western pine beetle and mountain pine beetle galleries mixed together in the mid portion of the tree.

Perhaps more important competitors for the phloem tissue are the large wood borers in the families Buprestidae (flatheaded borers) and Cerambycidae (roundheaded borers). These insects commonly lay their eggs on the outer bark of attacked and dying trees. The larvae hatch and work their way through the bark into the phloem tissue. The larvae may make irregular winding galleries through the phloem and since the larvae may be as long as 3 cm, they are capable of destroying large amounts of potential bark beetle food. Many of the large wood borers will bore into the xylem or wood tissue also. Coulson et al. (1976) demonstrated the existence of interspecific competition between a cerambycid, Monochamus titillator, and the southern pine beetle by defining the distribution of the southern pine beetle in bark samples where M. titillator was absent and comparing these with bark samples where M. titillator was present.

There is evidence that intraspecific competition is important to the success of several bark beetles. With the more aggressive bark beetles, such as some Dendroctonus, Ips, and Scolytus species that attack living trees, many attacking beetles need be present to effectively colonize the host. However, the mass attack phenomenon can also build densities that result in intense competition for the phloem tissue (Chap. 8). Coulson (1979) claims that the number of eggs laid by the southern pine beetle, D. frontalis, is controlled by a density-dependent compensatory feedback process, so that the higher the attacking density of females the fewer eggs each female would deposit. This type of relationship no doubt exists for a number of bark beetle species, but Coulson (1979) claims that it is not a universal phenomenon and cites Ips and Scolytus as genera in which no regulation occurs.

Intraspecific competition has been demonstrated experimentally in the laboratory for the mountain pine beetle (D. ponderosae) (Cole 1962) and the Douglas-fir beetle (D. pseudotsugae) (McMullen and Atkins 1961). In both cases, natural enemies were eliminated from the system, and it was found that egg galleries were shorter and that mortality was higher in the larval and pupal stages as attack densities increased. Cole (1962) found that the ratio of parent adults to new adults was highest at the lowest attack density. Beaver (1974) studied the effects of intraspecific competition among larvae on three bark beetles, Scolytus scolytus, S. multistriatus, and Tomicus piniperda, and found the major effect to be an increase in mortality and a decrease in mean adult weight. He felt that mortality was density-dependent with S. scolytus and T. piniperda, but not with S. multistriatus.

It appears that both intra- and interspecific competition are extremely important mortality factors and perhaps much of unaccounted for mortality is due to competition. The interaction of all mortality factors should be examined carefully in natural populations.

SEQUENCE OF ARRIVAL

A generalized picture of the sequence of associated species
of a representative bark beetle is presented in Fig. 5.8.

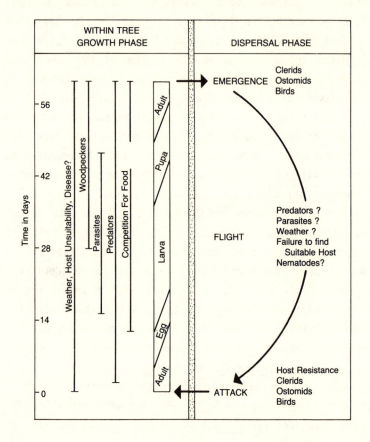

Figure 5.8. Life history of Dendroctonus brevicomis, showing
the sequence of mortality factors. Predators are insect and
arachnid predators of eggs and parent adults, and insect
predators of larval and pupal stages. The larval period is
extended by about 200 days in the overwintering generation.
The first or summer generation lasts 55-60 days. (From Stark
and Dahlsten 1970).

The sequence of events would begin with the primary attraction of the attacking beetles to a tree. The females initiate attack in the monogamous species (e.g., Dendroctonus spp.) and the males initiate the attack in the polygamous species (e.g., Ips spp.). Once these pioneer beetles are there, secondary attraction begins, resulting in a directed attack by many beetles. A number of predators (e.g., clerid beetles and trogostid beetles) use the bark beetle pheromone to find the attacked trees and are thus able to feed on attacking beetles. At the same time, predators lay eggs, and the larvae that emerge from these eggs will feed on bark beetle larvae beneath the bark. The attacking beetles have brought fungi that help them to overcome the tree and also phoretic mites and nematodes that may feed on the bark beetle eggs and early instar larvae. Mites that are predatory on other mites, nematodes and mycetophagous organisms may be introduced at this time as well.

Once egg gallery construction has begun and the bark beetles begin laying eggs, small histerid beetle egg predators begin to arrive along with a group of inquilines dominated by midges (cecidomyids and sciarids). These organisms have their own set of natural enemies that also begin to build up. Finally, after the bark beetle larvae have increased in size (about the third or fourth instar), the parasitic wasps become abundant on the bark and, in some cases, in the bark beetle galleries, searching for hosts and laying their eggs on or in the vicinity of the bark beetle larvae.

The sequence ends with the emergence of the bark beetle broods, the inquilines, predators and parasitoids, but it is only the beginning for a group of secondary insects that utilize the decomposing wood. This stage in succession is demonstrated by the large wood borers (e.g. cerambycids and buprestids) and this continues until the tree is decomposed and reenters the nutrient cycle of the forest ecosystem.

Insect control generally has been dominated by attempts to reduce target pest insect populations with little consideration of the organisms associated with the pest. This approach has led to many problems in forestry much like those in agriculture. However, the forest ecosystem is far more complex than an agroecosystem and a broader perspective is necessary.

With the advent of integrated pest management (Chap. 9), the single pest population approach to insect control is rapidly becoming a thing of the past. This is essentially an ecosystem or community approach to pest control. One need only to review the complex relationship of the bark beetles to trees and the community of organisms that have co-evolved through thousands of years with them to realize that trying to kill bark beetles with no concern for the other components in this complex system is futile. From the discussion above, it can be seen that the factors limiting beetle populations are not clearly understood. Hopefully, a community approach will lead to a better understanding of the factors limiting beetle populations and therefore to longlasting, less ecologically disruptive, and more economical pest management strategies.

6. Relationships between Bark Beetles and Symbiotic Organisms

H. S. WHITNEY

Bark beetles, like most macroorganisms, are not single organismic entities (Read 1970, Trager 1970). They exist together with a variety of microorganisms on and in their bodies (Fig. 1). Barras (1979) described the scolytid-microbe-woody host ecosystem as a "supra-organism." Early North American workers, such as Von Schrenk (1903), Craighead (1928), and Rumbold (1931) believed bark beetles carried fungi into trees, but it remained for Nelson and Beal (1929) and Leach et al. (1934) to present experimental evidence that Dendroctonus and Ips beetles vector blue stain fungi. In subsequent years, numerous other fungi, bacteria, and protozoans were discovered in and on bark beetle bodies. Many of the bacteria and fungi have been identified, but we know little about the protozoans. The factual evidence for symbiosis and its ecological and evolutionary consequences for bark beetles are the subjects of this chapter.

I will deal primarily with bark beetles that attack and kill coniferous trees in North America. These include all species of Dendroctonus, and major damaging species of Ips, Scolytus, and Dryocoetes (Furniss and Carolin 1977, Baker 1972). Although this assemblage of bark beetles is not a random sample of the family Scolytidae, their common characteristic of attacking living coniferous hosts probably has special significance in symbiology.

Symbiosis, as originally described by deBary (1879), referred to the living together of different species. He made no reference to numbers or to benefits or disadvantages of the relationship, only that the idea of living together implied a sense of continuity throughout the life of the partners. Organisms in symbiosis are more than casually associated in space and time. They share each other's physical and physiological properties and functions; their sum is greater than their parts. The organisms' togetherness

MICROORGANISMS AFFECTING BARK BEETLES

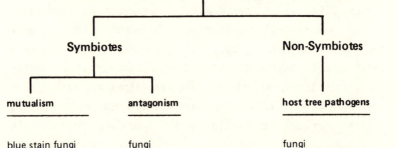

Symbiotes		Non-Symbiotes
mutualism	antagonism	host tree pathogens
blue stain fungi (prevent resinosis)	fungi (direct killers)	fungi (predispose tree to beetle attack, primarily by causing root diseases)
yeast (enhance food)	protozoans (chronic debilitators)	
	bacteria (conditioned diseases)	

Figure 6.1.   Relationships between bark beetles and their associated microorganisms.

is usually ensured by special mechanisms, and despite occasional enforced separations, togetherness prevails. Two sorts of symbiosis will be considered in this review of bark beetles and their associated microbes: mutualism, wherein both partners benefit, and antagonism, wherein one or both suffer injury. Bark beetles disease, treated here as an antigonistic symbiosis, has not been reviewed previously. Commensalism or neutralism, sometimes used to describe the situation wherein there are neither beneficial nor harmful effects, might better be considered as a lack of our ability to perceive effects. Symbiote refers to the smaller member(s) of the symbiotic unit, and host refers to the bark beetle (Figure 6.1). Hosts may have more than one symbiote at a time and the same symbiotes may have different hosts. Descriptions and reviews of the advantages of symbiotic mutualism for bark beetles and their symbiotes have been given by Graham (1967) and Francke-Grosmann (1967) and Barras and Perry (1975) have provided an annotated bibliography of symbiotic relationships for North American species.

No attempt is made here to review the taxonomy and nomenclature of bark beetle symbiotes. The reader is referred to Weijman and De Hoog (1975) and Upadhyay and Kendrick (1975) for entry to the literature on Ceratocystis and related fungi imperfecti. The other organisms, by comparison, are much less frequently represented in symbiosis with bark beetles and their taxonomy is best pursued from their specific references. Macrosymbiotic organisms associated with bark beetles will be dealt with in Chapter 5. It should perhaps be mentioned here that mites and nematodes, phoretic or parasitic on bark beetles, may affect microbial symbiotes by consuming them or by bringing or taking their spores to or from the beetles (Hetrick 1949). A sporogenous mycangium might serve as food source for mycophagous nematodes (Hunt and Poinar 1971) and mites.

SYMBIOTES OF BARK BEETLES

Nearly one hundred different species of microorganisms have been identified as associates of conifer-killing bark beetles in North America (Table 6.1). Of the 38 genera of fungi and 17 genera of bacteria only a few are known to be important symbiotically.   All four genera of protozoans thus far reported are symbiotic.   New discoveries of symbiotes continue to be made, especially when intensive searches are mounted (Marler and Barras 1978, Sikorowski et al. 1979). As expected, the largest number of known symbiotes have been identified on the most thoroughly studied species of bark beetles.   Because several species of North American bark beetles have not been studied extensively, we can therefore expect to find new symbiotes with time.   Perhaps if virologists began working on bark beetles, viruses would soon be added to the list.

MUTUALISM

Reviews of mutualism and ectosymbiosis in wood- and bark-inhabiting beetles by Graham (1967) and Francke-Grosmann (1967) substantially increased our understanding of the mutualism between tree-killing bark beetles and some of their associated fungi.   Mutualism has since been related to vectoring host-tree pathogens, eliciting responses from trees, overcoming tree resistance, enhancing beetle nutrition, and producing or modifying chemicals affecting beetle aggregation.   Perhaps the most significant contribution of the sapwood- and phloem-invading fungi vectored by attacking bark beetles is their killing of live host tissue.   This is accomplished primarily by blue stain

Table 6.1. Microorganisms Associated with the Major Bark Beetles Attacking Coniferous Trees in North America.

Microorganism	Bark Beetle	Source[1]	Reference
FUNGI			
Alternaria	Dendroctonus frontalis	beetle	Moore 1971
Aspergillus spp	D. frontalis; D. adjunctus	beetle; beetle	Moore 1971, Barras and Perry 1971b
Aspergillus flavus Link[a]	D. frontalis; Scolytus ventralis	beetle; beetle habitat	Moore 1971, Ashraff and Berryman 1969, Thomas and Poinar 1973
Basidiomycete, R593[m]	D. brevicomis	beetle	Whitney and Cobb 1972
Basidiomycete, SJB 122[m]	D. frontalis	beetle	Barras and Perry 1972
Beauveria bassiana	D. brevicomis; D. frontalis; D. ponderosae; D. pseudotsugae; D. rufipennis; Ips paraconfusus.	beetle	De Mars et al. 1970, Moore 1970, Whitney et al. 1979, Weiser 1970, Morris and Olsen 1970, Steinhaus and Marsh 1962
Candida curvata[m]	D. brevicomis, D. ponderosae	beetle	Shifrine and Phaff 1956
Candida parapsilosis, C. mycoderma	D. pseudotsugae	beetle	Lu et al. 1957
Candida rugosa Candida tenuis	D. ponderosae, D. frontalis; D. ponderosae	beetle habitat	Farmer 1965, Howe et al. 1971, Farmer 1965
Cephalosporium sp[a]	D. frontalis	beetle	Moore 1971

Table 6.1 (continued)

Microorganism	Bark Beetle	Source[1]	Reference
Ceratocystis[m] spp	D. frontalis;	beetle, habitat	Moore 1971
Ceratocystis abiocarpa	Dryocoetes confusus; Ips		Davidson 1966,
Ceratocystis bicolor	D. rufipennis, Ips; I. hunteri, I. pilifrons	beetle, habitat	Davidson 1955, 1958
Ceratocystis coerulescens	D. rufipennis	habitat	Davidson 1955
Ceratocystis dryocoetidis Molnar	D. confusus	beetle, habitat	
Ceratocystis gossypina	Dendroctonus spp., Ips spp.	habitat	Davidson 1966
Ceratocystis gossypina var robusta	D. adjunctus	habitat	Davidson 1966
Ceratocystis huntii Grinchenko	Dendroctonus; D. ponderosae; I. pini	beetles; habitat.	Davidson and Robinson-Jeffry 1965, Robinson-Jeffry and Grinchenko 1964
Ceratocystis ips	D. ponderosae, D. valens; I. avulsus; I. calligraphus, I. emarginatus I. grandicollis, I. intiger, I. lecontii, I. plastographus, I. pini.	beetles, habitat	Mathre 1964, Rumbold 1931, Yearian et al. 1972, Rumbold 1936 Ellis 1939
Ceratocystis leucocarpa	Dendroctonus spp.	habitat	Davidson 1966

Table 6.1 (continued)

Microorganism	Bark Beetle	Source[1]	Reference
Ceratocystis minor	D. brevicomis, D. frontalis; D. ponderosae, D. pseudotsugae	beetle, habitat	Rumbold 1931, Mathre 1964, Robinson 1962, Hunt 1956
	D. frontalis	beetle	Barras and Taylor 1973
Ceratocystis minor var. barrassi			
Ceratocystis minuta	D. ponderosae; I. emarginatus, I. paraconfusus	habitat	Robinson 1962, Mathre 1964
Ceratocystis minuta-bicolor	Ips sp., I. paraconfusus, I. pini	habitat	Davidson 1971
Ceratocystis montia	D. jeffreyi; D. ponderosae	habitat, beetle	Mathre 1964, Rumbold 1941, Robinson 1962
Ceratocystis nigrocarpa	D. breviconis	beetle, habitat	Davidson 1971, Whitney and Cobb 1972
Ceratocystis olivaceapini	Dendroctonus	habitat	Davidson 1971
Ceratocystis penicillata	D. confusus	habitat	Davidson 1958
Ceratocystis perfecta	I. pilifrons	beetle	Davidson 1958
Ceratocystis piceaperda	D. rufipennis	beetle	Rumbold 1936, Hunt 1956
Ceratocystis shrenkiana	D. ponderosae	habitat	Mathre 1964
Ceratocystis truncicola	D. rufipennis	habitat	Davidson 1958

Table 6.1 (continued)

Microorganism	Bark Beetle	Source[1]	Reference
Cladosporium	D. frontalis, D. ponderosae, Dendroctonus	beetle	Moore 1971, Whitney and Farris 1970
Cryptococcus diffluens, C. neoformans	Dendroctonus	beetle	Shifrine and Phaff 1956
Cryptoporus volvatus	D. brevicomis, D. ponderosae, D. pseudotsugae, D. frontalis	habitat	Borden and McClaren 1970
Cylindricolla dendroctonia	D. frontalis	beetle	Charles 1941
Europhium[m] aureum	Dendroctonus	beetle, habitat	Robinson-Jeffrey and Davidson 1968
Europhium clavigerum	D. ponderosae	beetle, habitat	Robinson-Jeffrey and Davidson 1968
Europhium robustom	Dendroctonus	beetle, habitat	Robinson-Jeffrey and Davidson 1968
Epicoccum	D. frontalis	beetle	Moore 1971
Endomycopsis scolyti[m]	D. ponderosae	beetle	Farmer 1965
Fomes annosus[m]	D. terebrans	habitat, beetle	Himes and Skelley 1972
Fusarium solani[a]	D. frontalis	beetle	Moore 1971
Gliocladium	D. frontalis	beetle	Moore 1971
Hansenula capsulata	D. ponderosae, D. jeffreyi, D. pseudotsugae	beetle, habitat	Robinson 1962, Mathre 1964, Farmer 1965, Shifrine and Phaff 1956, Lu et al. 1957
Hansenula holstii	D. ponderosae, D. rufipennis	habitat, beetle	Robinson 1962, Wickerham 1960
Isaria[a]	D. brevicomis	beetle	DeMars et al. 1970

Table 6.1 (continued)

Microorganism	Bark Beetle	Source[1]	Reference
Leptographium engelmanii	D. rufipennis	habitat	Davidson 1955
Leptographium terebrantis	D. terebrans	beetle, habitat	Barras and Perry 1971b
Metarrhizium anisopliae[a]	D. frontalis; D. terebrans	beetle	Moore 1971, Jouvenaz and Wilkinson 1970, Holt 1961
Mortierella	D. frontalis	beetle	Moore 1971
Oedocephalum	Scolytus ventralis	beetle, habitat	Stark and Borden 1965
Paecilomyces[a]	D. frontalis	beetle	Moore 1971
Penicillium	D. frontalis; D. ponderosae	beetle	Moore 1971, Whitney and Farris 1970
	D. adjunctus	beetle	Barras and Perry 1971b
Penicillium decumbens	D. terebrans	habitat	Barras 1969
Penicillium implicatum[a]	D. adjunctus	beetle	Barras and Perry 1971b
Penicillium spinulosum	D. murrayanae, D. ponderosae, D. rufipennis	habitat	Whitney and Funk 1977
Pezizella chapmanii	D. brevicomis	habitat	Whitney and Funk 1977
Pichia pini[m]	D. frontalis; D. ponderosae; D. jeffreyi	beetle	Holst 1937, Robinson 1962, Farmer 1965. Shifrine and Phaff 1956.
Rhodotorula crocea		beetle	
Saccharomyces bisporus[m]	D. frontalis	habitat, beetle	Howe et al. 1971

Table 6.1 (continued)

Microorganism	Bark Beetle	Source[1]	Reference
Saccharomyces pastori[a]	D. frontalis; D. pseudotsugae	habitat	Howe et al. 1971, Lu et al. 1957
Saccharomyces steineri	D. frontalis	habitat	Howe et al. 1971
Scopulariopsis[a]	D. frontalis	beetle	Moore 1971
Spicaria[d]	D. brevicomis	beetle	Steinhaus and Marsh 1962
Sporothrix[m]	D. frontalis	beetle	Barras 1972
Torulopsis ernobii[m]	D. frontalis	habitat	Howe et al. 1971
Torulopsis nitratophila	D. ponderosae	beetle	Shifrine and Phaff 1956
Torulopsis melibiosum	D. ponderosae	beetle	Shifrine and Phaff 1956
Torulopsis pinus	D. frontalis	habitat	Barras and Perry 1971a, Whitney and Farris 1970
Trichoderma	D. adjunctus; D. ponderosae	beetle	Wright 1935, Livingston and Berryman 1972
Trichosporium symbioticum	S. ventralis	beetle, habitat	Kendrick and Molnar 1965
Verticicladiella dryocoetidis[m]	D. confusus	beetle, habitat	Barras and Perry 1971a
Verticicladiella huntii	D. adjunctus	beetle	
Verticicladiella wagenerii	D. valens	habitat	Goheen and Cobb 1978

Table 6.1 (continued)

Microorganism	Bark Beetle	Source[1]	Reference
BACTERIA			
Actinotobacter calcoaceticus			
Aerobacter aerogenes			
Aeromonas hydrophilia			
Alcaligenes faecalis, metacaligenes, recti, viscolatis			
Bacillus cereus[a], megaterium subtilis, thuringiensis[a]	D. frontalis	beetle, habitat	Marler and Barras 1978
Brevibacterium acetylicum, incertum, lipolyticum, quale			Moore 1971
Enterobacter aerogenes, cloacae			Barras and Marler 1974
Flavobacterium aquale[a], ferrugineum, lutescens, sp.			Steinhaus and Marsh 1962
Kurthia bessonii			Jouvenaz and Wilkinson 1970
Micrococcus luteus, varians			Wood 1961
Mycobacterium			
Proteus rettgeri			
Pseudomonas aerugenosa[a], fluorescens[a]			
Sarcina flava, lutea			
Serratia marcescens[a],			
Staphylococcus epidermidis			
Xanthomonas			

Table 6.1 (continued)

Microorganism	Bark Beetle	Source[1]	Reference
PROTOZOANS[a]			
Microsporida	D. rufipennis	beetle	Morris and Olsen 1970
Chytridiopsis typographi	D. pseudotsugae	beetle	Weiser 1970
Nosema dendroctoni Weiser	D. pseudotsugae	beetle	Weiser 1970
Ophryocystis dendroctoni	D. pseudotsugae	beetle	Weiser 1970
Unikaryon minutum	D. frontalis	beetle	Knell and Allen 1978

[1]Source: beetle = microorganism on or in any or all life stages; habitat means microorganism on or in gallery walls and adjacent tissues including frass.
a means supposed or proven antagonistic symbiote (pathogen).
m means supposed or proven mutualistic symbiote.

fungi of the genus Ceratocystis. Living tree tissues prevent successful beetle reproduction by resinosis, the production of copious amounts of resin and callus tissue (Coulson 1979, Safranyik et al. 1975, Berryman 1972, Reid et al. 1967). Death of the tree results not only in absence of resinosus but also in changes in gas exchange and moisture that are beneficial to the beetle's brood and to the symbiotic fungi that have invaded the sapwood (Webb and Franklin 1978). Tree death following bark beetle attack is unique. It is not the same as death due to felling, debarking, or defoliation. It results from cumulative effects of fungal killing of live tissue in the close vicinity of beetle entrance holes. During the early stages of colonization these blue stain fungi are colorless; there is no hint of blue color at this time. Translocation no longer occurs in tissues killed this way. The impact of a bark beetle attack on tree vicinity depends on tree health and energy reserves, the density and circumferential distribution of attacks, and the quantity and quality of the introduced blue stain fungi. When a stem becomes girdled by fungus-killed tissue the crown soon becomes water stressed, vital functions decline throughout the tree, and it rapidly dies. Only when the tree is dying is there sufficient oxygen available for the symbiotes to produce abundant mycelium and spores, some of which become brown. This brown pigment in the mycelium impacts a characteristic blue stain to the sapwood it has colonized. Who is primarily responsible for the death of the tree -- the beetle, or the fungus that it carries? This question has generated considerable interest and controversy with some investigators declaring that blue stain fungi may not be essential for success of bark beetles and others demonstrating pathogenicity of blue stain fungi in the absence of the beetles (Hetrick 1949, Kulman 1964, Amman

1980, Mathre 1964, Nelson and Beal 1929). The culprit is not likely to be the beetle or the fungus exclusively; these organisms are tightly locked in a mutualistic association. On the one hand, the fungi would not be able to gain access to the phloem and sapwood if it were not for the bark-penetrating beetles.   On the other hand, the tree would continue producing resin and callus if its living tissues were not killed by the invading blue stain fungi.

The mutualistic Ceratocystis-type fungi are strongly pleomorphic.   That is, they can grow either as hyphae and mycelium or as spores.   As mycelia, they penetrate extensively into the sapwood and phloem adjacent to beetle galleries and thus are able to kill live tree tissues relatively far removed from the beetles.   By contrast, as spores growing on, in, or next to the beetle's body, the fungi can become concentrated in large numbers and be available as food or fit for transport in mycangia (see below).   Chemicals produced by larvae and/or pupae of the southern pine beetle, D. frontalis, may act as stimulants favoring the sporogenous (ambrosial) mode of growth of Sporothrix sp. and an unidentified basidiomycete (Happ et al. 1975, 1976).   These workers have hypothesized that the bark beetle not only provides protection for the mycangial symbiotes but also controls the growth form that the fungi will have when in contact with the insect.

A second major role postulated for symbiotic microorganisms is that they serve as a source of nutrition for tree-killing bark beetles.   Wild southern pine beetles, presumed to have a full complement of symbiotes, produced significantly more progeny in a shorter time than did beetles without their symbiotic fungi (Barras 1973) and, in a similar study, it was shown that wild I. avulsus benefited from yeast and bacterial symbiotes as well (Gouger 1972).

Population size of second-generation adult mountain pine beetles, Dendroctonus ponderosae, was directly correlated with the amount of yeast ingested in the larval stage of development (Farmer 1965), but in another study mountain pine beetle larvae developed into apparently normal pupae and young adults while mining on bolts in fresh unaltered axenic lodgepole pine phloem that was free of yeasts or other microbes (Whitney 1971). When Ips paraconfusus (the California five-spined ips) were reared axenically from egg to adult in autoclaved ground phloem without yeast supplement (Bedard 1961), development times were abnormally long and only a few axenic adult pine bark beetles (one Southern pine beetle, two six-spined ips, and ten Southern pine engraver beetles) were able to survive in autoclaved pine phloem sandwiches (Holst 1937). Colonization of pine phloem by the southern pine beetle lowers the ratio of carbon to nitrogen (Barras and Hodges 1969), but it is not clear how this might influence the production of brood. In natural bark beetle attack, the sequence of events is such that parent adults initiate and may even complete their reproductive activity before the tree tissues become extensively colonized by symbiotes, so that if microbial symbiotes are functioning as nutritional benefactors for bark beetles it is more likely that it is during the larval stage or during maturation feeding of young adults rather than during their specific reproductive activities. The foregoing examples of possible symbiotic involvement in bark beetle nutrition coupled with the fact that at least some progeny can be produced without fungus (Barras 1973) indicate that this symbiosis is quantitative. The actual symbiote benefit to the bark beetles is likely to depend on the inherent nutritional value of the phloem prior to its colonization by the symbiotes. Young, healthy, vigorous trees on a good site with adequate moisture and fully exposed crowns likely have more nutritious phloem than old decrepit trees on a poor site with inadequate moisture and shaded, broken, or defoliated crowns. A facultative nutritional symbiosis probably enlarges the range of host tree conditions in which bark beetles could successfully maintain themselves.

A third way in which fungal symbiotes may aid bark beetles is indirect. The debilitating activities of certain fungi on the host tree increase its susceptibility to subsequent attack by bark beetles. For example, D. pseudotsugae, the Douglas fir beetle, vectors the sapwood decay fungus Cryptoporus volvatus (Castello, Shaw, and Furniss 1976) and D. valens, the red turpentine beetle, possibly transmits the root pathogen Verticicladiella wagenerii (Goheen and Cobb 1978). The actual effect of these fungi on bark beetle survival is not known.

A major part of the symbiosis between bark beetles and their associated fungi is their joint action to overcome the defense of trees. This symbiotic action is certainly evident in mass attacks when the density of beetles is high (Chap. 8), but the potential of the symbiotes to weaken and predispose trees to subsequent attacks may be very important when beetle density is low (endemic populations) and there are not enough beetles to mount successful mass attacks. Although the first beetles to attack may be overwhelmed and killed by resins (Chap. 8), they may innoculate the tree with fungi that may linger for a long time. For example, two sapwood-invading symbiotes remained alive for at least two years in necrotic tisue of trees which survived bark beetle attack (Molnar 1965, Whitney et al. 1979). If these symbiotes exact a substantial energy toll from the tree, it may be highly susceptible to its next attack by beetles.

A fourth way in which symbiotes may benefit bark beetles involves bark beetle behavioral chemicals (Chap. 4). Gut bacteria of Ips confusus, the pinon ips, and several other bark beetles can produce verbenol from the host monoterpene alpha-pinene (Brand et al. 1975). Verbenol is an integral component of chemical complexes affecting aggregation of several bark beetles. Mycangial symbiotes of the southern pine beetle have the capacity to further oxidize verbenol to verbenone which has anti-aggregant properties (Brand et al. 1976). These symbiotes produce other substances, such as isoamyl alcohol, that enhance the action of aggregation pheromones (Brand and Barras 1977, Brand et al. 1977). This information was obtained from laboratory experiments, but the investigators hypothesize that similar effects probably occur in nature, and that symbiotes will have to be considered when developing an understanding of insect responses to volatile substances in the forest.

THE BLUE STAIN DECEPTION

The sapwood of coniferous trees that have produced bark beetle broods is usually more or less blue rather than the natural wood color. This observation, first published by von Schrenk almost eighty years ago, has given rise to a long-standing emphasis on blue stain which has unfortunately impeded our understanding of certain aspects of bark beetle symbiosis. Certainly, blue stained sapwood is strongly correlated with beetle-killed coniferous trees, and it is caused by mutualistic Ceratocystis spp. vectored to trees by bark beetles. However, the appearance of the blue stain in relation to the occurrence and timing of various events in bark beetle biology has largely been neglected. Not only is the blue color complex, being a combination of physical and pigmented color, but also several important mycangial symbiotes produce very little, if any, pigment (Barras and

Perry 1972, Whitney and Cobb 1972) and therefore cannot stain
the wood. The stain symbiotes themselves produce very little
stain when they are actively growing.   Production of the
brown pigment that imparts the blue color to fungus-colonized
sapwood indicates that the fungi have successfully colonized
the substrate, are maturing, and are becoming dormant. Thus
interpretation of observations relating blue stain per se to
events associated with bark beetle brood development and host
tree interactions must consider that, by the time stain is
evident, most primary events, such as gallery establishment,
fungus infection and colonization, mating, egg laying, and
some larval mining, will have taken place. Reliance on the
appearance of stain as an indicator of successful beetle
colonization of the host is all the more deceptive because
the stain fungi themselves are highly variable and may not
produce pigment.   Furthermore, sapwood may be stained by a
variety of nonsymbiotic fungi. Statements that blue stain is
not essential for bark beetle success (Hetrick 1949, Rudinsky
1961, McCowan and Rudinsky 1958, Kulman 1964, Amann 1980) or
that blue stain is inhibitory to bark beetle development
(Barras 1970, Franklin 1970) have not fully taken into
account the biology of blue stain production. Careful time
place culturing and microscopic studies must be done to
confidently ascribe stain or other effects to specific bark
beetle symbiotes.   Free intermingling of many symbiotes in
the developing brood habitat emphasizes the need for
diligence in making isolations and pure cultures (Robinson
1962, Molnar 1965, Whitney 1971).

ANTAGONISM

Disorders of insects caused by infectious agents have been
known for almost 150 years (Steinhaus 1949).   The earliest

report of an entomopathogen of a North American species of bark beetle attacking coniferous trees is of Cylindricolla dendroctoni on the southern pine beetle in West Virginia in 1896 (Charles 1941). Diseases of bark beetles are generally attributed to fungi, bacteria, or protozoans. Viruses are suspected of causing diseases in larvae of the southern pine beetle (Sikorowski et al. 1979), and although viral and rickettsial diseases of coleopterans are known (Nienhaus and Sikora 1979, Vaughn 1974, Tinsley and Harrap 1978), there are no confirmed reports that they cause diseases of bark beetles. Apart from a survey and laboratory testing of pathogens of the southern pine beetle (Moore 1970, 1972a, 1972b, 1973, Sikorowsky et al. 1979) and preliminary field experiments with Beauveria bassiana for direct control of the mountain pine beetle (Whitney et al. 1978), there are only occasional diagnostic records of entomopathogens in bark beetles in North America (Table 6.1).

.It is noteworthy that Koch's rules of proof (Thomas 1974), which require that the test organism produce the typical disease when inoculated into the host, have been applied to only a few host-pathogen combinations. If conducting the tests under natural conditions is added as a corollary to Koch's postulates, then in only one case in Table 6.1 has pathogenicity been rigorously tested (Whitney 1979). Satisfying Koch's postulates is often beset with difficulties. For example, in the case of certain parasitic protozoans requiring unknown specific insect host substances for their growth and reproduction, it may be necessary to undertake elaborate insect tissue culture procedures to produce a pure culture of the candidate parasite before itcan be tested for possible pathogenic effects. There is a strong need for reliable bioassay systems in which known and suspected pathogens can be studied to find out if and how such agents induce disease under natural conditions. Obviously, the study of bark beetle diseases and their effects on natural populations is in its infancy.

North American forest entomologists describing bark beetle population dynamics have long recognized disease as a possible contributing cause of "unexplained mortality" (Miller and Keen 1960, Bennett and Ciesla 1971, Moore and Thatcher 1973, Coulson et al. 1972). However, signs and symptoms of naturally occurring insect diseases are not always readily apparent (Moore 1970, Steinhaus 1963, Bucher 1973) and the evidence that pathogens are able to keep bark beetle populations at endemic levels is only indirect (Moore 1971, Sikorowski et al. 1979). Despite the fact that disease is a potential regulator of bark beetle population numbers, there are no documented reports that bark beetle diseases are responsible for major population changes.

There is also very little known of the mechanisms of pathogensesis in bark beetle diseases. Three sorts of antagonistic interaction are apparent: acute killing, chronic debilitation, and conditioned pathogenesis. Virulent aggresive fungal pathogens, such as Beauveria, Paecilomyces, and Metarrhizium, are fast acting and very lethal. They die along with their host, and consequently they are not usually transmitted from generation to generation and therefore occur only sporadically. By contrast, protozoan pathogens produce a reduction in vigor and reproductive ability. These organisms are often transmitted from one generation to the next, sometimes transovarially, and they may be widespread in endemic populations. The concept of potential pathogens causing conditioned pathogenesis was introduced by Lysenko (1959) and Bucher (1960). These pathogens, normally noninvasive gut-bacteria, produce fatal bacteremias when given entrance to the hemocoel. Such bacteria have been isolated from the California five-spined ips, the six-spined ips, the southern pine beetle, and the mountain pine beetle (Table 6.1) and are more likely present in most bark beetle species. Facultative or conditioned pathogens may have the potential to produce significant changes in the size of natural populations that have become predisposed to infection by environmental stress such as extremes of temperature and/or moisture. Additionally, other gut symbiotes, such as gregarine trophozoites (Weiser 1970, Massey 1974) and nematodes, may weaken gut walls and facilitate entrance of the pathogen to the hemocoel. The presence of "harmless commensals" such as Eugregarines, should be re-examined in light of their possibly predisposing beetles to conditioned diseases.

Some researchers have suggested that the apparent low incidence of disease in bark beetles reflects conditions in the host tree unsuitable to development of disease. Lack of sufficient moisture, especially low relative humidity, is believed to inhibit infection and spread of fungal diseases of insects (Roberts and Campbell 1977). However, low moisture should not be a limiting factor in the spread of microbial pathogens in the early stages of brood development because the phloem and sapwood would not yet have begun to dry. Considering the microenvironment in the vicinity of intersegmental membranes and the thin abdominal integument beneath the elytra where pathogens are likely to gain entry into the beetles, it seems unlikely that moisture would be limiting at these sites, even in the relatively dry ecosystem that many bark beetles inhabit (Ferron 1977).

Notwithstanding the lack of evidence for widespread disease outbreaks, it would be naive to believe that in nature bark beetles do not have diseases that significantly affect their population dynamics. There is a low probability that disease will be detected in cryptic insects, such as bark beetle, for diseases characteristically remain obscured until host organisms become very well known and understood (Bucher 1973). As more knowledge is gained of the biology of bark beetles and as improved techniques become available for testing Koch's postulates in reliable bioassays, we can expect that more diseases and disorders of these complex insects will be recognized and described.

In addition to the antagonism between beetles and symbionts, competition and antagonism among symbionts have been observed (Barras 1969, Wilkenson 1968). The aggressive sapwood-invading symbiote C. minor produces isocoumarins, recently suspected to be significant in phytopathological processes (Hemingway et al. 1977, McGraw and Hemingway 1977). Europhium clavigerum and Ceratocystis montia, two sapwood-invading symbiotes of the mountain pine beetle, are much more tolerant to antifungal substances, such as pinosylvin produced by the host tree, than are nonsymbiotic fungi.

MYCANGIA

Next to the appearance of blue stained sapwood in trees attacked by bark beetles, perhaps the presence of a mycangium is the most conspicuous physical manifestation of the symbiosis between bark beetles and fungi. Like the blue stain, however, mycangia are beguiling because they are highly developed in some species and apparently absent in others. Interestingly, all known types of mycangia in North American bark beetles may occur geographically within a few kilometers of one another. Representative examples of mycangia are illustrated in Chapter 3.

A mycangium was defined by Batra (1963) as "... a sac- or cup-shaped fungal repository commonly located on the exterior of the beetle, or rarely, in the oral cavity of wood inhabiting Scolytidae that use fungi as the main source of their food." He goes on to say that mycangia are found primarily in females, near glands, and are usually paired. Detailed descriptions of mycangia have been published for six North American bark beetles. The transverse elevated-callus mycangium on the pronotum of Dendroctonus adjunctus, the roundheaded pine beetle, D. brevicomis, the western pine beetle, and females of the southern pine beetle is by far the most conspicuous and most studied (Francke-Grosmann 1965a, Barras 1967, 1975, Barras and Perry 1971a, 1972, and Happ et al. 1971). Females of D. parallelacolis (=approximatus), the larger Mexican pine beetle, also have the pronotal callus ridge, and although there is no published confirmation, the structure probably contains fungal symbiotes and functions as a mycangium in this species as well. The reported pronotal mycangium in males of the roundheaded pine beetle (Francke-Grosmann 1965b) could not be substantiated. This much reduced structure, which also occurs in the males of the western pine beetle and the southern pine beetle, has been renamed a pseudomycangium because it lacks certain essential features of a mycangium as defined by Batra 1963 (Barras and Perry 1971). A maxilliary mycangium occurs near the proximal end of the cardines in the mountain pine beetle (Farris 1965, Whitney and Farris 1970), and a mandibular mycangium occurs at the base of the mandibles in the western balsam bark beetle, Dryocoetes confusus (Farris 1969). These paired pronotal and oral mycangia in species of Dendroctonus and Dryocoetes contrast with the non-paired multiple cup-shaped pit-mycangia on the top and sides of the head of the fir engraver, Scolytus ventralis (Livingston and Berryman 1972).

These nonpronotal mycangia occur in both sexes, whereas the paired pronotal mycangia are secondary female characteristics (Barras 1967).

Cultures have been made from the mycangia of the roundheaded pine beetle, the western pine beetle, the southern pine beetle, and the mountain pine beetle, and the known fungal symbiote for each bark beetle was recovered. In some cases, where more than one fungal symbiote had been recognized, both species were found co-habitating in the mycangium. Based on spore morphology, distribution of spores on the body, and correlative cultural evidence, it is suspected that the spores in mycangial pits of the fir engraver are, in fact, those of Trichosporium sumbioticum Wright, this beetle's known fungal symbiote.

Investigators of the pronotal mycangia of the southern pine beetle and the roundheaded pine beetle have revealed the presence of glandular structures which empty into the mycangia (Barras and Perry 1971, Happ et al. 1971). These investigators hypothesize that secretions from the gland cells regulate symbiote morphology, suppress their sexual reproduction, and prevent nonmycangial fungi from becoming established. Such regulation of symbiote growth and reproduction would tend to secure for the beetle a long-term, genetically stable symbiote adapted to rapidly colonize the host tree. Apparently these mycangia deliver symbiote spores passively as the females become involved in gallery construction and egg laying (Barras 1975). However, the outer mycangial wall is thin and forms a ridge on the pronotum so that pressure could easily be brought to bear on a spore-engorged mycangium by pushing against the host tissue. Whether ovipositing females do, in fact, make mycangial squeezing movements during gallery construction and egg deposition is not known. The precise mechanisms by which these mycangia become colonized by symbiotes and the specific physiological effects of glandular secretions are unknown.

Just as the beetles have adaptations to carry symbiotes, some symbiotes have adaptations that help them to travel with the beetle and to colonize trees. The yeasts Hansenula spp. and the blue stain fungi both produce adhesive material on their spores, thus improving their chances of remaining with dispersing bark beetles as they fly and bore into new host trees (Hawker and Madelin 1976, Wickerham and Burton 1961). The adhesiveness of these spores results in their sticking together in spore masses containing dozens to thousands of spores. Interestingly, masses of asexual spores of the blue stain fungi disperse in water, whereas masses of sexual spores do not (Whitney and Blauel 1972). However, these workers found that the sexual spores readily separated from one another in conifer resin. Fresh liquid resin is likely to be encountered by attacking bark beetles and the resulting dispersal of the sexual spores would not only enhance inoculation of the new habitat but also promote genetic diversity in the fungi.

Mycangia are not required for successful transmission of symbiotes. There are several well-documented symbioses that do not involve any apparent adaptive modification of beetle morphology. Castello et al. (1976) did not locate a mycangium in the Douglas fir beetle but showed that the fungus Cryptoporus volvatus was carried as mycelial fragments lodged in intersegmental areas. Moreover, although specific blue stain fungi are associated with the spruce beetle (D. rufipennis) and the Douglas fir beetle (Davidson 1955, 1958), no mycangia were discovered in young adults of these beetles (Farris 1965). Similarly, a consistent association with certain microbes is maintained by Ips avulsus in the absence of any apparent physical adaptation that would serve as a means of ensuring a specific association (Gouger et al. 1975).

It will be unfortunate if the term mycangium becomes rigidly defined before sufficient information on the various modes of microbial association and transmission is obtained for these insects. The word itself means simply "Fungus container," and whether it must be furnished with elaborate glands, occur in pairs, be sex related, or need be present at all to ensure successful insect-microbe symbiosis is not clear. Indeed, although it is attractive to imagine that the elaborate pronotal mycangium was developed specifically for preserving and propagating pure cultures of selected microbes, fungi may be more adapted to adhering to the beetles than the beetles are adapted to carrying the fungus about (Leach 1940, Gouger et al. 1975). Francke-Grosmann's (1965) view that mycangia are either modified body parts or highly developed glands of uncertain morphology, together with the extreme range of variability seen in mycangial morphology, suggests that a simple all-inclusive definition of a mycangium may not be possible.

Furthermore, fungi are fully capable of pleomorphism in the absence of bark beetles and their antibiotic-producing capabilities to keep other microbes, including other fungi, at bay are well known (Romano 1966, Brian 1957). Thus there is no necessity to ascribe pleomorphism induction and antibiosis functions to mycangial gland secretions. Also, physical disruption of mycelia is known to induce sporulation in certain fungi (Hawker 1957, Leonard and Dick 1973) and it may be that larval movements, and particularly abdominal flexing of pupae (Whitney 1971), also induce production of nutritionally rich ambrosial growth that immature stages of bark beetles feast on. Perhaps mycangial fungi are merely exploring and exploiting the surface and subsurface of bark beetle bodies using their own resources to colonize these ecological niches the same way as they do on a multitude of other naturally occurring substrates. Until more is known about physiological and biochemical activity of specific glandular secretions of insect origin and experiments have been done challenging axenic mycangia with pure cultures of various mycangial and nonmycangial candidate fungi, the precise mechanisms ensuring successful symbiosis between bark beetles and fungi are likely to remain speculative.

It is interesting to wonder what the function of the pronotal glandular "mycangium" of the roundheaded pine beetle and other allied species might be if it were not colonized by fungi. Barras and Perry (1971a) noted the similarity of mycangial gland cells to defensive gland cells of certain tenebrionid beetles. In some tenebrionids, the integument is provisioned with cuticular sacs that serve as reservoirs for the toxicants produced by defensive glands (Happ 1968, Roth and Eisner 1962). It is tempting to speculate that glandular mycangia in certain Dendroctonus species may have originated as a result of an adaptation by microorganisms to tolerate toxicants in integumentary receptables of what were once defense glands. The mycangium may be truly an organ of symbiotic origin if an evolutionary compromise between aggression (pathogenecity) on the part of the original symbiotes and a defensive reaction (antibiosis) on the part of the insect host has been reached.

SUMMARY

In mutualism between bark beetles and certain of their symbiotes, the beetles benefit by obtaining help in location of hosts, aggregation, overcoming host resistance, obtaining adequate nutrition, and conditioning hosts for reproduction. The symbiotes benefit by being transported to new host trees by the beetles.   Thus, they are aided in dispersal, inoculation, and penetration into fresh substrate.   The classical question in mutualism, "What's in it for the participants?" should be secondary to the question "What's in it for the partnership?"   To this extent symbiology challenges certain concepts of natural selection.   In the present case, suppressed sexual and enhanced asexual reproduction of fungal symbiotes, a disadvantage perhaps for the fungus, provides advantageous stability to the partnership.   Similarly, the symbiotic units' commitment to a nutritionally inefficient food (high carbon-to-nitrogen ratio) is successful in part because mutualistic microbes concentrate and convert host-tree nitrogen into food stuffs for the beetles.

Considering the large number and diversity of organisms in and on various stages of bark beetles in host tree tissues of varying degrees of physiological deterioration, the task of identifying and understanding the important relationships of the organisms to one another seems almost insurmountable. Perhaps it is axiomatic that in symbiology a preoccupation with simple hypotheses is counterproductive (Graham 1967). Occam's razor may not be sharp enough to slice through the jungle of information because the simplest hypothesis may be merely the simplest in a complex series of hypotheses. Historically, the approach to the problem has been to tear the community apart, describing all the components in detail, and then to put it back together bit by bit to find out how it works.   Presently, we seem to have more bits than we can handle. A possible way out of this dilemma is to follow the lead of Bedard (1961, 1966) and Barras (1971, 1972) and use axenically reared bark beetles to establish standards of their biological capabilities, that is, to first gain a knowledge of what the beetles alone are potentially capable of doing and then to begin reconstruction of the natural system. Questions of symbiote involvement in bark beetle nutrition, pathogenesis, and pheromone synthesis could be approached directly with rigorous bioassays in axenic systems. Direct approaches to investigating the role of symbiotes in overcoming host tree resistance and tree changes that optimize bark beetle brood production are hampered by an inability to induce bark beetle attack experimentally in specific trees in the forest. This problem has been partially overcome recently by use of pheromones that will direct or guide beetles to certain trees. However, primary host finding and establishment cannot be fully studied using this means alone. Techniques that would modify trees so that searching beetles would recognize and attack them would be helpful.

Finally, the effects of symbiotes may be studied in highly sensitive biochemical analyses that are now becoming commonplace in modern laboratories. Isozyme analysis has recently been used to identify the prey that insect predators have consumed. The prey isozymes were readily detectable in the body of the predator (Murry and Solomon 1978, Wood et al. 1978). Clearly, diseased bark beetles containing a high titre of inter- or intra-cellular pathogenic symbiotes, or beetles with one or more large mycangia filled with one or more mutualistic fungi, can be expected to differ chemically, either qualitatively or quantitatively, from beetles that are not carrying these symbiotes.

Because of its complexity, the "supra organism" -- bark beetle plus micro and macrosymbiotes plus host-tree -- should be especially vulnerable to biological manipulations. As such, it may offer unique opportunities for integrated pest management. However, insights gained thus far give a clear indication that harnessing bark beetle symbiotes for man's specific purposes is a sizeable challenge.

7. Host Resistance and Susceptibility

R. G. CATES and H. ALEXANDER

Foresters have long recognized that healthy, vigorously growing forest trees show differential resistance to forest pest attack, whether these pests be defoliators, tree killing bark beetles and their associated fungi, or fungal pathogens (McDonald 1979, Hanover 1975, 1980, Berryman 1972, Smith 1972, Bingham 1966, Callaham 1966a,b, Hopkins 1902). Since the early 1900's, entomologists have pursued the effect of conifer resin on bark beetles and their associated fungi (Smith 1966a,b, Hanover 1980). However, it has been only since the mid to late 1960's that any significant progress has been made in the understanding of bark beetle-fungus-host plant interactions. Even to date, it appears that breeding trees for resistance to forest pests and pathogens is not an integrated part of forest management (National Academy of Sciences 1975, Hanover 1980, Belanger 1981) even though considerable success has been found in the resistance of western white pine (Pinus monticola) to blister rust (Cronartium ribicola) (McDonald 1979, Bingham 1966). Hanover (1980) remarked that few rigorously demonstrated cases of genetic resistance to forest pests exist. He cited about 10 examples, only one of which included interactions between the American bark beetles (Dendroctonus, Ips, and Scolytus) and their conifer hosts which have been major subjects of research for about 80 years.

Most of the papers published prior to 1960 and several recent papers do not contain a concise definition of resistance of trees to bark beetles. Several definitions exist in the more recent literature ranging from genetic to phenotypic to pseudoresistance (for reviews see Hanover 1975, Maxwell 1972, Beck 1965, Painter 1951). Host plant resistance, as defined in this treatment, includes "the collective heritable characteristics by which a plant species, race, clone, or individual may reduce the probability of successful utilization of that plant as a host by an insect species, race, biotype, or individual" (Beck 1965). The complexity of this phenomenon is readily apparent and several authors have delineated various mechanisms and types of resistance apparent among host plants and their phytophagous insects (Hanover 1975, Maxwell 1972, Beck 1965, Painter 1951). It is clear from the agricultural literature that many resistant characteristics of a plant are genetically controlled and heritable, and that varying degrees of resistance of host plants to phytophagous insects or pathogens are useful in management programs. Genetic resistance can be modified by environmental factors and "phenotypic resistance" is often a good indicator of a high level of genetically controlled resistance in trees (Bingham 1966).

PLANT RESISTANCE MECHANISMS

Physical Systems

Mechanisms in the resistance of plants to phytophagous insects include morphological characteristics, nutritional imbalances, chemical or natural plant products, and asynchrony between the emergence of host plant tissues and of the insect.

Morphological Systems

The information concerning the effects that various structural and morphological characteristics have on reducing the impact of phytophagous insects has come predominantly from other disciplines, most notably the agriculture literature. It is, therefore, important to discuss this information before considering the effect of structural and morphological characteristics in the bark beetle-host interaction. There have been few well designed experimental investigations of the role of morphological characteristics in conferring resistance to a plant. Feeny (1970) suggested that winter moth emergence and survival is highest during the rapid flush of young, nutritious oak leaves, when moths can avoid the deleterious effects of both protein-complexing tannins and also the more lignified, mature leaves. It is well known that cattle will not eat several very palatable cacti until the thorns are removed from them by burning.

Rice plants containing high concentrations of silica in stems appear to be resistant to larvae of the rice stem borer (Sasamoto 1958), and some varieties of cucumbers are evidently resistant to the squash borer due to their woody stems and close arrangement of vascular bundles that reduces larval attack and feeding (Howe 1949, 1950). Some pine species are less susceptible to attack by the pine needle miner, Exoteleia, and white pine weevils, Pissodes, depending on the number and size of the resin ducts (Bennet 1954, Stroh and Gerhold 1965). Other examples of morphological characteristics or physical aspects of resin that confer resistance to a plant are reviewed by Hanover (1975), Beck (1965), and Painter (1951).

Chemical Systems

The role of natural plant products in biological systems in general is just becoming evident, although their significance in conferring resistance to plants has been suspected for more than 90 years. As long ago as 1888, Stahl suggested that plant chemical systems may be important components in the resistance of plants to phytophagous insects as well as being used as cues in the selection of host plants by insects (Rhoades 1979, Rhoades and Cates 1976, and Fraenkel 1959). In addition, chemicals are now known to be important in host plant-pathogen interactions, in the allelopathic or plant-plant interactions (Rice 1977, 1974), and in ecosystem dynamics. In fact, individual chemicals or certain compounds within a class may function in several different ways in a plant, not only as agents conferring resistance or increased competitive ability, but also may play important roles in the primary metabolism of plants (Rhoades 1977, Rhoades and Cates 1976, Seigler and Price 1976).

That many of these chemicals are important in resistance of plants to phytophagous insects is well verified in the literature (Rosenthal and Janzen 1979, Cates and Rhoades 1977, Feeny 1976, Rhoades and Cates 1976, Wallace and Mansell 1976, Harborne 1972, Van Emden 1973). Many biochemically resistant varieties of plants are known and used to reduce the adverse effects of insect pests and pathogens in the management of agricultural systems. For example, maize (Zea mays) genotypes differ in their degree of resistance to the European corn borer, Ostrinia nubilalis. Differential resistance is attributed to differing concentrations of 2,4-dihydroxy 7-methoxy (2H)-benzoxazin-3 (4H)-1 (DIMBOA) and other unknown, presumably chemical, plant factors. The sesquiterpenoid gossypol is important in the degree of utilization of different varieties of cotton which differ in

the amount of this compound produced as well as in the degree
of pubescence (Maxwell 1972). The percent survival to the
adult stage of the cotton jassid Amrasca devastans is more
than 50% higher on susceptible varieties of cotton versus
that on resistant varieties, and the duration of the nymphal
stages is significantly shorter on the susceptible varieties
(Agarwal and Krishnananda 1976). Various alkaloids inhibit
the feeding of the Colorado potato beetle or are toxic to the
larvae (Beck and Reese 1976). Sinigrin, a mustard oil, is a
growth deterrent to Papilio larvae, and causes severe
mortality at concentrations normally found in crucifer plants
(Erickson and Feeny 1974). Over 60 different plant chemicals
have been shown to have an adverse effect on the growth,
development, or reproduction of numerous species of insects
(Cates 1980, Beck and Reese 1976).

Six wheat varieties are resistant to the wheat stem sawfly
(Cephus cinctus) and numerous varieties are resistant to the
Hessian fly (National Academy of Sciences 1969, Painter
1951). Fecundity of herbivores is often greatly reduced by
resistant varieties (Lawton and McNeill 1979, Huffaker 1974,
van Emden and May 1973, Starks et al. 1972, Wyatt 1970).
Presumably this resistance is due to the chemical assemblages
within the plants.

The above examples include primarily annual and herbaceous
perennial crop plants. Until recently there has been a lack
of a general framework of the natural product chemistry of
woody perennials and how this chemistry may affect
phytophagous insects and pathogens. Theoretical treatments
by Cates and Rhoades (1977), Feeny (1976), and Rhoades and
Cates (1976) provide ideas and basic information on the types
of chemicals produced in different tissues of plants and how
these chemicals may influence the feeding patterns, growth
rates, and survival of larval and adult phytophagous

insects. In this chapter, we will discuss and provide evidence for the importance of natural plant products as agents conferring resistance to woody perennials against phytophagous insects. For a discussion of the defensive chemistry of herbaceous plants, the reader is referred to reviews by Cates (1980), Janzen and Rosenthal (1979), Cates and Rhoades (1977), Feeny (1976), and Rhoades and Cates (1976).

Woody perennials, especially forest trees, are known to produce a multi-faceted natural product chemistry consisting of two major types of defensive chemistries that are important in ameliorating the effect of phytophagous insects and pathogens. These are qualitative defenses or toxins and quantitative defenses or protein-complexing systems. These two types are delineated by their mechanism of action against insects and pathogens. Furthermore, the above authors suggest that monophagous, oligophagous, and polyphagous phytophagous insects (herbivores) and pathogens represent at least one of the major selection pressures producing the observed patterns of distribution of these two types of defenses among tissues of any given plant, among individuals within a species, and among plant species.

Since the defensive systems of members of the Pinaceae are included in the following discussion, it is appropriate to define the components of these systems at this time. The oleoresin, especially in pines, is defined as "pine gum, the nonaqueous secretion of resin acids dissolved in terpene hydrocarbon oil which is (a) produced in or exuded from the intercellular resin ducts of a living tree, and (b) accumulated, together with oxidation products, in the dead wood of weathered limbs or stumps" (Stark 1965). The defensive chemistry of conifers is much more complex than the above definition suggests, not only from a chemical but also from a functional perspective. Klement and Goodman (1967) describe the defensive systems of most plants as being either a preformed system or an induced response that occurs in surrounding tissues upon wounding or infection.   In the former system, the attacking bark beetle-fungal complex would encounter a defensive system that exists in the tree prior to infection. This system is exemplified by Pinus ponderosa and usually consists of a short-lived flow of resin. The induced defensive system is divided into either a premunity or hypersensitivity response (Klement and Goodman 1967). Premunity is regarded as a general, nonspecific immunity against pathogens that is acquired after one pathogen infects a plant. Hypersensitivity is a rapidly induced reaction that is developed by an incompatible host upon wounding or infection.   The result of this reaction is cellular necrosis accompanied by the synthesis and/or release of terpenes, phenolics, proteinase inhibitors, and other compounds (Ryan 1979, Hadwiger and Schwochau 1969, Kosuge 1969).

There are, however, some problems in the use of these terms.   The components and dynamics of the premunity and hypersensitivity responses are not at all well known.   In addition, evidence is now accumulating that some plants produce more defensive compounds as the level and intensity of herbivory increases.   Finally, the term hypersensitivity resistance, as defined by Klement and Goodman (1967), primarily describes the necrotic process and not the induced chemical responses of a plant to wounding or infection. As a result, we prefer to use the term "induced system" to describe the necrosis, increased metabolic activity, and the synthesis and/or production of new compounds at the wound or infection site (Ryan 1979, Hadwiger and Schwochau 1969, Kosuge 1969). The result of this system with regard to the bark beetle-fungus complex is to restrict their growth and survival. It is our contention that most plants that are resistant to herbivores and pathogens possess some type of induced system. In many cases, such as in some of the conifers (Berryman 1972), this is a complex system requiring considerable energy expenditure that may confer a high level of fitness on the tree. Several terms have been used by other authors to describe the induced response. Included are terms such as wound response (Shrimpton 1978), hypersensitive reaction (Berryman 1972 and others), secondary resinosis (Reid et al. 1967), and dynamic physiological response in the reaction zone (Shain 1967). We equate these terms with the "induced system" and, occasionally, use these terms interchangeably. The preformed and induced systems of conifers include both toxin or qualitative defenses and the protein-complexing or quantitative defenses.

More specifically, the qualitative or toxin defensive systems of woody perennials represent that portion of the natural product chemistry that primarily affects the metabolic processes occurring within the herbivore. Classes of compounds included in this defensive mode are monoterpenes, acetates, alcohols, phenols, alkaloids, cyanogenic glycosides, isothiocyanates, pyrethrins, and many others. In general, they are present in small quantities, often less than 2% dry weight of plant tissue, gain easy entry into the herbivore's target cells, are highly active against the physiological systems of animals in small concentrations, and are often characteristic of the ephemeral tissues of woody plants. These have been shown to be major defensive systems of flowers, young leaves, and other ephemeral or rapidly growing tissues of woody perennials. This class of compounds is characterized by an extremely diverse group of substances, both from a physiochemical and

functional viewpoint.   For example, pyrrolizidine alkaloids
and oxalates interfere with liver and kidney function, other
alkaloids and amines attack the central nervous system,
saponins destroy erythrocytes by lysis, cardenolides act upon
the muscle systems of animals, cyanide acts by inhibiting the
action of the porphyrin enzyme cytochrome oxidase, and many
other adverse effects of toxins on animal systems are known
(Kingsbury 1964).

   In the preformed and induced resin system of conifers,
toxins take the form of low molecular weight volatile
monoterpenes, acetates, alcohols, juvenile hormone analogs,
resin acids, low molecular weight phenols, and possibly other
compounds (Mabry and Gill 1979, Ikeda et al. 1977, Hadwiger
and Schwochau 1969, Kosuge 1969).   In general, the
concentration of toxins decreases sharply among the more
mature tissues of all plant growth forms (Rhoades and Cates
1976) and this may be part of the cause of increased
susceptibility of older trees to pathogen and insect attack.
For example, bark beetles are known to prefer trees of older
age classes (Shrimpton 1978, Kozlowski 1969).

   Quantitative defenses or the digestibility-reducing,
protein-complexing substances have properties that are very
different from those of plant toxins (Rhoades and Cates
1976).   These compounds are primarily generalized
protein-complexing and carbohydrate-complexing agents, are
often present in large quantities, usually ranging from
10-50% dry weight of tissue, and are often characteristic of
the mature plant tissues of woody perennials such as bark,
wood and mature leaves.   In this class are found resins,
protein-complexing phenolics, condensed tannins, and the
hydrolyzable ellagi- and gallotannins. The complexing nature
of tannins is accomplished through the mechanism of phenolic
hydroxyl-hydrogen bonding or covalent bonding with the proper

substrate forming a complex that is not easily digested (Swain 1979, Ribereau-Gayon 1972). Rhoades and Cates (1976) have calculated that about 80% of dicotyledonous woody perennial plant species contain tannins, while only 15% of annual and herbaceous perennial dicot species contain tannins in their tissues. Protein-complexing systems are known from the foliage of conifers, but it has not yet been established conclusively for the oleoresin system. The induced oleoresin system is known to contain phenolics (Kosuge 1969), some of which may complex with proteins, carbohydrates, and other compounds. Detailed chemical analyses of the preformed and induced resin systems are greatly needed. These analyses must be followed by studies designed to determine the effects of these compounds against phytophagous insects and pathogens.

In order to understand which of the two classes of compounds will offer the most effective defense against any particular phytophagous insect, we must understand the types of selection pressure exerted by different insects. Phytophagous insects and pathogens are usually categorized as monophagous-oligophagous and polyphagous herbivores or pathogens, or specialists and generalists, respectively. The division is arbitrary and it is fully realized that a gradation of herbivore-pathogen types from strict specialism on one plant species to extreme generalism on many host plant species exists, and thus we are referring to opposite ends of a continuum. Most bark beetles are of the more specialized mode. The degree of specialization of the fungus complex associated with bark beetles, however, is not well documented. This fungal complex infests a tissue (the xylem or sapwood) that is composed primarily of cells that are dead at maturity. Among conifers it is assumed that the mature xylem cells per se (tracheids in conifers) transport rather innocuous materials (water and nutrients), and have no intrinsic defensive system. The defensive system is apparently a preformed system found in the oleoresin ducts and/or an induced system. The interaction between the fungal complex and the xylem tissue is in need of further work.

This specialization of the bark beetle and the characteristics of the tissue that is invaded by the fungal complex have important consequences as to the types of chemicals that may be most effective in conferring resistance to trees (Cates and Rhoades 1977, Rhoades and Cates 1976). In the extreme case, where each plant species is experiencing attack by its own specialist herbivore or pathogen, the defense evolved in any one plant species (or tissue) will not be affected by the defenses evolved in other plant species (or tissue) in the community. Parallelism or convergence of chemical defenses can be expected, therefore, for predictable plants and tissues such as forest trees and their contained sapwood, cambium, and phloem (Berryman 1972). We and others propose that the defensive system that has been converged upon is the quantitative mode of defense. Support for this supposition comes from the fact that the diversity of known quantitative defensive systems is much lower in comparison to the plethora of toxins, and of those known, they function in a very similar way. These defensive systems, as compared to the toxin systems, are thought to be effective against all herbivores and pathogens because of their very non-specific, complexing mode of action. In particular, however, quantitative defenses are suspected to be especially well adapted against monophagous-oligophagous herbivores and pathogens. The mode of action of these generalized quantitative substances is to immediately disrupt the securing of nutrients by binding proteins and other plant nutrients when the cells are disrupted, and to reduce their utility as a food resource. Chemical reactions begin before the plant tissue enters the gut or before the fungus and/or bark beetle reach the tissues (Rhoades and Cates 1976). Consequently, there are fewer pathways by which even the specialized herbivore can interfere with these systems as compared to those available against the toxic system. For example, tannins are very general and nonspecific protein-complexing agents, and compounds in creosote bush resin possess similar properties. Tannins complex with and denature a wide variety of digestive and non-digestive enzymes. Chemically, a plant tannin extract consists of a heterogeneous mixture of polymeric phenols of varying chain lengths. It is the very heterogeneous and nonspecific nature of tannin action which may render herbivore digestive enzyme adaptations against these compounds difficult. Rhoades (1977) has found that the proteolytic enzymes of the grasshopper, Astroma quadrilobatum, a creosote bush herbivore, are inhibited by creosote bush resin no less than are porcine enzymes lending weight to the above arguments. For nutrients possessing low intrinsic digestibility, e.g., refractory polysaccharides, the system is in operation before the arrival of the herbivore or the pathogen.

For many plant species, constraints may limit the type of defensive system evolved by the plant or plant tissue. These constraints may be due to the mode of action of quantitative defenses, growth form, phenological condition and activity of tissues (such as rapidly growing tissue with mitotic division among cells vs. a mature tissue), or the allocation of energy to other high priority physiological processes. For example, annuals, many herbaceous plants, flowers, buds, young rapidly growing leaf tissues, and meristematic tissues are notably free of quantitative defenses and are typically characterized by the qualitative or toxin defensive system (Cates and

Rhoades 1977, Feeny 1976, Rhoades and Cates 1976). Quantitative defenses are most common in woody perennials, and in such tissues as long lived green fruits (Dement and Mooney 1974), mature leaf tissues, bark, and wood (Cates and Rhoades 1977, Feeny 1976, Rhoades and Cates 1976). The defensive systems of conifers, <u>Larrea</u> (creosote bush), and other woody plants producing resin systems often contain both qualitative and quantitative systems together. Quantitative defenses may be 'compartmentalized' in some manner, such as by producing resin ducts, specialized storage glands, or by placing the chemicals externally on the tissue. Compartmentalization is often required because these compounds may interfere with plant cell metabolic processes, and in fact, their mode of action is often to bind with cell contents and to disrupt cell metabolism, thereby disrupting their utility as a food source to an herbivore or pathogen (Rhoades and Cates 1976, Berryman 1972).

There is little doubt that chemical defensive systems are under genetic control and are highly heritable (Forde 1964, Peters 1971, Smith 1964, Squillace 1971). Studies dealing with the resistance of white pine to white pine blister rust have verified the genetic control and heritability of this resistance (McDonald 1979, Bingham 1966). Monoterpene composition and concentration are under strong genetic control (Squillace 1976, 1971, Mergen et al. 1955). Hanover (1966 a, b, c) suggests that the quantitative variation in some monoterpenes may be controlled by single genes. It is encouraging that resistance due to phenotypic characteristics is usually an important indicator of a high level of genetic, heritable resistance suggesting that pursuit of phenotypically resistant trees may be useful in the management of forest pathogens and pests (McDonald 1979, Belanger 1981, Bingham 1966). A need exists for the determination of the extent of the variation in the chemical genotypes among trees within populations over the geographic range of the species, the nature of the interaction between endemic bark beetle populations and host chemistry, and the ways in which the environment may modify the defensive chemistry of trees in a stand. The effect of modifying the natural variation among trees by the present silvicultural management methods, or by natural phenomena such as fire, may also provide clues to the dynamics of bark beetle populations both at endemic and outbreak levels. Detailed analyses of the chemical variation among trees within species and among sympatric species is beginning to be addressed in several laboratories (Smith 1977, Squillace 1976, Zavarin 1975, 1965, Squillace et al. 1971, von Rudloff and Rehfeldt, in press) and is an area that warrants detailed investigation. Knowledge of the biosynthetic pathway involved in the formation of terpenes in plants (Mabry and Gill 1979, Nicholas 1973, Tinus 1965) will help determine how habitat and genetic variation affects chemical variation.

Within tree variation of oleoresin composition is not well understood. Hanover (1966 b,c, 1971) found very little influence from non-genetic factors such as season of the year, tree age, vigor, and position of the tree for Pinus monticola. Hodges and Lorio (1975) on the other hand, demonstrated that moisture stress can cause changes in both the monoterpene and resin acid contents of P. taedea. The fact that at least some conifers have more than one resin system containing different compounds serves to complicate matters even more (Russell and Berryman 1976, Rockwood 1973). The authors believe that, as a general rule, within a tree the number and types of chemicals do not change radically, but that amounts and concentrations will vary significantly over time.

What kind of defenses might be expected to have evolved in conifers (and other trees) in response to attack by the bark beetle-fungal complexes? Since the individual beetle species attacking conifers show a high degree of host specificity we would expect the toxin component of the defensive repetoire to be less effective than the quantitative type of defense that interferes with nutrient availability. Bark beetles should not be greatly affected by toxins in the preformed resin systems since, for many generations, they have encountered some level of the toxic components in every tree attacked. Due to constant exposure to oleoresin, beetles that have the capacity to detoxify toxins are selected, survive at a greater level than those without the appropriate detoxification system, and produce progeny that have a higher resistance to the toxins. Any mechanism that would reduce the constant exposure of the beetles to the resin system would be expected to increase the effectiveness of the toxin systems. Consequently, the toxins in the induced system of certain conifers may confer a significant level of resistance against the monophagous-oligophagous beetles.

Some believe that the fungi accompanying the bark beetles are the main tree killing agents. If this is true, the defensive chemistry may have evolved primarily against the fungal complex and not the beetle. A considerable amount, but not all, of the research on tree resistance has been directed toward the bark beetle. Phytopathologists have contributed to our limited understanding of the interaction between the fungi and their host tree (Chap. 6). The relationship of the fungal complex, however, to tree resistance-susceptibility characteristics is an area in need of clarification.

The microsomal detoxification systems of beetles, and animals in general, appear to be very important in detoxifying qualitative defensive chemicals (Brattsten 1979, LaDu et al. 1971, Wilkinson 1968). These systems do have a finite response and can be overloaded. Thus, we would expect that trees with rare genotypes, producing certain unique qualitative or quantitative combinations of compounds, might enjoy reduced levels of attack. The rare genotype would, therefore, increase in abundance, and as the chances that beetles would encounter the resistant genotypes increased, an intense selection pressure would be exerted on the beetles to detoxify the chemicals in this no longer rare genotype. Those beetles with the capacity to detoxify the toxic components of this new abundant tree, in concert with fungi that developed a tolerance for the new chemistry, would survive and reproduce at a higher rate than other beetle-fungal complexes that lacked these capacities. Eventually, the resistance conferred by the genotype would be greatly reduced (Cates and Rhoades 1977, Rhoades and Cates 1976, Callaham 1966a). If the responses of the beetle and the host species is as suggested above, we would expect that one toxin or combinations of toxins that confer resistance against bark beetles might vary over the range of the host plant. This scenario is similar to that experienced in the use of chlorinated hydrocarbons against agricultural pests where eventually the pests become resistant to even the most potent pesticides, resulting in the need to develop an even more toxic compound (Ehrlich et al. 1977).

On the basis of the above considerations, the best defense against any herbivore or pathogen appears to be some system that renders the attacked tissue less utilizable as a food resource, i.e. some mechanism that will decrease the digestibility or the availability of nutrients to the beetle-fungal complex. Such a system has been described in Abies grandis where the beetle and the fungus are confined by

Table 7.1.   The effect of conifer oleoresin, oleoresin
components, and phenolics on bark beetles.

Chemical System	Host Tree	Beetle Taxon
Resin vapor	Pinus ponderosa	Dendroctonus brevi-comis
	P. jeffreyi	D. jeffreyi
Limonene 3-carene myrcene alpha-pinene= beta-pinene inert control	P. ponderosa	D. brevicomis
Limonene plus myrcene	P. ponderosa	D. brevicomis
Limonene 3-carene beta-pinene= standard	P. ponderosa	D. brevicomis
Alpha-pinene beta-pinene camphene limonene terpineol geraniol	Pseudotsuga menziesii	D. pseudosugae
Myrcene	Agar diets done in laboratory	---------------
Monoterpenes in cortical blister resin vs. those in induced resin	Abies grandis	Scolytus ventralis

Proposed Effect	Type of Analysis	Reference
Each beetle species tolerated oleoresin components of host better than that of non-host	Resin vapor	Smith 1961
Limonene most toxic to beetles. Inert control least toxic.	% of total monoterpenes	Smith 1965
Beetles not successful in attacking trees that produce high amounts of both in the oleoresin	% of total monoterpenes	Smith 1966
Trees producing high amounts of limonene were more resistant to beetle attack.	% of total monoterpenes	Smith 1969
Beetles were repelled by Douglas-fir resin and its fractions when tested at close range in the laboratory.	Douglas-fir resin and resin fractions used.	Rudinsky 1966
Inhibits growth of pathogenic fungi found in bark beetles.	Components purified by fractional distillation of oleoresin	Cobb et al. 1968a
Induced resin qualitatively and quantitatively different than cortical resin and more repellent to bark beetles. All monoterpenes were repellent to some degree.	Monoterpenes from commercial supplies and whole resin were used.	Bordasch and Berryman 1977

Table 7.1.   (cont.)

Chemical System	Host Tree	Beetle Taxon
Induced oleoresin phenolics, etc.	A. grandis	S. ventralis
alpha-pinene beta-pinene camphene 3-carene isobornyl acetate limonene myrcene	P. ponderosa	D. ponderosae
Total monoterpene concentration	A. grandis	S. ventralis
Limonene alpha-pinene beta-pinene myrcene 3-carene camphene phellandrene	P. echinata	D. frontalis
Resinosis	A. grandis	S. ventralis
Cortical resin system	A. grandis	S. ventralis

Proposed Effect	Type of Analysis	Reference
Lesions formed due to bark beetle and fungal infections, caused degenerative metabolism of tree tissues followed by production of resin, phenolics, etc.	Resin and phenolics produced in response to wounding and fungal inoculations.	Wong and Berryman 1977
No significant difference in monoterpene content or composition between attacked trees (n = 19) and unattacked trees (n = 167).	Amount monoterpene/unit resin.	Alexander, this paper and Ph.D Dissertation
Trees which produced the least amounts of monoterpenes were the ones successfully attacked.	Amount monoterpene/lesion sample.	Wright et al. 1979
Limonene substances were usually the most toxic. Beetles often responded differently to isomers of the same compound.	Used commercially supplied monoterpenes.	Coyne and Lott 1976
Increased resinosis inversely correlated with brood mortality. 68% of 60 trees received at least 1 attack; 88% of 25 trees attacked, but none of the attacks were successful.	Field observations.	Berryman and Ashraf 1970
Variables influencing tree suitability were (in order from most influential to least): phloem thickness, cortical resin canals, host resistance and predation by woodpeckers.	Amount of beetle egg gallery surrounded by resin impregnated tissues and extent and intensity of the distribution of cortical resin canals.	Berryman 1976

Table 7.1.   (cont.)

Chemical System	Host Tree	Beetle Taxon
Cortical blister resin vs. induced system	A. grandis	S. ventralis
Alpha-pinene beta-pinene limonene and myrcene	P. ponderosa	D. brevicomis
Alpha-pinene beta-pinene 3-carene myrcene and limonene	P. ponderosa	D. brevicomis
Oleoresin and fractions	P. ponderosa	D. brevicomis

Note: Compounds listed in the first column are arranged in descending order of toxicity for each study.

Proposed Effect	Type of Analysis	Reference
Cortical blister resin not as effective as the induced system in conferring resistance to fungus that is associated with S. ventralis.	% of total monoterpenes	Russell and Berryman 1976
Limonene more common in oleoresin of trees growing in sites with a history of bark beetle epidemic. Beta-pinene, the least toxic terpene, and myrcene, which the beetle requires to synergize its pheromone, were the least variable among trees and populations, and were in highest concentration where the beetle is abundant and most destructive. Beetles appear to prefer trees low in limonene and high in alpha-pinene.	% of total monoterpenes	Sturgeon 1979
Trees high in limonene content in oleoresin appear to be more resistant than those low in limonene. Where beetles most destructive and abundant, trees have most balanced monoterpene composition and highest concentration of alpha-pinene. Most susceptible trees appear to have high concentrations of myrcene.	% of total monoterpenes	Smith 1977
Whole oleoresins were repellent. Less volatile oils were less repellent or neutral.	Whole oleoresin and distillation of oleoresin.	Person 1931

the tree to an area characterized by the formation of a necrotic lesion lacking nutrients and flooded with resin (Wong and Berryman 1977). While these quantitative defenses are most efficacious, they also may exact a greater demand on the energy budget and physiology of the plant.

The fungal pathogens appear to be more generalized than the beetle. But even if they are not generalists, the sapwood among conifer species is very similar, being composed of tracheids that are non-living at maturity. Consequently, tracheids do not have the innate capacity to produce a defensive system. They transport primarily water and nutrients, both of which are essential but innocuous to the pathogen. Presumably the hyphae can select among xylem cells and avoid the preformed resin systems of conifers, whereas a beetle or the beetle larvae will be exposed eventually to the preformed resin system or the cortical blister resin. We would expect the quantitative defenses to be effective against the beetle and its associated pathogens, and also suggest that the qualitative compounds will be more effective against the fungal pathogens than against the beetle or the beetle larvae (Cobb et al. 1968a, Shrimpton and Whitney 1968).

To date, comparisons of the differential effectiveness of the various defensive systems against the beetle and fungus are not available. Russell and Berryman (1976) have shown the induced system to be more effective than the preformed resin system against Trichosporium symbioticum. Until recently, the monoterpene, acetate, and alcohol toxin components have been the easiest to analyze and, hence, the most frequently discussed and examined (Table 7.1). The above discussion, however, suggests this approach may not be as fruitful as will a total analysis of the complete defensive system using the beetle larvae, beetle adults, and fungal pathogens as the bioassay of the degree of effectiveness.

HOST RESISTANCE TO ATTACK BY BARK BEETLES

Physical Systems

The woody habit of plants is often cited as a mechanism conferring resistance to some phytophagous insects, but this is not likely for either bark beetles that are well equipped to bore through bark and wood or for the associated fungus that thrives on sapwood.   The thickness of phloem is a primary influence on the number of beetles attacking a tree. There is also a strong positive relationship between phloem thickness and larval growth and brood development (Shrimpton 1978, Amman 1978, 1977, 1972, Berryman 1975, Cole 1973).   The production of minimal phloem in large trees to confer resistance against bark beetle attack appears to not be feasible in view of the supply and demand for water, nutrients, and carbohydrates among tissues of a tree.

In 1902, Hopkins noted that "when they [bark beetles] enter the inner living bark, or bast, the tree commences to exert its resistance by throwing out pitch to fill and heal the fresh wounds in the living tissue."   Since then, 'pitching out' of adult bark beetles and fungal spores has often been considered a physical mechanism that confers resistance to forest trees having high oleoresin exudation pressures or yields (Shrimpton 1978, Smith 1977, 1972, Callaham 1966b, Smith 1966b, Blackman 1931).

Squillace (1971), Mason (1969), and Mergen et al. (1955) infer that this mechanism is, at least partly, genetically determined.   Hodges et al. (1979) point out that of four pine species attacked by D. frontalis, the two species which contain low yields of oleoresin (shortleaf and loblolly) are also more susceptible than the species with high yields (slash and longleaf).   Smith (1977) suggests that 5% of the ponderosa pine in western United States are highly susceptible to bark beetle attack because of a lack of resin flow and that another 5-10% of all ponderosa pine trees may be highly susceptible to bark beetles on the basis of lack of sufficient oleoresin pressure. Smith estimates, however, that 20-30% may be very resistant to bark beetle attack because of good to excellent oleoresin pressures.

Considering the ponderosa pine trees we measured at the San Isabel National Forest in southern Colorado, oleoresin exudation pressure ranged from '0' to greater than 150 psi. Measurements were taken just prior to the mountain pine beetle flight period. Trees successfully attacked by the mountain pine beetle had significantly lower average oleoresin pressure than unattacked trees (Table 7.2). While oleoresin pressure varies both seasonally and diurnally among these trees, the ranking of trees by oleoresin pressure or flow rate did not vary significantly over the flight period of the beetle (July-Sept.).

Vite and Wood (1961) found that mass attacks by D. monticolae and D. brevicomis were significantly correlated with low oleoresin exudation pressures in the trees they measured. In an experiment by Wood (1962), Ips confusus was not able to exploit ponderosa pine trees until the oleoresin pressure of these trees was reduced to zero. Rudinsky (1966), Vite (1961) and Stark et al. (1968) also present data suggesting that trees with high oleoresin pressures are more resistant to bark beetle attack. Hodges et al. (1979) found physical properties of resin (total flow, flow rate, viscosity, time to crystallization) to be more accurate in predicting resistant trees than other chemical properties.

Oleoresin pressure is only a first line of defense and is not always absolute in its adverse effect upon the beetle-fungus complex (Shrimpton 1978, Stark 1965). Oleoresin pressure depends greatly upon the water balance of

Table 7.2. Oleoresin exudation pressure (OEP) and resin flow rates of non-attacked and attacked Pinus ponderosa trees in the San Isabel National Forest, Colorado.

Attack Status	OEP (psi)	Flow Rate (ml/24 h)	N
1979			
Successfully attacked trees	4.9 ± 15.7*	9.3 ± 16.0*	18
Non-attacked trees	37.9 ± 47.0	25.7 ± 24.3	62
1980			
Successfully attacked trees	3.8 ± 9.1*	11.1 ± 3.9*	10
Non-attacked trees	46.8 ± 45.7	33.5 ± 27.8	34

* Indicates significant (P < 0.02) differences between attacked and non-attacked trees. Data presented are means ± standard deviations. Non-attacked trees were within 30 m of an infected tree.

the tree, is of limited duration, and is specific to only the resin canal and to the cells around the entrance tunnel (Shrimpton 1978, Berryman 1972, Vite 1961). Viscosity, rate of crystallization, and the quantity of oleoresin produced may be important in resistance, but their effects have not been well documented. The trees which we observed under attack in southern Colorado, that produced enough resin for an analysis of viscosity, were significantly less viscous than those that were not attacked (p < 0.02). Smith (1975) suggests that ponderosa pines with greater resin flow tend to be more resistant than trees of similar resin composition which have lesser resin flow. Eggs of some bark beetle species do not hatch when covered with resin (Amman 1975, Reid and Gates 1970). Rudinsky (1966) suggests that

Douglas-fir resin exuding from trees with high oleoresin pressures results in death of adult Douglas-fir beetles by suffocation. This effect may not be true for all bark beetles, for we have observed D. ponderosae working for almost six hours while covered with resin, and a similar phenomenon has been reported by Blackman (1931). Berryman (pers. comm.) has observed D. ponderosae for 7 days or more in liquid resin, and the beetles displayed no adverse effects. An adverse effect is observed, however, if the resin crystallizes rapidly (Raffa 1980, Berryman, pers. comm.).

Chemical Systems

One of the greatest limitations in the quest to investigate the plant resistance to phytophagous pests or pathogens is the general lack of knowledge of the chemical defensive systems of plants; the Pinaceae are no exception. Resin systems are present in all genera of the Pinaceae but they differ in whether they are largely a preformed or an induced system. In the genera Pinus, Larix, Picea, and Pseudotsuga, the xylem oleoresin system is suggested to be separate from the needle and root systems and is primarily a preformed system that is continually active and present in the stem tissues (Shrimpton 1978, Berryman 1972). Whether species in these genera also contain induced systems is not known and is an area in great need of clarification. The resin system in lodgepole pine, P. contorta is composed not only of a preformed system, representing the first chemical barrier to insect and pathogen attack, but also an induced system that is activated shortly after the wound occurs, but before the preformed system ceases flowing. The genera Abies, Tsuga, Cedrus, and Pseudolarix appear to lack a preformed resin duct system, the primary defense being an induced resin system in the wood, cambium, and phloem (Berryman 1972). The only preformed system in these genera may be the resin-containing blisters or structures in the outer bark that vary in number, size, and position along and in the stem (Berryman 1969).

Several lines of evidence, other than that presented earlier, suggest that the conifer chemical systems represent a major line of defense against phytophagous insects and pathogens. First, various types of abnormal stresses increase the chances that a tree will be successfully attacked (Hodges and Lorio 1975, Berryman 1975, Hanover 1975, Ferrell 1971, Stark et al. 1968, Rudinsky 1966). It is hypothesized that these stresses adversely influence the chemical defensive systems, thereby lowering the tree's ability to resist the beetle-fungal attack (Rhoades 1979, White 1974, 1969). Secondly, components of resins vary qualitatively among species and non-host resins are more toxic than are the components from the host species (Callaham 1966a, Smith 1963, 1961, Mirov 1961, Mirov and Iloff 1958). This suggests that the beetles possess detoxification mechanisms that permit them to cope with the chemical regimes specific to their host trees but not to those possessed by non-host species (Cates and Rhoades 1977, Baker et al. 1971).

Literature on the resin systems of genera in the Pinaceae is not extensive, but reviews of current knowledge are found in Shrimpton (1978), Berryman (1972), Srivastava (1963), and Bannen (1936). In those species with a preformed system, i.e. one that exists prior to attack by pathogens or insects, the bark beetles are usually inundated by a copious flow of resin. This is a result of well-developed resin ducts that extend both vertically and horizontally in the outer bark, phloem, and sapwood that are severed by the beetle while boring (Shrimpton 1978). The size, distribution, and production of vertical or horizontal ducts in the wood and bark varies among species with a preformed system (Cabrera 1978). The efficacy of this system in conferring resistance to a tree depends on the health and general vigor of the tree, tree age, the water relations of the tree at the time of attack, and the chemical composition, viscosity, and rate of crystallization of the resin (Hodges et al. 1979, Shrimpton 1978, Smith 1977, Amman 1975, Berryman 1972, Shrimpton 1973a, Cobb et al. 1968b, Stark et al. 1968, Rudinsky 1966). Although many of these characteristics have been investigated in preformed resin systems, the complete chemical composition of the oleoresin, the seasonal and developmental changes within a tree, the variability in the qualitative and quantitative composition of the resin among trees in a population and the qualitative and quantitative changes during an attack are in need of greater investigation. Studies in progress are showing tremendous quantitative (Table 7.3) and some qualitative variation in the composition of the foliage resin among individuals of Douglas-fir and the oleoresin of ponderosa pine (Table 7.4) (Sturgeon 1979, Smith 1977). We, among others, believe that this variation is the result, in part, of herbivore and pathogen selection pressures.

Lodgepole pine responds to bark beetle and pathogen attack with a preformed resin system followed by an induced system that becomes active before the former system ceases to flow (Shrimpton 1978). Shrimpton (1978), Reid et al. (1967), and Srivastava (1963) suggest that the induced system originates in lodgepole pine from the ray parenchyma and other cambial parenchyma cells near the wound. In loblolly pine, Pinus taeda, it is believed that the resin in the 'reaction zone' tissue is similar in composition to the original oleoresin. The accumulation of oleoresin in the wound vicinity is due to

Table 7.3. Variation among Douglas-fir trees in the chemical constituents of the current years foliage. Deer Lodge National Forest, Boulder, Montana. (n = 62 trees).

Constituent	Mean	Standard deviation	Range Min.	Max.
Terpenes*				
Alpha-pinene	176	224	0.0	1001.0
Beta-pinene	135	94	0.0	429.0
Camphene	63	63	0.0	211.0
Myrcene	43	64	0.0	311.0
Limonene	85	69	0.0	403.0
Bornyl acetate	63	38	6.0	189.0
Citronellyl acetate	8	8.0	0.0	40.0
Cadinene isomer (C_{15})	7	5.0	0.0	33.0
Tannin**	0.13	0.07	0.03	0.27
Polyphenols	0.41	0.17	0.10	0.96
Total nitrogen ***	1.8	0.46	1.27	3.5

*   counts per 40 mg tissue.

**   Tannins and polyphenols = % fresh weight.

*** % dry weight.

Table 7.4.   Variation among ponderosa pine trees in
monoterpene content (microliter per gram resin) of the
oleoresin. San Isabel National Forest, Colorado.

Compound	Mean	Standard Deviation	Minimum	Maximum
alpha-pinene	17.5	18.8	0	113.1
beta-pinene	90.6	72.7	3.3	350.0
3-carene	30.0	36.3	0	350.0
camphene	6.6	6.0	0	28.9
myrcene	8.8	9.1	0	63.0
limonene	6.4	4.3	0	26.3
isobornyl acetate	11.0	7.5	0	71.8

n = 186 trees sampled except for camphene where 71 trees were
sampled.

pressure deficits that arise in the resin canals upon death
of the parenchyma cells that line these canals (Shain 1967).
Shrimpton (1975) and Berryman (1969) have also shown that
parenchyma cells in phloem and sapwood rays undergo changes
in form and content, and expell resinous materials under
pressure into the wound. The resolution of these conflicting
findings are important to our understanding of plant defenses
and warrant further study.

Although minor differences may exist in the induced
response of cells in the phloem or inner bark and those in
the sapwood, the basic process appears to be similar in both
tissues, at least based upon our present knowledge (Shrimpton
1978, Berryman 1972). Within the bark and sapwood, tissues

surrounding the wound undergo rapid cellular dessication and necrosis. This is accompanied by the loss of starch and the accumulation of oleoresin, phenols, other phenolics, resin acids, lignans, and possibly other compounds (Shrimpton 1978, Russell and Berryman 1976, Berryman 1972, Shain and Hillis 1971, Shain 1967). From the live cambium, a callus is developed well into the undamaged tissue, and finally a periderm is formed surrounding the resin-soaked, chemically diverse, necrotic area, thereby compartmentalizing the fungus and the beetle (Figure 7.1).

At the other end of the spectrum are the genera that have very little or no preformed defensive system, relying on an induced system except for localized cortical resin blisters. In this group, grand fir, Abies grandis, has been studied in depth by Berryman and coworkers, and Struble (cited in Smith 1972) discusses a similar phenomenon in white fir, A. concolor. The induced systems in grand fir and white fir are similar to that of lodgepole pine. Wounding, whether mechanical, pathogen, or insect in origin, induces a reaction characterized by cellular dessication and necrosis that occurs in advance of the pathogen and beetle, thereby confining them to a localized area. Upon necrosis, nutrients and water essential for the growth of the pathogen and beetle are removed, broken down and/or transported out of the area, or in some other way made unavailable to the beetle and pathogen. An area in advance of the site of beetle and fungal attack is flooded with toxic, inhibitory, and possibly digestibility-reducing compounds. Callus tissue, consisting of parenchyma cells produced by the vascular cambium, surrounds the pathogen and seals the wound off from living tissue. In other areas, a periderm develops. The callus and the periderm may contain toxic and inhibitory compounds. The most resistant trees appear to be those that quickly restrict

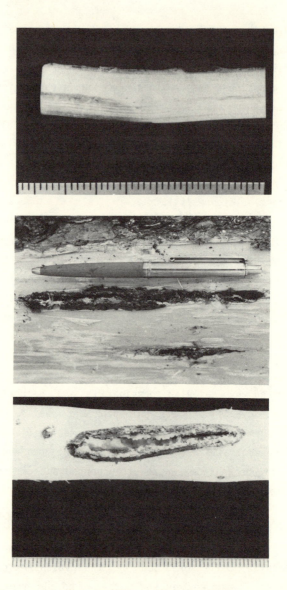

Figure 7.1.   (a) The stem of a six year-old lodgepole pine
seedling 60 days after inoculation with a fungus. (b) Radial
section of a six year-old pine seedling 60 days after
inoculation showing resin soaked tissue.   (c) The induced
response in a mature lodgepole pine tree (about 35 years old)
near Mac's Inn, Idaho. Photographs by R.G. Cates, 1981.

the growth of the fungi through the production of defensive compounds, removing nutrients and water essential to pathogen and beetle growth. Cells at the site of infection and in advance of the pathogen die and form a restricted lesion, resulting in compartmentalization of the pathogen and beetle (Bordasch and Berryman 1977, Wong and Berryman 1977, Russell and Berryman 1976, Berryman 1972, 1969).

The chemical components of the preformed oleoresin systems that have received the most attention are the monoterpenes, acetates, alcohols, resin acids, and fatty acids. Colony growth of the fungi, Fomes annosus (a root rot) and Ceratocystis pilifera (one fungus species associated with bark beetles) was significantly reduced in the presence of vapors of heptane, and myrcene reduced the growth of C. ips by 36% over the controls but its effect was not significantly different from those of phellandrene, limonene or beta-pinene (Cobb et al. 1968a). 3-carene significantly reduced growth of C. ips. Limonene and myrcene had a greater inhibitory effect on C. minor than C. ips. C. schrenkiana was not affected as much as the above were, but still the most inhibitory compounds were limonene, myrcene, and beta-pinene. C. pilifera was adversely affected by most of the volatile compounds more than any other Ceratocystis species (Cobb et al. 1968a). F. annosus growth was reduced by 72% by myrcene, greater than 60% by beta-pinene, limonene, and phellandrene, and by 54% by alpha-pinene over the controls. Sporulation of C. minor was reduced by all volatiles except alpha-pinene, and that of C. pilifera was adversely affected by limonene, phellandrene, and beta-pinene. Myrcene, limonene, beta-pinene, 3-carene, and undecane significantly reduced conidia production in F. annosus (Cobb et al. 1968a). Loblolly pine and pitch pine produce the fungitoxic phenols, pinosylvin, and pinosylvin monomethyl ether, along with pinobanksin and pinocembrin as well as an accumulation of oleoresin monoterpenes, acetates and resin acids as a response to wounding (Shain 1967, Jorgensen 1961). Sinclair and Dymond (1973) report similar compounds, resin acids, and a large quantity of unknown compounds in the acetone extracts of wood of jack pine. In Picea abies the lignans, matairesinol, hydroxymatairesinol, conidendrin, and liovil increased significantly in the 'reaction zone' (Shain and Hillis 1971). In addition, there was quantitative variation in the production of these chemicals among trees, and matairesinol and hydroxymatairesinol significantly inhibited mycelial growth of Fomes annosus. The most striking increase was in hydroxymatairesinol which was also the most inhibitory (Shain and Hillis 1971). Quantitative variation among trees in the resin production upon wounding was also noted for lodgepole pine (Reid et al. 1967). It is important to note that the tree's response to cambial injury results in different compounds being produced than found in the rest of the cells in the 'reaction zone,' such as pinoresinol, ipipinoresinol, isolariciresinol, and secoisolariciresinol (Shain and Hillis 1971, and references therein). Proteinase inhibitors are also known to be induced upon wounding in western white pine (Pinus monticola), ponderosa pine (P. ponderosa), and western larch (Larix occidentalis) (Ryan 1979). These rapidly accumulate upon wounding in sapwood and may adversely affect the digestion of nitrogen containing compounds by the adult beetle and possibly beetle larvae. The action of the proteinaceous fungal enzymes might also be inhibited. In grand fir, the wound resin is different in its qualitative and quantitative composition (Bordasch and Berryman 1977, Russell and Berryman 1976). Wound resin lacks tricyclene, camphene, and bornyl acetate, and contains large quantities

of alpha-pinene, myrcene, and 3-carene, which are also not
present in the cortical blister resin. The sapwood of grand
fir is known also to produce several leucoanthocyanidins
(Puritich 1977). Puritch suggested that aphid infestations
cause alteration in sapwood chemistry, resulting in the
production of at least one different phenolic compound.
Smith (1965) and Cobb et al. (1968a) suggested that 3-carene
and myrcene are toxic or inhibitory to bark beetles and their
associated fungi. In laboratory and field tests, Smith
(1972, 1966a, 1965) showed that, in ponderosa pine xylem
resin, limonene is most toxic or inhibitory to the western
pine beetle, followed by 3-carene. Although considerable
information is known about the preformed and induced resin
systems, substantially more is needed. The types of
chemicals produced, the ways in which they vary both
quantitatively and qualitatively among trees, the mechanism
of their formation, the nature of their adverse effects on
bark beetle adults, eggs and larvae, and on their associated
fungi are not well understood.

HOST SELECTION

No definitive studies have been made that delineate the
complete basis for host selection by bark beetles or, for
that matter, any other herbivores on forest trees. Many
factors are involved in the complex interactions between
phytophagous insects and the selection of their host plants
(Hanover 1975). For any phytophagous insect, the most
important factors in determining evolutionary success are the
selection of a resource providing suitable food and shelter
from the physical environment, the abundance of the resource,
the quality, quantity, and variation in nutrition and the
defensive chemistry of the host, and, finally, the insect
attack density. Investigation into the importance of these

factors has led to two main schools of thought on the most important factors that determine successful host selection. These are phloem thickness, or an adequate food supply, versus chemical resistance, or the insects ability to cope with defensive chemicals.

There appears to be little question that some bark beetles prefer large diameter host trees with thick phloem that will provide suitable food for the adult and the developing larvae, thereby permitting the production of a large number of large offspring (Amman 1978, 1977, 1973, 1972, Cabrera 1978, Amman and Pace 1976, Berryman 1975, Cole 1975, 1973, Cole and Cahill 1976). Climate, however can be an overriding factor. At high elevations or northern latitudes where cold temperatures are common, high mortality rates can occur even in the presence of an adequate food supply (Amman 1978, Safranyik 1978, Safranyik et al. 1975). Otherwise, the productivity of an individual bark beetle, in terms of the number of offspring surviving per given unit of tree diameter or phloem thickness, is related in almost a linear fashion to food abundance, all other factors being equal (Amman 1973, Powell 1966). Rarely, if ever, are the other variables equal among individual trees nor among species that are hosts for the various beetle species. For example, lodgepole pine trees that are 60 years of age or younger appear not only to have smaller diameters and thinner phloem, but also produce a spongy and more resinous phloem tissue. Shrimpton (1973) found that 90% of such trees were resistant when inoculated with a blue stain fungus (but see Peterman 1977). After 60 years of age, resistance to artificial fungal inoculation and fungal establishment decreased significantly. Only 30% of the trees in the 111-140 age class were resistant. Roe and Amman (1970) found that for trees infested by the mountain pine beetle, the average age was 104 years and the average

size was 33 cm in diameter at breast height (dbh). The level
of stand resistance, which is determined by individual tree
resistance, increases up to 60 years of age and then declines
(Safranyik et al. 1975). In other cases, however, stands
much older are not attacked (Mahoney 1978, Berryman 1976).
In western Canada, outbreaks in stands younger than 60 years
have not been reported. Outbreaks are rarely reported from
stands in the 60-80 year age class, but are regularly
reported from stands in the greater than 90 year old age
classes. Even though a general relationship between phloem
thickness and preference by the mountain pine beetle for
trees with thicker phloem was established, phloem thickness
accounted for only 25% of the variance between attacked and
non-attacked trees (Amman 1972). This suggests that some
other variables are important in host selection.

Clearly trees vary in their degree of resistance to bark
beetles and their associated fungi. Table 7.1 summarizes
many of the studies dealing with host resistance systems to
bark beetles. Almost all have dealt with the effects of
monoterpenes which are known to represent only a small
portion of the chemicals found in tree oleoresin. A large
number of the studies cited in Table 1 are laboratory
studies. Comprehensive studies investigating the effects of
differential resistance among trees in a field population on
bark beetle-fungal dynamics are lacking.

It is interesting to note that interspecific differences
and geographical differences within a species are common
(Smith 1977). An interesting question is "What are the
selection pressures maintaining these differences"? We
suggest that the differences are at least in part due to the
bark beetle-fungal interactions with their host plants. One
problem that is pointed out in this table is the need to
standardize methodology. Some studies have simply used total

resin or extracts with the implicit assumption that all resins are alike. Others have considered only the qualitative aspects of monoterpene analysis. Where quantitative differences among resins have been investigated, some measured the amount of monoterpenes per unit resin while others measured the amount of monoterpene per unit phloem. Hodges et al. (1979) have the only published data dealing with differences between resin components other than monoterpenes and how these differences might affect host resistance (Table 7.1).

Berryman and coworkers (Bordasch and Berryman 1977, Wong and Berryman 1977, Berryman 1976, Russell and Berryman 1976, Berryman and Ashraf 1970) have estimated the degree of resistance and susceptibility of grand fir to the fir-engraver beetle and its associated fungus, and are continuing to investigate the mechanisms causing differential resistance levels among host trees. Coyne and Lott (1976) showed that limonene compounds were most toxic to the southern pine beetle, D. frontalis, and speculated that "limonene may contribute to host-tree resistance to beetle attack". Sturgeon (1979) and Smith (1966, 1965) both found that high limonene content in ponderosa pine trees in California may confer resistance to the western pine beetle. They also presented evidence that suggests that western pine beetle infestations are the most destructive where the percentage of monoterpenes are more evenly balanced. In addition, they suggested that areas where this beetle has been most destructive are where the oleoresin is high in myrcene, a compound required by the beetle to synergize pheromones (Chap. 4), and also where the oleoresin is high in beta-pinene. Beta-pinene is the least toxic monoterpene to the beetle (Table 7.1).

Heikken and Hrutfiord (1965), on the other hand, found that D. pseudosugae is repelled by high concentrations of beta-pinene. Rudinsky (1966) indicated that D. pseudosugae is repelled by high concentrations of any of six monoterpenes, including beta-pinene.

Alexander (Table 7.5) could find no difference in monoterpene and acetate concentration or composition between 17 ponderosa pine trees attacked by the mountain pine beetle, versus 49 non-attacked trees in southern Colorado. Hodges et al. (1979) come to similar conclusions for both monoterpenes and resin acids in four Pinus species attacked by the Southern pine beetle. Both studies did determine that physical properties, especially the amount of resin the trees put out in response to wounding, were important. In a slightly different approach, Waring and Pitman (1979) and Wright et al. (1979) found a significant relationship between carbohydrate reserves and beetle attack. High amounts of carbohydrate reserves indicated that the tree was healthy and not likely to be attacked. It would be very interesting to know the relationship between carbohydrate reserves and defensive chemistry.

All of these studies represent important contributions to our knowledge of the mechanisms of resistance of trees to bark beetles. But until more comprehensive studies of the effects of the resin system on bark beetles and their associated fungi are undertaken, the use of host resistance as a management tool is severely limited

ENVIRONMENTAL FACTORS MODIFYING RESISTANCE

Even though resin quantity and composition are under genetic control, their expression is dependent also upon the presence of pathogens, disease, competition, prior levels of insect

Table 7.5.   Monoterpene content of the oleoresin from attacked (n = 17) and non-attacked (n = 49) ponderosa pine trees. No significant differences were found.

	Non-attacked Trees	Attacked Trees
alpha-pinene	15.3 ± 16.5	15.9 ± 11.9
3-carene	23.3 ± 21.4	25.6 ± 17.2
beta-pinene	83.9 ± 45.7	79.6 ± 32.9
myrcene	9.0 ± 9.6	6.9 ± 6.4
limonene	7.8 ± 3.2	6.7 ± 5.0
bornyl acetate	8.9 ± 4.14	9.1 ± 3.07
Total monoterpenes	149.3 ± 65.6	141.4 ± 73.8

Note: Data are presented as the mean ± the standard deviation.   Total monoterpenes are presented as microliters per gram of resin.

attack, drought, pollution, fire, logging, lightning, and any other factor which can affect the physiological condition of a tree.

Resistance is a function of age, with outbreaks occurring in lodgepole pine stands older than 90 years (Safranyik et al. 1975, Shrimpton 1973b, Roe and Amman 1970, Reid 1963, Mogren 1955). Often these outbreaks develop in stands that appear to be occupied by vigorously growing trees with the thickest phloem (Berryman 1976, Cole 1973, Amman 1972) or bark (Kushmaul 1979). Wherever this statement has appeared in the literature, there has been no assessment in the stand of the level of the resistance of the trees, the age of the trees, nor the physiological condition of the trees. Shrimpton (1978) presents data suggesting that beetles tend to prefer trees that have grown at a faster rate during the first 60-80 years, but after that grow at a slower than average rate. Other data suggest also that lodgepole pine may have reached physiological maturity as evidenced by a low current rate of incremental growth combined with a trend in decreasing phloem thickness with increasing age (Shrimpton 1978). Conversely, Ferrell (1973) determined that the densities of scars left by the fir-engraver beetle on white fir did not increase with tree age.

As pointed out by Mahoney (1978), Peterman (1978), Shrimpton (1978), Amman et al. (1977), Cates and Rhoades (1977), Rhoades and Cates (1976), Safranyik et al. (1975), and others, resistance in the form of defensive chemistry is energetically and biochemically costly. For example, the maintenance of such a defensive repetoire in lodgepole pine, as described in earlier sections of this paper, may represent too great a cost for trees greater than 90-100 years of age. Decreasing vigor as trees advance in age may be one of the major reasons for beetle populations reaching epidemic proportions in areas where there appears to be no abnormal stress from the physical environment.

The reasons for insect outbreaks have become an important area of research among foresters, entomologists, and many others working on plant-herbivore problems (Rhoades 1979, Cates 1978, Krebs 1978, Safranyik 1978, Stoszek et al. 1977, Kalkstein 1976, White 1976, 1974, 1969, Berryman 1975, Batzli and Pitelka 1975, Hodges and Lorio 1975, Johnson and Denton 1975, Mattson and Addy 1975, Freeland 1974, Kozlowski 1969, Lack 1966, Chitty 1960, Andrewartha and Birch 1954).

Terminology for the effects of climate or weather on insect populations vary, but all have centered around the idea of a 'climatic release' of an endemic population due to several stressful factors which include drought, air pollution, excessive moisture, hot and dry periods, trees growing on poor soils, ridgetops, or other hazard rated sites. White (1969) suggested that when plants undergo unusual stress there are significant changes in the nutrition of the plant tissues. It is clear from numerous studies (McNeil and Southwood 1978, Dixon 1973, Southwood 1973) that the nutritional quality of plant tissues in healthy trees is only marginally adequate for an insect and that nitrogen is especially limiting for herbivorous insects. Any factor(s) that may change nitrogen to a more soluble form or bring about significant increases in soluble nitrogen in the tissues of a host is proposed to result in an increased larval growth rate, survival, and adult fecundity (White 1969).

In addition to its effects on nutritional quality, stress affects the defensive chemistry of plants. Mattson and Addy (1975) suggested that, while the causal mechanisms eliciting outbreaks of insects under stressful conditions are not fully understood, they are related to changes in both the defensive systems of plants and their nutrients. Rhoades (1979) reviewed the literature on plant defenses and concluded that toxins increase during periods of stress and that digestibility-reducing substances (quantitative defenses) may decrease. He suggested that the production of large quantities of digestibility-reducing substances represents a significant drain on the energy budget of a plant, whereas the production of small molecular weight toxins are less of a commitment (Cates and Rhoades 1977, Feeny 1976, Rhoades and Cates 1976, Cates 1975). If toxins do provide some degree of protection against generalist pathogens or herbivores, then natural selection might favor the increased production of toxins to reduce the adverse effect of these generalists. However, in plants growing under unusually stressful conditions energy must be transferred from one system to another. It is proposed that energy may be shifted from the large commitment to quantitative defenses resulting in a reduction in their production (Hodges and Lorio 1975).

The net postulated changes in plant tissue quality due to unusually stressful conditions are the following: (1) an increase in nitrogen to a more available form to help ameliorate the stress by producing new roots, leaves, or other tissues, (2) an increase in toxin concentration to at least provide some chemical protection against pathogens or herbivores, and (3) a decrease in the production of the large molecular   weight,   generalized   digestibility-reducing compounds   or   other   quantitative   defenses   which   are physiologically expensive. The result of increased nutrition and a simultaneous decrease in the anti-monophagous or oligophagous defensive system is predicted to be an increase in insect and pathogen growth, survival, and fecundity. The degree of the outbreak also depends upon the number of trees affected by stressful conditions, the amount of resource available to the herbivore-pathogen complex, the age of the trees, and other factors.

To our knowledge, these ideas have not been tested using bark beetles, but there is strong evidence supporting such an interaction between these insects and various kinds of stress (Coulson 1981, Hicks 1981, Wright et al. 1979, Alexander et al. 1978, Ferrell 1978, 1973, 1971, Lorio and Hodges 1977, Reid and Gates 1972, Helms et al. 1971, Smith 1969, Vite 1961, Mogren 1956, Caird 1935, Hopkins 1901). Cates and others have tested some of the ideas above using phytophagous insects and their respective host plants. Various levels of water stress were induced using polyethylene glycol 6000 on four   year-old   white   firs   grown   hydroponically   in   the laboratory. The percent weight gain of Douglas-fir tussock moth larvae reared on white fir saplings grown at -22 bars (xylem water potential) was 144.2 ± 41% versus 95.2 ± 16% for the larvae reared on the control trees that were grown at -10 bars. This represents a significant (p < 0.05) increase in larval weight gain on the experimentally stressed trees. In other studies, drought stress (and possibly nutrient stress) was conferred on Douglas-fir trees growing on a south facing slope in the Jemez Mountains of New Mexico by digging 0.5 m trenches 0.5 m from the bole of the tree and then covering the area from the trench to the tree with plastic. Non-trenched trees growing on a north facing slope served as controls. Alpha-pinene and soluble nitrogen concentrations were higher in the foliage of stressed trees as compared to the controls, and beta-pinene, bornyl acetate, and 2 unknown terpenes were lower in the stressed trees as compared to the controls (Table 7.6). Spruce budworm larvae were reared on both sets of trees and the above changes in the foliage quality of the stressed trees resulted in a 31% increase in female adult budworm weight and a 16% increase in male adult weights. This is the first time that the stress phenomenon described above has been shown to benefit the physiology of herbivorous insects. These data suggest that changes in tree physiology and tissue quality due to stress may have a significant effect on phytophagous insect population dynamics and that this interaction may be part of the reason for outbreaks of herbivorous insects, including bark beetles. It is well verified that ponderosa pine trees exhibiting advanced symptoms of pollution injury are heavily infested by bark beetles (Stark et al. 1968). As the severity of the pollution increased, oleoresin yield, rate of flow, and pressure decreased (Cobb et al. 1968b, c). Fir-engraver

Table 7.6. The effect of drought stress on the defensive chemicals and soluble nitrogen content of Douglas-fir foliage 43 days after trenching.

Foliage Chemicals that change with drought stress	Concentration	
	Higher	Lower
Young Needles		
Soluble nitrogen	*	
Alpha-pinene	*	
Beta-pinene		*
Bornyl acetate		*
Terpene 7		*
Terpene 10		*

beetles often are very abundant on their hosts after these trees have been defoliated (Wright et al. 1979, Wickman 1978, Dewey et al. 1974) and other Scolytus species have been shown to be attracted to injured or abnormally stressed hosts (Graham 1968, Meyer and Norris 1967, Rudinsky 1966, Goeden and Norris 1964).

The ability of a tree to resist beetle attacks is also a function of the number of beetles attacking it. After a pioneer beetle 'decides' to settle on a tree, it releases a pheromone which will start the 'mass attack' phase of colonization. This pheromone, plus volatile components of oleoresin from the tree, will attract other beetles to the tree (Chap. 4, Chap. 8, Billings et al. 1976, Hughes 1973, Pitman 1971). Payne (1981) and the authors of this chapter believe that a vigorous, healthy host tree can be killed only if beetle density is high. McCambridge (1969), Smith (1969), and Schmid (1972) have produced evidence that as the number of beetles attacking a tree increases, the greater are the chances that the attack will be successful. Wagner et al. (1979) compare their data with those produced by Fargo et al. (1978) and conclude that low attack densities cause gallery construction to be slower and increase the time between attack and oviposition.

As beetle density goes up, the amount of food reserves per beetle goes down. Beetle densities that are too large result in insufficient food to allow high levels of brood production and survival (Wagner et al. 1980, Amman 1977, Berryman 1976, Coulson et al. 1976, McMullen and Atkins 1961). Thus, the highest rate of beetle survival will be at some intermediate density (Berryman 1974). Hodges et al. (1979) found that for the "average" pine tree, about 100 attacks were needed. Raffa (1981) found that an attack density of around 62 attacks per m^2, for the western pine beetle attacking

lodgepole pine, resulted in maximum reproduction of the
beetle. We hypothesize that this optimum density should
fluctuate, decreasing in less resistant trees and increasing
as tree resistance increases (Figure 7.2). These
considerations may have been important in the evolution of
the attractant and repellant pheromone systems (Chap. 4).

The attack density is influenced also by the distance from
a tree with emerging beetles to a tree that is potentially
colonizable. Knight and Yasinski (1956) claim that only 16%
of an emerging brood will travel more than about 100 meters
from the tree in which they developed.

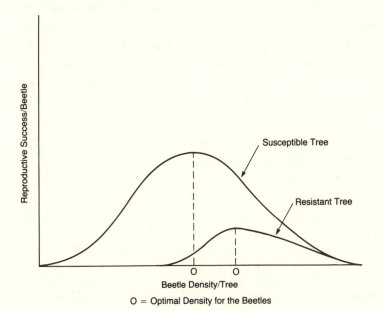

Figure 7.2. Hypothetical graph of beetle success as a
function of beetle density and tree resistance.

McCambridge (1967) showed that trees close to an infested tree were attacked and colonized sooner and more often than trees farther away. Of the 31 trees we observed which were attacked in 1978-1980, all of them were well within 200 feet of an infested tree. In another study, we attempted to induce mass attacks on healthy trees by placing a small number of beetles within a screen attached to the trees. We were successful only if the tree was near an infested tree (Table 7.7).

Long distance travel reduces the probability of a beetle's success in colonizing a tree. Coulson (1981) showed that survival probability in the laboratory decreased drastically after a few days, especially at the high temperatures beetles encounter during their mid-summer emergence. Rain also may be detrimental to exposed beetles (Gauge et al. 1980). In addition, we suppose that a beetle flying or exposed on a tree trunk is more likely to be taken by a predator than one inside a tree. If a long distance flight causes a beetle to be relatively far away from an infested tree, recruitment may become difficult. Finally, since the female's fat body and flight muscles are reabsorbed after a suitable tree has been

Table 7.7.   The effect of distance on the likelihood of beetles successfully establishing in ponderosa pine trees in the San Isabel National Forest, Colorado.

Distance from nearest infected tree	Number of trees not attacked	Number of trees surviving attack	Number of trees overcome by attack
less than 75 m	16	1	16
more than 75 m	15	0	0

located (Reid 1958), a relatively short flight should leave a female with more energy reserves to put into reproduction.

Thus, trees within a stand become more likely to be attacked if there are just a few susceptible "brood" trees within that stand. Thinning, a highly recommended way of controlling beetle infestations (Cole and Cahill 1976, Sartwell and Dolph 1976, Sartwell and Stevens 1975, Stevens et al. 1974), will reduce crowding stresses on the trees and increase the distance beetles must travel to colonize a tree. If the trees that are cut are of low resistance and are the largest trees with the most phloem, the benefits might be even greater.

SUMMARY

The data presented in this paper strongly suggest that genetic resistance of conifers to bark beetles and their associated fungi is determined by a multifaceted chemical defensive system consisting of toxic terpenes, fungitoxic phenols, fungitoxic lignans, leucoanthocyanidins, proteinase inhibitors, resin acids, and possibly other unknown toxins or digestibility-reducing substances. It is very likely that all conifer species have some sort of defensive system that is induced upon wounding or attack by bark beetles and fungi. For instance, ponderosa pine, which has a well developed preformed resin system, is now known to produce proteinase inhibitors upon wounding (Ryan 1979). Although much is known of the general preformed and induced systems of lodgepole pine and grand fir, considerably more information is needed to describe the types of chemicals produced by other species, the quantitative and qualitative chemical variation among trees in a population, and the adverse effect of these chemicals on bark beetle adults, eggs, larvae, and their associated fungi.

An analysis of current plant-herbivore theories suggests that conifers, as well as other woody perennials, have evolved some type of digestibility-reducing or quantitative system as the primary defense against monophagous-oligophagous herbivores such as bark beetles (Rhoades and Cates 1976). Toxin defenses may adversely influence monophagous-oligophagous herbivores, but are proposed to have the greatest effect on polyphagous herbivores. The induced reaction in conifers due to wounding is characterized by cellular dessication, removal of nutrients essential for the growth of the beetle progeny and the fungi, and flooding of the general area with both toxins and digestibility-reducing compounds (Wong and Berryman 1977). Simultaneously, necrosis occurs in tissues well in advance of the spread of the attacking organisms. The physical texture of the resin may also be an important component in the quantitative defensive system.

Although monoterpenes have received the greatest attention of all the chemicals in the conifer defensive system (Table 7.1), digestibility-reducing mechanisms or quantitative defenses may be the most important in conferring resistance of host trees against bark beetles and their associated fungi. This is an area that deserves much greater attention.

The selection of a tree by attacking beetles is clearly dependent upon several factors, but two of the most important are the quantity of food available (phloem thickness and number of susceptible trees in a given area), and the physiological condition and level of chemical resistance of the host. Resistance, in turn, may be modified by tree age and vigor. Clearly, the interaction between the genotype of a tree and its environment must be considered before resistance can be understood. In addition, bark beetle and fungal bioassays are needed to determine how variations in these factors may influence beetle and fungal growth and population dynamics.   Finally, it will be instructive to analyze beetle and fungal behavior in natural stands that vary in their chemistry, age, vigor, density, species composition, and site characteristics.   These types of stands will be important in assessing the potential of host resistance as a management tool.

8. Population Dynamics of Bark Beetles

A. A. BERRYMAN

The study of population dynamics concerns change, specifically changes in the state of a population in space and time and the causes of these changes. The population state in which we are usually most interested is the number or density of individuals. However, we may also be interested in the distribution of individuals, the frequency of age classes, and the frequencies of particular genes or any other variable which describes the state of a population. The variables we observe in order to characterize the state of a population are called, reasonably, state variables. Changes in state variables over space and time are caused by the operation of the processes of natality (births), mortality (deaths), and migration (movements into and out of a particular region). These processes, in turn, are controlled by the individual properties of the organisms making up the population and by the properties of the environment in which the population lives (Figure 8.1). Of course, neither the environment nor the individuals in a population are static entities, for they change continuously in time and space. Environmental factors such as weather, food, predators, and competitors, may change independently of the population, or may change in response to the population state variables. Weather, for instance, usually operates as an independent external variable to affect birth, death and migration rates within the population. Predators, however, often respond to the density of their prey population because they migrate into regions where their prey is abundant and reproduce more offspring when food is plentiful. When a population influences the properties of its environment, which in turn feed back to affect the population, we have what is called a feedback loop (Figure 8.1). Feedback loops have important qualitative effects on the dynamic behavior of population systems, positive feedback giving rise to growth or decay dynamics and negative feedback to steady-states, oscillations and population cycles (Berryman 1981). When the feedback involves changes in the genetic properties of biotic environmental components, such as food organisms, predators and competitors, these organisms are said to coevolve with the population in question. Coevolution, in turn, has important qualitative effects on characteristics of the population, particularly its stability and persistence (Pimentel 1971).

Feedback also occurs when population state variables influence the properties of individuals in the population (Figure 8.1). For example, when populations become very dense the "normal" behavior of individuals may be disrupted, causing reduced fertility and fecundity and high death and emigration rates. These changes in behavior may be physiological or genetic in origin. Physiological changes in individuals result from stresses induced under crowded conditions, while genetic changes in the population occur when different genotypes are selected for under different population densities (Christian and Davis 1971, Chitty 1971). With this, brief general look at population systems and their functioning, we will now examine one particular system, that of bark beetles.

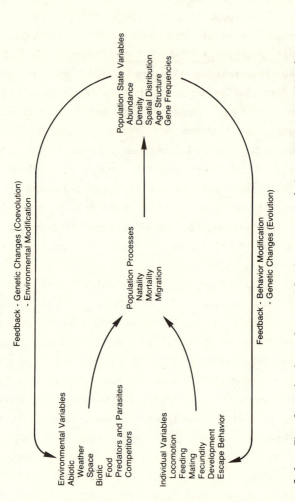

Figure 8.1.   The functioning of a population system (after Berryman 1981a).

STABILITY, PERSISTENCE AND COEVOLUTION

The bark beetles are a unique and highly specialized group of insects. Unlike most plant feeders, which browse on or extract sap from their hosts, bark beetles live for much of their lives within the tissues of their host which they have to kill, or at least kill part of, in order to reproduce successfully. In this way they resemble some mammalian predators which attack and kill prey much larger than themselves. Such predators often work together in cooperative groups or packs in order to overcome their powerful prey. Similarly, bark beetles have evolved complex cooperative interactions, mediated by chemical pheromones (Chap. 4), which enable them to overwhelm the defenses of their coniferous hosts (Chap. 7). In addition, mutualistic associations between bark beetles and plant fungi have evolved to play a critical role in killing the attacked tree (Chap. 6). These two specializations, chemical communication and symbiosis, contribute to the evolutionary success of bark beetles and form the key to understanding their population dynamics.

Like all species which cause the death of their host, bark beetle-fungus associations are faced with the problem of conservation; that is, if the beetle-fungus populations become too efficient at finding and attacking their hosts, they may reduce their food supply to such a low level that their own extinction becomes a distinct possibility. This problem is particularly acute for bark beetles because their hosts must grow for many years to reach a size which is favorable for beetle reproduction. Simultaneously, the host trees have evolved defenses against the beetle-fungus association which have served to maximize their own survival and persistence. These reciprocal interactions between the beetle-fungus complex and their hosts have lead to a system that has persisted in time and space. This process is called coevolution.

Perhaps we can examine the coevolutionary process by looking first at a system in which the organisms have not had the opportunity to adapt to one another. The smaller European elm bark beetle (Scolytus multistriatus) and its fungal symbiont (Ceratocystis ulmi), the pathogen which causes Dutch elm disease, were introduced into the United States in the early 1900's. In the 80 years since its introduction, the Dutch elm disease has spread throughout the eastern and midwestern United States causing catastrophic losses to this valuable shade and forest tree, and it is now beginning to make inroads into the western States. The beetle feeds on healthy elm twigs prior to boring into the trunks and larger branches of weakened trees to lay its eggs. During this preliminary feeding activity the fungus may be inoculated into the tree and, if infection occurs, the tree is weakened and eventually dies. The weakened and dying trees then form suitable breeding sites for more beetles, producing more fungus infections and more weakened elms. This type of unstable positive feedback loop, or "vicious circle", will eventually lead to the death of all the trees and the extinction of elm, beetle and fungus (Figure 8.2). Will the American elm become extinct because of this introduced pest complex? It will if the interacting species are unable to coevolve and form a stable community.

A likely way for elm trees to adapt to this challenge is to evolve resistance to invasion by the fungal pathogen. There may be a few resistant genotypes in the population, and their offspring could comprise the majority of the next generation

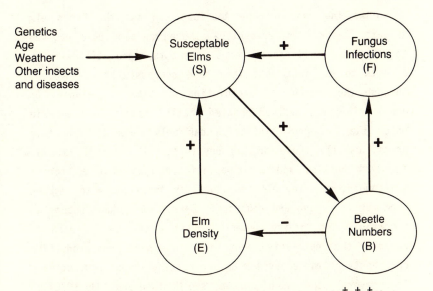

Figure 8.2. The unstable positive feedback loop ($S\overset{+}{\to}B\overset{+}{\to}F\overset{+}{\to}S$) in the interaction between the smaller European elm beetle, a pathogenic fungus, and the American elm.

of American elms. Other possibilities for adaptation include the evolution of less virulent strains of the fungus which cannot colonize healthy trees, or a change in beetle feeding behavior prohibiting the spread of the fungus to healthy hosts. Whatever mechanisms are involved, it is necessary for the positive feedback loop in Figure 8.2 to be broken before the system can reach equilibrium and thereby persist in time; e.g. the link F-S of Figure 8.2 will be broken if healthy elms are resistant to the fungus, if the fungus is avirulent when inoculated into healthy trees, or if the beetle does not feed in healthy hosts.

Unlike the case of the American elm and the Dutch elm disease, native conifer-beetle-pathogen systems have evolved together over millions of years. That they are still with us attests to the fact that they have successfully co-adapted. Through the process of coevolution a state of balance, or equilibrium, has evolved between beetle, fungus, and tree in which the pathogenic infection can only succeed in certain hosts, usually those of low physiological vigor (Chap. 7). The bark beetles and their associated fungi act as forest scavengers, removing old and unhealthy individuals and making room for the young and vigorous. Bark beetles have become an important part of the dynamic forest community, helping to recycle the nutrients contained in these non-productive components of the ecosystem and increasing forest productivity (for several other such examples see Mattson and Addy 1975).

Perhaps the general properties of the coevolved beetle-fungus-tree system are best illustrated in a feedback diagram similar to that of Figure 8.2. In the simplified system shown in Figure 8.3, we see that the vigor of individual trees is negatively affected by increased stand density, because of increased competition for light, water and nutrients; the beetle population is negatively affected by tree vigor, because beetles can only reproduce in weakened individuals; and stand density is negatively affected by beetle population density, because trees are removed by the activity of the beetle and its fungal symbionts. This interaction forms the negative feedback loop D→V→B→D Figure 8.3, because an odd number of negative interactions (3 in this case) gives a total negative effect. Negative feedback loops are intrinsically stabilizing (Berryman 1981a) in that they tend to create equilibrium conditions in which the variables

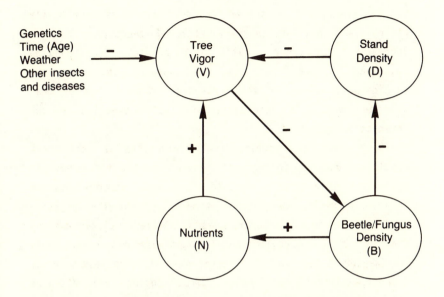

Figure 8.3. Potentially stable negative feedback loops
(D→V→B→D and V→B→N→V) in the interaction between the
smaller European elm beetle, a pathogenic fungus, and the
American elm.

attain a steady state; that is, stand density, tree vigor and
beetle population density will eventually reach relatively
constant values. However, if there are time-delays in the
negative feedback structure then we may find oscillations or
cycles in the steady-state behavior. As it may take some time
for tree vigor to respond to stands being thinned by bark
beetles, we might expect beetle population and stand densities
to cycle around some mean value. The second feedback loop in

8.3 ($B \overset{+}{\rightarrow} N \overset{+}{\rightarrow} V \overset{-}{\rightarrow} B$) involves the recycling of nutrients from beetle killed trees back into the residual stand, thereby increasing its vigor.   This also creates a stabilizing negative feedback loop but, once again, there may be considerable delays in the breakdown and recycling of dead wood and this will accentuate the cyclic dynamics of the interacting system.

This elementary qualitative evaluation of the bark beetle-fungus-tree interaction indicates that the coevolved system is intrinsically stable, but that we may observe population cycles because of time-delays in the negative feedback structure.   However, this simplified analysis assumes that all external environmental components remain constant. In Chapter 5 we saw that environmental stresses such as drought, nutrient deficiencies, fire, logging, root diseases and insect defoliation may alter the vigor of individual trees or stands.   In addition, time itself acts as a stress factor because trees become less vigorous as they age.   If we include the concept of external environmental stress, then the behavior of the system becomes much more complicated.   When stands come under permanent stresses, such as old age (overmaturity), bark beetle populations will slowly increase and may eventually explode to outbreak levels.   Devastating large-scale outbreaks are, therefore, much more likely to occur when the tree species grows in extensive, even-aged, pure stands.   For instance, lodgepole pine often regenerates in even-aged pure stands following forest fires.   Because these stands reach susceptible age simultaneously, dramatic outbreaks of the mountain pine beetle often erupt, killing most of the larger trees in the stand (Safranyik et al. 1974;

Amman et al. 1977). However, such infestations, which are devasting from man's point of view, may play a central role in a coevolved system. As Peterman (1978) has pointed out, the likelihood of forest fires is increased in areas where there are large numbers of dead trees and fire, in turn, creates conditions favorable for lodgepole pine regeneration. Peterman showed (via simulation modeling) that the beetles attack stands at an age when the accumulation of seeds in their serotinous cones is optimal for regeneration of the new stand. In other words, the beetles and the pines interact in such as way as to optimize the fitnesses of both species. This is surely a remarkable example of co-adaptation!

Environmental stresses which permanently weaken individual trees or small groups of trees (root diseases, fire, logging, etc.), or which temporarily weaken whole stands (fires, insect defoliation, droughts, etc.), create conditions for local or short-term beetle outbreaks which usually kill the least vigorous trees in the stands. For example, Figure 8.4 illustrates a situation where a fire weakened large numbers of ponderosa pines and precipitated a 2-year explosion of the western pine beetle. Beetle caused pine mortality also increased substantially in uninjured stands, probably because large numbers of beetles migrating from the outbreak area were able to overcome more resistant trees.

Having examined the general nature of the interaction between forest trees, bark beetles and their associated fungi, it is now time to narrow our scope to get a more intimate perspective of bark beetle population dynamics.

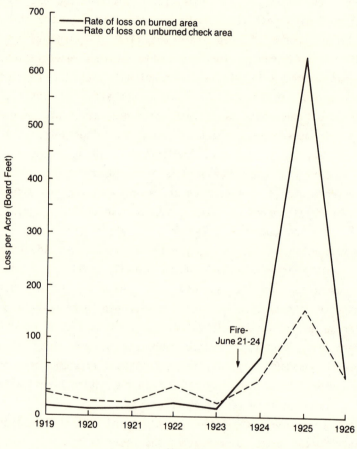

Figure 8.4. Volume of ponderosa pine killed annually by the western pine beetle on an area (5,460 acres) burned on June 21, 1924, and on an adjoining unburned area (1,880 acres) (redrawn from Miller and Keen 1960).

POPULATION SYSTEMS

All bark beetle population systems can be divided into three basic components--dispersal, host colonization and production (Figure 2.2). Each of these components of the system can be subdivided into a series of behavioral or physiological processes which are characteristic of a particular bark beetle species (i.e., the individual properties of Figure 8.1). To understand why bark beetle populations behave as they do we need to examine these components and their processes in some detail.

Dispersal

When mature bark beetle adults emerge from the dead trees in which they have been reared, they fly in search of living hosts to attack. Emergence and flight are governed, to a large degree, by weather factors, particularly air temperature and light (Graham 1959, Shepherd 1966, Atkins 1966). Each species of beetle flies within a certain temperature range and newly emerged individuals are usually photopositive, flying towards the light. Their behavior during the initial stages of the flight period appears to be an evolutionary adaptation to the problem of locating rare and diffuse food sources, i.e., weakened trees widely scattered throughout the forest. For instance, newly emerged beetles of some bark beetle species seem to require a period of flight exercise before they will actively search for hosts to attack (Graham 1959, Atkins 1966, Shepherd 1966). This initial flight behavior tends to disperse individuals widely and increases the probability that some individuals will encounter susceptible hosts (Raffa and Berryman 1980). However, there is some evidence to suggest that, at least in some species like the southern pine beetle, the primary dispersal urge may be overridden if the beetle flies into a pheromone plume emanating from an attacked tree (Gara and Coster 1968). This behavior is adaptive because a pheromone plume identifies a suitable breeding site.

After their initial flight, adult beetles become physiologically conditioned to search for feeding sites. Some species, particularly among the genus Scolytus, go through a maturation stage, where they feed in the crotches of twigs and branches before searching for permanent hosts. The question of how bark beetles select their hosts has generated an ongoing controversy. Some authors have concluded that they are guided by odors emanating from weakened trees (Person 1931, Rudinsky 1966, Werner 1972, Anderson 1977, Heikkenen 1977). However, most of these experiments were conducted on logs from cut trees and, in some of these studies, beetles producing pheromones may have been present in the logs. Inferences from such studies are difficult or impossible to extrapolate to standing, living trees. Most of the experiments which have been conducted on living hosts support the null hypothesis; i.e., that flying beetles land at random and are not guided by hosts volatiles (Berryman and Ashraf 1970, Moek 1975, Hynum and Berryman 1980, Raffa and Berryman 1980). There is evidence that some bark beetles (e.g., the mountain pine beetle and southern pine beetle) orient by sight to large, dark, vertical objects (Gara et al. 1965, Shepherd 1966), which means that they would land with greater frequency on larger trees, but other evidence suggests that landing rate per unit area of tree surface is independent of tree size and species (Hynum and Berryman 1980).

At first sight, this haphazard behavior of indiscriminate landing on hosts and non-host trees appears to be an inefficient strategy. However, to gain insight into its evolutionary significance we must examine its place in the full spectrum of host selection behavior. Once a beetle has landed on a tree it reaches a point where a crucial decision must be made--to bore or not to bore. Some beetles appear to bore indiscriminately into any member of their host species; e.g., the fir engraver beetle bores into any host on which it lands, but will only establish galleries in those hosts which are unable to respond by producing resins and other toxic or repellent materials (Berryman and Ashraf 1970, Berryman 1972, Bordasch and Berryman 1977). In this case, if the tree proves unsuitable, the beetle will leave and attack another tree. Other bark beetle species (e.g. many Dendroctonus) are not repelled by resin flow, per se, but seem to make their decision after initially feeding, or "tasting" the bark of the tree (Baker and Norris 1967, Hynum and Berryman 1980). Apparently these species are able to sense chemicals present in the outer bark or phloem, perhaps to assess the levels of toxins (Chap. 7) and pheromone precursors (Chap. 4) before committing themselves to the attack. For these species, initial flight and random landings, coupled with gustatory stimulation, is a simple yet effective adaptive strategy for an insect searching for a rare and widely scattered suitable host, for it ensures that a large proportion of the host population will be "tasted" and that few susceptible hosts will escape detection. This process of population dispersal is summarized in the upper part of Figure 8.5. Notice that the host plays a passive role in this interpretation of the dispersal process.

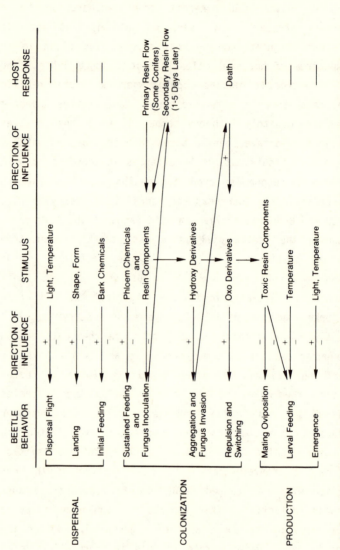

Figure 8.5. Stimuli and host responses governing the attack behavior of bark beetles; + = direct relationship, − = inverse relationship (modified from Raffa and Berryman 1980).

Host Colonization

Colonization of the host is a critical phase of the bark beetle's life cycle, for it is here that they meet face to face with the defensive systems of the tree (Figure 8.5). As beetles enter the inhospitable environment of the tree's living tissues, pathogenic fungi are deposited and their growing hyphae penetrate the cells of phloem and sapwood (Chap. 6). In healthy hosts, this pathogenic invasion triggers a series of defensive reactions in the adjacent cells (Reid et al. 1967, Berryman 1969, 1972). Terpenoid and phenolic compounds released by intense metabolic reactions in the phloem and sapwood parenchyma cells soak the tissue surrounding the infection and inhibit fungal penetration and beetle boring activities (Wong and Berryman 1977). However, the precess of active defense takes time and, if the tree is colonized very rapidly (within one or two days), this defense system can be circumvented. It is this race between beetle attack and tree defense which sets the stage for the dynamic life and death struggle between beetle and tree. From the beetle's point of view, its success in this struggle is critically dependent upon production of pheromones which attract other beetles to help overcome the defenses of the host.

Pheromone Production. Bark beetle pheromone attractants are hydroxy derivatives of terpenoid compounds which were originally produced by the tree and were released when it was wounded (Chap. 4). Thus, pheromone production is directly linked to the defense system of the tree. There is evidence to suggest that the attractant chemicals are actually synthesized from resin precursors by bacteria living in the gut of the beetle (Brand et al. 1975). This is an interesting evolutionary sequence of events--the metabolic waste products of bacterial terpene metabolism now serve the beetles as pheromones which coordinate and direct the mass attack. In effect, the successful production of pheromones advertises the presence of a tree which has a high probability of being sucessfully detoxified (Raffa and Berryman 1980).

It has been demonstrated by several workers that pheromones are produced in the gut of beetles after mere exposure to the vapor of conifer resins (Vite et al. 1972, Hughes 1973). This suggests that pheromone production occurs as soon as the first attacking, or "pioneer" beetle, enters the living tissues of the host. However, our studies with the fir engraver beetle and mountain pine beetle indicate that pheromone release can be prevented by heavy and persistent resin flow (Berryman and Ashraf 1970, Raffa and Berryman 1980). We hypothesize that heavy host secretions may prevent the beetle from feeding and expelling the pheromones in its faeces, or it may mask or denature the attractive chemicals.

Some coniferous trees, particularly the pines, have a well developed system of resin ducts in the sapwood and phloem. This system seems to be particularly important during the early phases of colonization. Pioneer beetles boring into the phloem sever the radial ray ducts, and resin immediately flows into the open wound. The beetle deals with this flow by pushing the liquid resin out of its tunnel, forming a "pitch tube". It appears that pheromone production is prevented, or its effectiveness is greatly reduced, while the insect is engaged in fighting a heavy resin flow (Raffa 1980).

In some bark beetle species (e.g., the fir engraver beetle), the production of attractive pheromones appears to be inversely related to host resistance so that attraction is greatest to the least resistant trees (Figure 8.6). In others, like the mountain pine beetle, attraction to cut logs (dead trees) is considerably less than to living trees of intermediate resistance (Raffa 1980). In these more "aggressive" species, which are able to attack more resistant hosts, pheromone production increases directly with host resistance, up to a point, after which it is inhibited or masked by heavy resin flow (Figure 8.6).

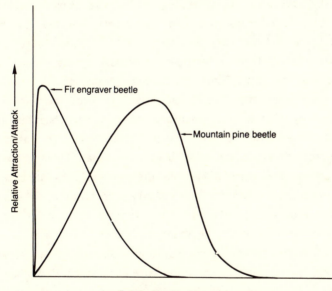

Figure 8.6. Hypothetical relationship between the relative attraction created by pioneer beetles and host resistance for a non-aggressive species (fir engraver beetle) and an aggressive species (mountain pine beetle).

Although aggregation for the purpose of overwhelming the defenses of living hosts is an effective evolutionary strategy, it poses some conflicting problems for the bark beetle. If too many beetles attack a living tree, then their offspring will find themselves short of food or space and many will die of starvation or cannibalism (Berryman and Pienaar 1973). Thus termination of the attack as soon as sufficient attacks have accumulated to kill the tree would be advantageous to the beetles already in the tree. Attack termination may be brought about by the cessation of attractive pheromone production or, in some beetles, by the production of anti-aggregating or "masking" pheromones (Rudinsky 1968). If this strategy is to work, the production of masking substances must be intimately related to the physiological status of the host (Raffa 1980, Raffa and Berryman 1980). There is evidence that masking pheromones can be produced by some fungal symbionts (Brand et al. 1976). These pheromones signal that the tree has been successfully colonized by fungi, and, therefore, that no more beetles are needed to overcome its resistance. In fact, the rate of attack for the mountain pine beetle declines linearly with attack density (Figure 8.7), and the maximum rate of attack and final attack density are highly correlated. Because attraction to susceptible hosts is directly related to their resistance (Figure 8.6), we would expect the maximum rate of attack (beetles attacking per day), and thus the final attack density, to be greater on the more resistant individuals (Figure 8.7), at least up to the point where excessive resin flow begins to interfere with pheromone effectiveness.

Bark beetles have evolved complex aggregation and anti-aggregation systems which are closely tied to the

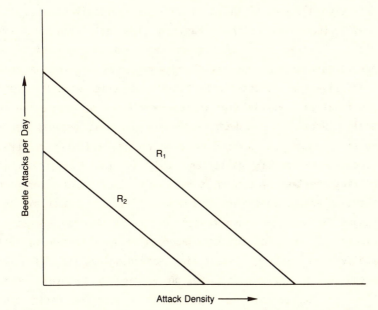

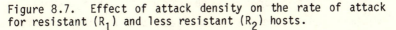

Figure 8.7. Effect of attack density on the rate of attack for resistant (R_1) and less resistant (R_2) hosts.

defensive responses of their hosts. Some species are adapted to attack hosts of greater resistance than are other species, providing a degree of niche displacement between species attacking the same host.

Production

Once bark beetle adults have successfully colonized their host they bore galleries in the phloem-cambium region and deposit, their eggs in niches cut in the lateral walls of these tunnels. The orientation of these egg galleries varies between species, but appears to be associated with the defensive systems of the tree and the aggressiveness of the beetle (Berryman 1972). Many species attacking conifers without a resin duct system, or dead and dying hosts, bore horizontal (across the grain) egg galleries, while the more aggressive species often orient their galleries with the grain (vertically). Vertical galleries seem to be an adaptation to deal with the resin duct system of the host because fewer vertical ducts are severed by a vertically oriented gallery. However, horizontal galleries would be more effective in killing the tree by acting as as girdle; they would sever more sieve tubes and tracheids, and more efficiently innoculate the fungus, which grows much faster vertically than in any other direction. Some of the most aggressive species may in fact modify their gallery orientation according to the defensive responses of their hosts, boring vertically to avoid resin flow and horizontally or downwards to increase resin flow which is needed for pheromone production. For example, the western and southern pine beetles bore winding or sinuous galleries with varying orientation, and the turpentine beetles make cave-like galleries, boring up to decrease and down to increase resin flow (Wood 1963, Berryman 1972).

As was mentioned previously, one of the greatest problems facing species which aggregate on their food supply is overpopulation, and the resulting shortage of food for their offspring. Bark beetles have evolved several mechanisms for minimizing the overpopulation problem. The first, pheromone-mediated termination of the attack when the host has been overwhelmed, has already been mentioned. A second adaptation is their utilization of the outer bark as an additional source of nourishment, thereby creating a three-dimensional food resource. This adaptation has been carried to its extreme by species such as the western pine beetle whose larvae feed for only a short time in the living phloem before boring out into the dry outer bark of its thick-barked host, ponderosa pine. A third adaptation is their regulation of the number of eggs laid in any one tree. In several species, the number of eggs deposited by each female decreases as attack density rises (Figure 8.8).

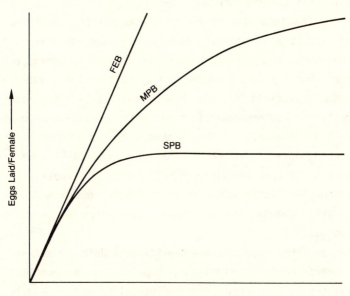

Figure 8.8. Relationship between female oviposition and attack density for the fir engraver beetle (FEB) (Berryman 1968b), mountain pine beetle (MPB) (Berrryman, unpublished data), and southern pine beetle (SPB) (Coulson et al. 1976a).

Females which have not laid their full complement of eggs may re-emerge from that tree, and attack a new host. The southern pine beetle is an excellent example of this type of oviposition behavior (Coulson et al. 1976a). Species like the fir engraver beetle, which do not regulate oviposition, or the mountain pine beetle, which only partly control egg depostion (Figure 8.8), are subjected to higher mortality as attack density increases. Most of this density-dependent mortality is due to intraspecific competition for food and space. Although the exact cause of death has been difficult to identify, it is usually attributed to starvation or inadvertent cannibalism (Berryman 1974, Berryman and Pienaar 1973). The important aspect of intraspecific competition is that is acts as a negative feedback loop regulating the number of beetles emerging from infested trees. Intraspecific competition may also be moderated by host factors, especially by the dimensions of the food resources. For instance, competition between mountain pine beetle broods is reduced in direct relationship to the thickness of the phloem layer in which the larvae feed (Amman 1972, Berryman 1976). We might expect other bark beetle species, particularly those like the western pine beetle which feed on the dry outer bark, to be similarly affected by the radial dimensions of their food resources.

In addition to the strong density-dependent control exerted by intraspecific competition, immature bark beetles are subjected to an array of other mortality agents, including parasites and predators (Chap. 5) and interspecific competition with other phloem feeding insects (Berryman 1973, Coulson et al. 1976b). Interspecific competiton amongst different species of bark beetles is minimized, to a certain degree, by preferences for different dimensions of host material. For instance, western pine beetles usually attack the mid-section of their host trees, red turpentine beetles attack the basal portions and various engraver beetles their tops (Miller and Keen 1960). This behavior provides a degree of niche displacement in space, and reduces the intensity of competition. The aggressive nature of the particular species also provides a degree of niche displacement in time, the more aggresive species being first to colonize living trees. However, even with thee differences in behavior, competition between different bark beetle species is often seen and may even lead to the competitive displacement of one species (Berryman 1981).

Weather, in particular temperature, may also have important effects on bark beetle production. Excessively cold temperatures during winter months can kill a high percentage of the overwintering broods (Berryman 1970), and excessively hot temperatures may have the similar effects in summer (Miller and Keen 1960). Temperature also determines the rate of larval growth the development and, therefore, the number of beetle generations produced during the year (Miller and Keen 1960). Species such as the western pine beetle and southern pine beetle, when living in warmer climates, may produce several generations a year. The rapidity of bark beetle population buildup is, therefore, closely tied to weather in these species. In more northerly temperate regions, most bark beetle species have but a single generation per year. However, warmer or colder than normal temperatures may cause them to overwinter in a stage which is poorly adapted to cold exposure (Safranyik 1978). Some species are able to minimize this problem by entering a dormant state (diapause) prior to the onset of cold weather.

POPULATION DYNAMICS

The dynamics of bark beetle populations are governed by the various processes operating during dispersal, colonization and production.   During each of these phases, the distribution and abundance of the beetle population changes in space and time.   We will examine these dynamics and their causes below.

Dynamics of Dispersal

The numerical dynamics of dispersing bark beetle populations is governed, to a considerable degree, by processes operating during the production phase, and by climatic factors.   These processes determine the number of beetles emerging from infested trees over time, and they vary considerably among species of beetles and from one geographic region to another. For instance, the emergence of mountain pine beetles in the northwestern United States is restricted to a very short two or three week period at the end of July and beginning of August, while the fir engraver emerges over a much longer period from June to September.   In California, western pine beetle emergence often occurs in two or three distinct peaks, depending on the number of generations per year, while in the southern states southern pine beetles may emerge more or less continuously over the entire year, although more slowly in winter.   Aside from these differences in the duration of the dispersal period, the dynamics of dispersal are quite similar for all bark beetle species (Figure 8.9).   In this figure, we see that the cumulative number of emerging beetles rises in a sigmoid fashion to a maximum determined by the number of trees previously infested and the beetle productivity within those trees.   The cumulative number of beetles which succeed in attacking susceptible hosts rises in a similar manner but reaches a lower maximum.   The difference between the maxima of

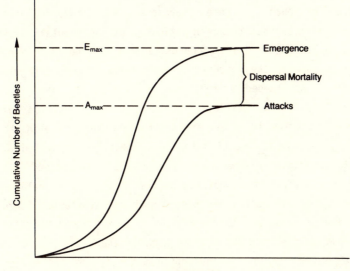

Figure 8.9. Expected changes in the cumulative number of beetles emerging and attacking over the period of dispersal.

these curves, then, represents the number of beetles which die during the dispersal phase.

In the absence of pheromones, emerging beetles go through a period of flight exercise before beginning their search for a suitable host. This behavior tends to distribute them evenly through the forest environment. In other words, their distribution in space is more or less uniform during the early stages of the flight period. As beetles are exercised by flight they begin to land indiscriminately on susceptible and resistant hosts, as well a non-host species, but will tend to land with greater frequency on larger trees because of their greater surface area. Pioneer beetles may then bore into hosts trees indiscriminately, or in response to olfactory stimuli and, depending on the resistance of these trees, attractive pheromones may be produced, precipitating the colonization phase.   Once pheromones are produced, the dynamics of dispersal change drastically as the population of flying beetles is drawn towards attractive hosts.   The distribution of the population in space becomes highly aggregated, or clumped.

Mortality is a critical process in the dynamics of dispersal, and yet it is extremely difficult to measure in the field.   There is no question that many beetles are killed by predators, unfavorable weather, and by host resistance.   In addition, survival of dispersing beetles must also be related to the number of susceptible hosts which are available within a particular area, and to the number of beetles searching for them.   Mortality is likely to be high if there are a lot of beetles searching for few susceptible trees, whereas flight losses should be small if few beetles are searching for a large number of susceptible hosts (Berryman 1979).   Thus, beetle survival should be inversely related to the emerging population size and directly related to the amount of susceptible hosts material available for attack (Figure 8.10).

Dynamics of Colonization

As we have seen, the spatial distribution of the beetle population changes drastically during the colonization phase as flying beetles are drawn towards attractive trees. Colonization begins when aggregation pheromones are produced by pioneer beetles boring into susceptible hosts.   The dynamics of the following "mass attack" is dependent on two important variables, the quantity and quality (i.e., the complex mixture) of aggregating and anti-aggregating pheromones as well as host synergists, and the density of

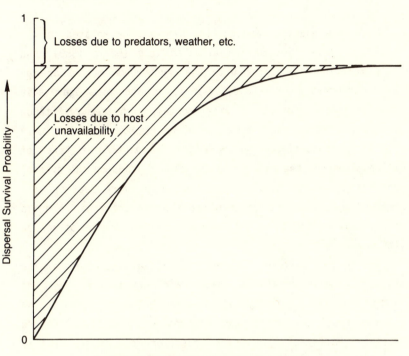

Figure 8.10.   Relationship between beetle survival during
dispersal and the relative abundance of host material per
emerging beetle (after Berryman 1979).

beetles flying in the vicinity of the pheromone plume. These two variables govern the rate at which the susceptible host is colonized.

Pheromone quantity and quality seems to be related to the resistant properties of the attacked tree (Raffa 1980), and this relationship varies with the aggressive nature of the beetle species (Figure 8.6). Thus, in the more aggressive species the rate of attack will increase with tree resistance up to the point where resin flow begins to interfere with pheromone production. However, given a particular level of attractiveness, the rate of attack also depends on the size of the flying beetle population. Obviously, more beetles will respond to a given pheromone plume if there are more flying in the vicinity, and this will create a higher rate of attack on the tree.

The spatial pattern of colonization usually develops in the following manner: First, newly arriving beetles tend to attack in the vicinity of pioneer beetles, establishing "beach-heads" in particular regions of the tree (Figure 8.11b) (Raffa 1980). The position of these initial "beach-heads" seems to vary with beetle and tree species; e.g., in the midbole region for western pine beetles attacking ponderosa pine and in the lower bole for mountain pine beetle attacking lodgepole pine. Thus, the original random distribution of pioneer beetles becomes highly clumped during the initial establishment of the infestation (Figure 8.11a and b). As more and more beetles arrive these initial "beach-heads" expand in all directions, and the attack pattern changes to a more random distribution (Figure 8.11c). Finally, as the tree fills in, and the attack terminates, the distribution of attacks becomes more uniform (Figure 8.11d). This uniformity of the final attack seems to be associated

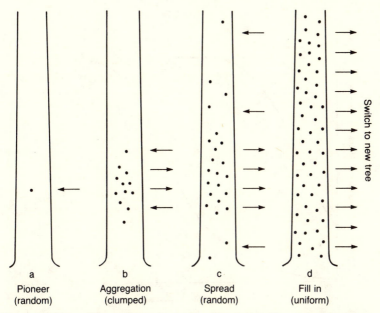

a	b	c	d
Pioneer	Aggregation	Spread	Fill in
(random)	(clumped)	(random)	(uniform)

Figure 8.11.   Space dynamics of a bark beetle colonization; incoming arrows indicate attraction and outgoing arrows repulsion from a particular area, or the relative dominance of aggregating and anti-aggregating pheromones (reconstructed from Raffa 1980).

with a spacing behavior, or territoriality, which may be controlled by anti-aggregation pheromones, sound, or phloem deterioration around established attacks, and acts to decrease competition between neighboring galleries (Barr 1969, Renwick and Vite 1970, Rudinsky and Michael 1973).

The numerical dynamics of colonization can often be described by an S-shaped, or sigmoid, response curve (Figure 8.12a) (McCambridge 1967, Gara and Coster 1968, Ashraf and Berryman 1969). The response curve can be characterized by three phases which correspond roughly to the spatial

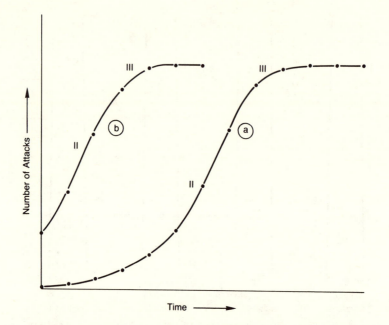

Figure 8.12. Numerical dynamics of bark beetle colonization; I = establishment phase, II = mass attack phase, III = termination phase (see text for references to data on which these curves were based).

patterns illustrated in Figure 8.11. In the first phase, the attack "beach-head" is established; in the second phase, there is a massive and rapid invasion and spread of the attack (Figure 8.11c); and in the final phase, the attack dwindles and terminates as the tree is "filled in" (Figure 8.11d). A second pattern of attack is frequently observed in which the first phase is absent (Figure 8.12b) (Vite and Crozier 1968). These trees usually start with a much higher "pioneer" attack and are frequently close to another tree on which the attack is in the termination phase. This brings us to the subject of infestations which expand to include groups of killed trees, a characteristic that is common to many bark beetle species.

Dynamics of Group Killing

Aggregation pheromones are generally involved in long-range communication, so that beetles are drawn to the tree under attack from considerable distances. On the other hand, anti-aggregation pheromones seem to operate over relatively short distances. Beetles responding to aggregation pheromones are primed and ready to attack and, when they are repelled from a tree which has "filled-up", they tend to immediately attack adjacent trees in large numbers. This gives rise to a process where the attack "switches", or "spills over", from tree to tree, and results in the group kills characteristic of may aggressive bark beetles (Gara and Coster 1968). Because the switch to a new tree may involve large numbers of beetles, which may be primed to produce aggregation pheromones without having fed (Rudinsky et al. 1974), the mass attack occurs extremely rapidly (Figure 8.12b) and even resistant hosts may be easily overwhelmed.

During switching, the choice of a new tree for attack seems to depend on its size and distance from the original focus tree, larger and closer hosts being more susceptible to switching attacks (Geiszler and Gara 1978). Hence, the direction of spot growth depends, to a large degree, on the spatial and size distributions of the trees in the stand (Figure 2 in Coulson 1979). Of course, the growth of a bark beetle group infestation also depends on the number of beetles in flight. Groups will expand much faster when large numbers of beetles are present but will terminate at the end of the flight period. However, some species, like the southern pine beetle, have an extended and more or less continuous flight period owing to their short, overlapping generations and re-emergence (i.e., parent beetles may attack one tree and then re-emerge to attack another). In these cases, group infestations may grow continuously throughout the year to encompass large areas of forest. Even in these species, however, expansion will slow down in the cooler season, and may terminate before warm weather returns.

Attack Thresholds. The success of a bark beetle attack is dependent on sufficient beetles being drawn to the tree to overwhelm its defense systems. There is, in effect, a dynamic interaction between the defenses of the trees and the beetle attack. Each attack drains resin from the duct system as well as energy in the form of energy-expensive hypersensitive responses (Wright et al. 1979), so that the tree's ability to defend itself against subsequent attacks is reduced. As the resistance of the tree declines, the quality and/or quantity of the pheromones changes and the attack rate slows down and eventually stops at some characteristic attack density (Figure 8.7). The characteristic attack density varies considerably between trees, and is related to the differences in resistance expressed by individual trees (Smith 1975, Berryman 1976, Waring and Pitman 1980). The concept of an attack threshold has emerged from these findings and the observation that trees frequently survive light attacks (Thalenhorst 1956, Berryman 1978). The threshold represents the density of attacks which must be attained to insure the death of the tree and the successful reproduction of the beetles. Because attack thresholds describe the transient states separating beetles success and failure, they can be empirically defined by observing the density of attacks on trees with differing levels of resistance and the success of these attacks (Figure 8.13A). The density of attacks required for successful colonization rises with hosts resistance.

Population Thresholds. The concept of attack thresholds provides a basic theoretical model for explaining the interaction between bark beetles and their hosts. When extrapolated to the population level, it enables us to explain, in a qualitative manner, the dynamic behavior observed in many bark beetle infested forests, and to develop risk decision models and control strategies (Berryman 1978, 1982). It is extremely important, therefore, that we understand the extrapolation of this theory to the population level.

Suppose we have a tree of a given resistance which is attacked by pioneer beetles. Provided that resistance is not too high, pheromones will be produced which have fixed quantitative and qualitative properties, and the pheromones will attract beetles from the surrounding area. If sufficient beetles are attracted to exceed the attack threshold, then the tree will be killed. However, if the flying beetle population is too small, the attack threshold will not be exceeded and the tree will survive. Hence, for any tree there exists a theoretical beetle population threshold which is sufficient to overcome its intrinsic resistance (Figure 8.13B). We can see that the outcome of the interaction between beetle and tree is dependent on both beetle population size and host resistance.

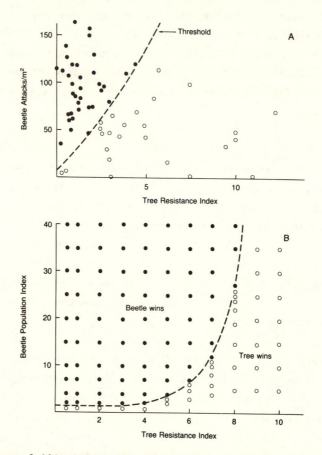

Figure 8.13A.   Identification of the mountain pine beetle
attack threshold required to kill lodgepole pines with
different levels of resistance; resistance index measured by
basal-area growth over sapwood basal-area (see Waring and
Pitman 1980 for original data and rationale for the resistance
index). ● = successful attack, tree dead; o = unsuccessful
attack, tree living. B. Results of a simulation experiment
with a model describing the interaction between beetle
population size and host resistance (see Raffa 1980 for a
description of the model; the beetle population index is a
relative measure of beetle number per unit area of forest).

These factors are overriding in their significance, and this elemental model forms a basic theoretical framework for evaluating bark beetle population behavior. For instance, the model predicts that trees adjacent to an attacked tree are at greater risk than more distant individuals. In effect, pheromones emanating from the infested tree cause the local beetle population to increase towards the population threshold which puts nearby trees at risk (Figure 8.14). For example,

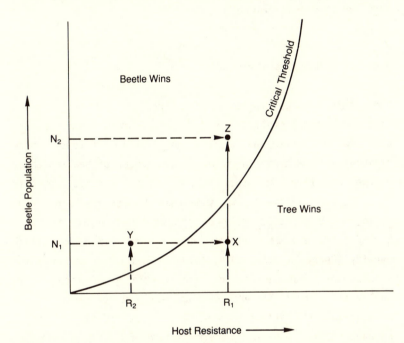

Figure 8.14. Theoretical relationship between host resistance and the beetle population required to overcome that resistance.

if the local population is originally at N_1, then a tree of resistance R_1 can not be successfully attacked because Y is above the threshold. However, an adjacent tree of resistance

R_2 can be successfully attacked because Y is above the threshold.   If beetles responding to this tree cause an increase of the local population, say to N_2, then the tree of resistance R_1 can be killed when the attack on the first tree terminates.   Thus, the threshold model accurately predicts the characteristic "switching" behavior of bark beetle populations and the growth of group infestations.   As we will see later it also forms the basis for understanding bark beetle epidemiology.

Dynamics of Production

After bark beetles reproduce, their progeny feed and develop within the successfully infested trees.   In effect, the tree can now be viewed as a sort of "beetle factory" whose efficiency is measured in terms of the number of beetles which emerge relative to the number which entered.   This reproductive efficiency is sometimes termed the production ratio, or the productivity of the beetle population; i.e., $P = E/A$, where E is the emergence density and A the attack density (Berryman 1974, 1976, 1979).   This ratio measures the success of reproduction of the beetle population.   In qualitative terms it provides an estimate of whether the population will increase (P>1), decrease (P<1) or remain unchanged (P=1).   However, the magnitude of P also provides an estimate of the maximum potential rate of increase of the population over a generation.

    A large number of variables affect the productivity of beetle infested trees but they can be conveniently classified into two groups (Figure 2.2); (1) habitat variables which determine the maximum productivity from a particular tree (e.g., tree size, phloem thickness and water content, weather, density-independent responses of predators, parasites and competitors), (2) variables which respond to the density of the beetle population and act as feedback regulators of productivity (e.g., oviposition behavior, competition for food and space, density-dependent predation and parasitism, host resinous responses).

Regulation of bark beetle productivity within infested trees is governed by two counteracting feedback effects. First, cooperation between attacking beetles helps to overcome the resistance of their host and to pre-condition the phloem for larval development (e.g., inoculating fungi and controlling moisture). This causes brood mortality to decrease, and productivity to increase, in direct relationship to attack density (Figure 8.15--cooperation curve). Second,

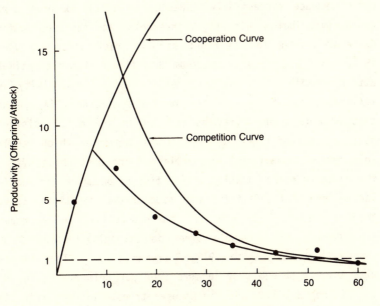

Figure 8.15. Productivity of the fir engraver beetle showing cooperation and competition effects and resulting productivity curve. Points indicate fit to field data (after Berryman 1979).

competition between attacking beetles and their offspring causes oviposition to decrease and brood mortality to increase with attack density (Figure 8.15--competition curve). The interaction of the positive (cooperation) and negative (competition) feedback effects results in the unimodal productivity function illustrated in Figure 8.15 (Berryman 1974, 1976, 1979).

The components of bark beetle productivity are strongly influenced by host factors, in particular host resistance and phloem or bark quantity and quality. As we have seen previously, there exists for a tree of given resistance a particular attack density, or threshold, which must be attained before the tree can be killed and the beetles can reproduce successfully. Hence, we should observe zero productivity below this attack threshold, and the cooperation curve will move to the right as tree resistance increases (Figure 8.16A). We also see from this figure that the optimal attack density (A_o), or that density where productivity is maximized, is somewhat higher than the threshold (A_c), and that maximum productivity is reduced in the more resistant hosts because of higher optimal attack densities which result in greater competition. Properties of the host which affect the quantity and/or quality of the tissues on which the beetle larvae feed will affect the amplitude of the competition curve (Figure 8.16B). For example, the productivity of mountain pine beetles in lodgepole pines is strongly influenced by phloem thickness (Amman 1972, Berryman 1976). Thus, trees with thicker phloem provide more food per unit area (e.g. H_3 in Figure 8.16B) and, in these trees, productivity is maximized at higher optimal attack densities.

A number of studies have demonstrated that the dynamic

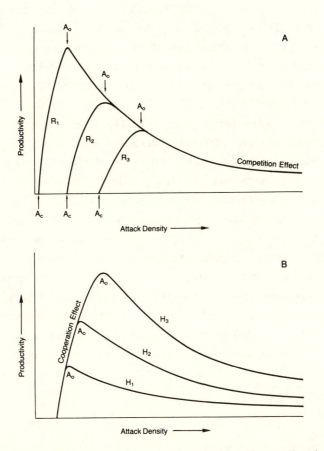

Figure 8.16.   Effects of host factors on beetle productivity:
Increasing host resistance (R_1 to R_3) causes the attack
threshold ($A_{c,1}$) to increase (to $A_{c,3}$) and maximum
productivity to decrease, where maximum productivity occurs at
the optimal attack density (A_o).   B.   Increasing food
quantity and/or quality (H_1 to H_3) causes increased
productivity, and also increases the optimal attack density
$A_{o,1}$ to $A_{o,3}$ (Berryman 1976 and unpublished).

interactions between beetle and host vary considerably with the vertical level in the host (Stark and Dahlsten 1970, Coulson 1979). In particular, the maximum suitability of the habitat (phloem thickness, moisture, etc.) varies with height in the tree, as does the density of attacks and the activity of parasites and predators (Stark and Dahlsten 1970, Shepherd 1965, Berryman 1968a, 1976). In addition, because of the dynamics of colonization (Figures 8.11, 8.12) and differences in temperature at different levels and aspects of the tree, the immature broods are often in different stages of development throughout the tree. These variation, together with the behavior of re-emergence in some species, may have considerable influence on the dynamics of beetle emergence and, consequently, on the ability of the population to invade resistant hosts. Some of the more aggressive species, such as the mountain pine beetle, seem to have evolved mechanisms which synchronize their emergence period over a fairly short interval of time. This adaptation tends to maximize the flying beetle population and increases the number of hosts which can be successfully colonized; i.e. more resistant trees can be killed because the flying beetle population is larger (Figure 8.13).

BARK BEETLE EPIDEMIOLOGY

Endemic Behavior

Bark beetles, with their pheromone communication systems and fungal symbionts, are highly adapted for seeking out and breeding in individual weakened hosts which are widely scattered throughout the forest. In a normal healthy forest, breeding sites are restricted in space and time to those trees which have been severely affected by various environmental stress factors. Under these conditions a beetle population will remain at very low levels and is said to be in an endemic state. Endemic beetle populations must suffer extremely high mortality during dispersal and their search for susceptible hosts, and this mortality probably plays a key role in population regulation (Figure 8.10). In other words, the endemic beetle populations are probably regulated primarily by the scarcity and spatial distribution of susceptible hosts; i.e., their high reproductive potential is not realized because of high mortality during their search for and attacks on rare hosts. Thus, we can visualize the endemic beetle population as being in a state of dynamic equilibrium with the host population, and that timber losses will be economically bearable or even beneficial to man; e.g., the beetles may act as thinning agents. By removing unhealthy trees whose decomposition recycles valuable nutrients, the beetles effectively increase the growth rates of the remaining healthy trees (Figure 8.3). Because of their dependence on rare susceptible hosts, the endemic distribution of bark beetle infested trees is reflected in the distribution of the various stress factors. For example, root pathogens and lightning strikes tend to weaken individual trees, or small groups of trees, at a particular place and time (Stark and Cobb 1969, Hertert el al. 1975, Hodges and Prickard 1971, Schmitz and Taylor 1969). Endemic beetle populations will, therefore, be distributed in a similar manner.

We can visualize the endemic system, and its relationship to the population threshold (Figures 8.13, 8.14), by considering the case of a vigorous stand composed of trees with similar genetic characteristics growing in a uniform environment. In other words, all the trees in the stand have similar defensive abilities or resistance. Given a high level of resistance, R_1, and a low beetle population, N_1, then the normal healthy trees in this stand will be immune from beetle colonization, because they are remote from the critical threshold. For example, the system may be at the point X in Figure 8.17A. However, should the beetle population rise substantially, or the resistance of the stand decline, so that the system is carried over the threshold, then all the trees will become susceptible to attack and the beetle population will increase rapidly. The population threshold, therefore, also separates the beetle/conifer system into endemic and epidemic patterns of behavior, and can also be considered as an epidemic threshold.

Epidemic Behavior

Bark beetle outbreaks, or epidemics, often erupt when whole stands or forests are weakened by widespread stress factors. For instance, outbreaks of the western pine beetle have followed droughts which placed large numbers of trees under water stress (Miller and Keen 1960) and nutrient stress resulting from insect defoliation has precipitated fir engraver outbreaks (Berryman 1973, Wright et al. 1979). In other words the system is moved from X to Y (Figure 8.17A) by the action of environmental stress. Widespread stresses not only increase the number of weakened trees but also cause them to be more evenly distributed in space. As a result of these sudden changes in the distribution and abundance of susceptible hosts, the beetle population suffers lower mortality during flight because individuals only need to fly

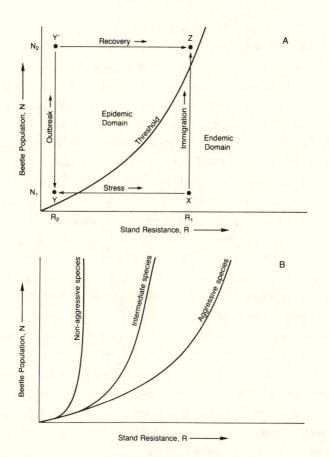

Figure 8.17A. Bark beetle population dynamics predicted by the threshold model; X = endemic population, N_1, sustained in severely weakened hosts but unable to colonize other healthy trees with resistance R_1; Y = system state in the epidemic domain following severe stress on the stand which reduces resistance to R_2.   Y' = system state following beetle population growth to N_2 in the stressed stand. Z = system state, still in epidemic domain, following the recovery of stand resistance to R_1, and Z also shows system state following a large immigration of beetles into a stand of resistance R_1. B. Hypothetical threshold functions for beetle species with different aggressive tendencies; aggressive species are usually more tolerant of host defensive chemicals.

short distances to find susceptible trees. This further accentuates the rate of population growth (Figure 8.10).

Stresses from drought or defoliation are usually temporary and the forests may recover quite rapidly when conditions return to normal. Thus, outbreaks of the less aggressive species, such as the fir engraver, often return to their endemic levels soon after the environmental stress is removed (Berryman and Wright 1978), while those of the more aggressive species (e.g. mountain, southern, and western pine beetles) may proceed for many years following the initial period of stress. Once again, these phenomena are anticipated by the threshold model (Figure 8.17A). Suppose that the population grows to N_2 during the time that the trees are under environmental stress. Then, even if the trees recover their original resistance (i.e. to R) the system will still be in the epidemic domain (i.e., the point Z remains above the threshold) and the outbreak will proceed. This result suggests that the less aggressive species, such as the fir engraver, will have more sharply rising threshold functions so that the system will enter the endemic domain as soon as the stress is removed. Thus, the aggressive characteristics of bark beetle species are reflected in the slope of their threshold functions (Figure 8.17B).

Although widespread stresses such as drought or defoliation are often associated with bark beetle outbreaks, they are not the only factors which can precipitate epidemics. In the more aggressive beetle species, outbreaks may sometimes be initiated by local stress acting in a relatively small number of trees. In these cases beetles produced from susceptible trees in a particular locality will migrate into surrounding stands, and may raise the population level above the epidemic threshold (e.g. from X to Z in Figure 8.17A).

Thus we may see the beetle population spreading in a wave-like fashion from a stressed epicenter, to create a general and widespread outbreak (Figure 8.18). In general, the greater

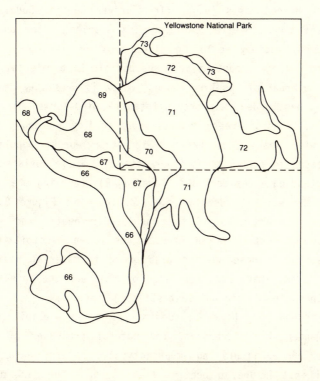

Figure 8.18. Spread of a mountain pine beetle outbreak in lodgepole pine stands into Yellowstone National Park. The epidemic started in 1958 in a relatively small area in the lower left corner of the map (redrawn from US Forest Service Survey Reports).

the stress under which the trees are placed, the greater is the beetle population buildup, because trees of very low resistance can be attacked at low densities where productivity is highest (Figure 8.16A). Thus, populations of beetles which attack downed trees (e.g., fir engraver beetle, Douglas-fir beetle, spruce beetle, and many Ips species) often build-up extremely rapidly following wind, snow or ice storms which create a large amount of very susceptible host material. The hugh outbreak of Ips typographus, which is presently ravaging Norway and Sweden, was precipitated by windstorms which blew down millions of trees, and was aggravated by a drought in the following years (Nov 1979). In other words, populations increased in fallen trees (increasing N), while resistence of the standing trees declined (lowering R), carrying the system into the epidemic domain (e.g., X to Y in Figure 8.17A). Similarly, outbreaks of the Douglas-fir beetle and spruce beetle in western North America are often precipitated by windstorms. These species also breed in material remaining after stands have been harvested by man, and the infestation may then spread into adjacent standing trees.

Although bark beetle epidemics are often precipitated by environmental disturbances, the natural processes of stand growth and aging play an important role. A very young stand is all but immune to beetle attack because the bark of the trees is too thin for beetles to successfully colonize (e.g., the point A in Figure 8.19A). As the trees grow in size, the bark becomes more and more suitable for beetle development and survival, but a young vigorous stand will remain resistant and beetles will only be able to infest a few scattered trees which have been weakened by lightning, root diseases, etc., placing the stand at point B in Figure 8.19A. As time progresses, and depending on tree growth rates and initial

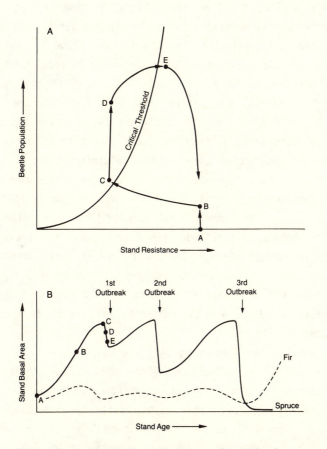

Figure 8.19A. Hypothetical trajectory of population states as a stand grows and matures; the threshold separates endemic from epidemic population behavior. B. Hypothetical time-series trajectory of a spruce-fir stand through three spruce beetle outbreaks (modified from Schmid and Hinds 1974). Basal area = cross sectional area of all stems per acre.

density levels, the trees will begin to compete with each other for water, nutrients and/or light, and stress will begin to increase, lowering the resistance of the stand. Along with this decline in resistance, the endemic beetle population will rise and eventually an outbreak may occur if the beetle population exceeds the critical threshold for the stand (B to C in Figure 8.19A). As the beetle population moves the stand to population state D, it will remove a portion of the stand, relieving competition between the remaining trees so that resistance slowly improves. Eventually, the system may again cross the critical threshold at E, and the population will then return to its endemic level B. The thinned out stand now resumes growth, competition may resume, and the cycle may occur repeatedly until the host species is completely eliminated and replaced by other species. In these successive outbreaks the debilitating effects of old age will become increasingly severe and the outbreaks are likely to increase in intensity (Figure 8.19B).

It should now be apparent that the behavior of bark beetle populations is a reflection of the dynamics of their host populations. Stands growing in different ways will create different conditions for bark beetle reproduction and survival, and will suffer different patterns of mortality. When forests grow in extensive, single-species, even-aged stands they are likely to experience devastating and widespread outbreaks when they come under stress from old age, competition and/or environmental disturbances. Stands with these characteristics are frequently formed following large forest fires which favor the reproduction of single species, such as lodgepole pine. It is not surprising, therefore, that lodgepole pine forests are frequently devastated by mountain pine beetle outbreaks (Figure 8.18). However, where lodgepole pine grows in uneven-aged stands in central Oregon, mortality from the mountain pine beetle is more restricted to individual older trees and groups of trees. In these conditions, the mountain pine beetle behaves very much like the western pine beetle in uneven-aged ponderosa pine stands.

Although our analysis of the threshold model has been conducted, for simplicity, under an assumption of uniform stand resistance, there is no reason why the system cannot be evaluated under conditions of variable stand resistance and variable endemic beetle population levels. Under these conditions, the state of a particular stand can be described by a two-dimensional distribution about the mean; i.e., the point X in Figure 8.17A will be the center of an ellipse whose horizontal and vertical axes reflect the variation in host resistance and beetle population, respectively. Analysis of such a system shows that beetle populations tend to remove the less resistant trees, reducing the variation in host resistance, and maintaining a more vigorous stand. Large scale outbreaks may still occur following severe environmental disturbances, but they will rarely be as destructive as in pure, uniformly resistant, stands.

I hope that I have convinced the reader that understanding bark beetle population dynamics rests on a sound knowledge of the interaction between beetle populations and the defense systems of their hosts. Other factors, such as parasitism, predation, and competing species may modify this behavior, largely through their effect on beetle population size, but they will not change the general qualitative properties of the system. The elemental threshold model seems to offer a general and robust approach for defining the risk of bark beetle epidemics, and for evaluating forest management

policies (Berryman 1978, 1981b).   However, its application
rests on our ability to measure both beetle populations and
the   resistance   of   forest   stands.   Unfortunately,   our
understanding of the nature of conifer resistance and, hence,
our ability to measure this complex variable, has not kept
pace with our ability to sample beetle numbers.   Hopefully
this will change in the future.

9. Integrated Management of Bark Beetles

R. N. COULSON and R. W. STARK

Most bark beetles are not considered to be forest pests. However, a number of species have gained notoriety because of their role as mortality agents of various age classes of forest and shade trees. Intensive research in North America on the more destructive of these scolytids, particularly those in the genus Dendroctonus, has been instrumental in solidifying concepts of integrated pest management as applied in forest ecosystems. This research has firmly established the need for understanding the functional role(s) of insects in forest ecosystems as a requisite for managing their population numbers and for minimizing tree mortality.

The goal of this chapter is to examine contemporary knowledge of management of bark beetles in forest ecosystems. Our specific objectives are (1) to identify the circumstances where bark beetles are considered to be pests, (2) to present the concept of integrated pest management as a component of forest management, (3) to investigate behavioral mechanisms characteristic of the population systems of bark beetles that have evolved to enhance survival in an ephemeral environment, (4) to review the various tactics and strategies that have been employed to suppress populations of bark beetles, (5) to consider the ecological roles of bark beetles in relation to forest management values, and (6) to provide a case history example of a management plan for a representative bark beetle pest.

BARK BEETLES AS PESTS

The term "pest" is an anthropocentric designation applied to certain bark beetle species when they impact on economic and social values associated with forest and shade trees. The term has no foundation in ecological principle. There are three general circumstances where bark beetles routinely occur as pests: (1) in forest ecosystems; (2) in specialized forest settings such as seed orchards, seed production areas, and turpentine plantations; and (3) in "urban forests". Our focus in this chapter will be directed primarily to pest management in forest ecosystems.

Forest Ecosystems

Although there are scolytid representatives associated with each developmental stage of managed forests, generally only mature and over mature stands are sufficiently threatened so as to warrant suppression activities. Furthermore, outbreaks of bark beetles are usually associated with trees in poor vigor.   A number of different conditions related to individual trees, sites, and stands can impair vigor. The following are representative of the conditions most often associated with bark beetle outbreaks: senescent trees, nutrient stress, water imbalance (too much or too little), defoliation, presence of disease, high stand density, stands composed of single species, windthrow, cultural disturbances (logging and urbanization), and smog (Coulson 1979). Severe bark beetle outbreaks usually occur as a result of poor site and stand conditions in combination with weather conditions favorable for insect development. The large scale of these outbreaks, caused primarily by Dendroctonus and occasionally by Ips, prohibits their direct suppression. Instead, forest managers attempt to prevent site and stand conditions that are favorable for rapid beetle population growth.

Because forest ecosystems are managed for multiple use and sustained yield, bark beetles have a number of different types of impacts on resource values. Historically, the greatest emphasis has been placed on the impact of the tree-killing species on wood and fiber production. In addition to volume of wood and fiber destroyed, there are unmeasured costs incurred by disruption of management schedules. Bark beetles also have influences on recreational uses of forests, fish and wildlife, grazing, and hydrology. Quantification of impacts on these values has proven to be a challenging task that has only recently been addressed, e.g., White (1976), Michalson and Findeis (1979), and Leuschner et al. (1978).

Specialized Forest Settings: Seed Orchards, Seed Production Areas, and Turpentine Plantations

Scolytid depredations in specialized forest settings such as seed orchards, seed production areas, and turpentine plantations can be extremely important because of the inordinate value of the products sought relative to traditional forest values.

In the case of seed orchards and seed production areas, cone beetles of the genus Conophthorus are the only scolytids causing significant damage. Since seed orchards and seed production areas are intensively managed and, therefore, costly to operate, only minimal damage by Conophthorus spp. and other insects is acceptable. Pest management practices are similar to those used in traditional agriculture and they

usually involve attempts to control the insects directly,
such as by the use of pesticides. In the case of turpentine
plantations (gum naval stores), trees are purposefully
damaged in the process of turpentine extraction. This damage
weakens the trees and creates some of the conditions
identified above that favor bark beetle attack. Dendroctonus
terebrans, D. valens, D. frontalis and several species of Ips
are often associated with turpentine plantations and their
control can be quite costly.

Urban Forests

The third circumstance where scolytids are considered pests
is in urban forests. Urban forests can be residential areas,
city parks, city streets, etc. Both conifers and hardwood
species are affected.

In conifers, mortality generally occurs in weakened
trees. In effect, the same abuses that contributed to host
susceptibility observed in forest stands have been
extrapolated to the urban environment. Cultural disturbance
associated with building construction, road maintenance,
power-line clearance, site preparation, nutrient and water
stress, have all been associated with tree mortality induced
by bark beetles. In most cases, the disturbances could be
avoided and the problems with bark beetles prevented. Most
bark beetle pest conditions in urban forests result in
suppression attempts after the problem has occurred. Again,
Dendroctonus and Ips species are the most serious pests.

In urban forests, the problem is limited primarily to elm
mortality resulting from the Dutch elm disease fungus
(Ceratocystis ulmi). This disease is vectored by either the
European elm bark beetle, Scolytus multistriatus, or the

native elm bark beetle, Hylurgopinus rufipes. This pest problem is not as directly attributable to cultural disturbances as that observed for bark beetles in conifers. The European elm bark beetle is an introduced pest to which native American elms have not had the opportunity to evolve resistance.

In both conifers and hardwood species, the social and economic values applied to tree mortality resulting from bark beetles in the "urban forest" are staggering, relative to traditional forest values. Therefore, bark beetle outbreaks in urban settings are, indeed, serious problems.

THE CONCEPT OF INTEGRATED PEST MANAGEMENT AS A COMPONENT OF FOREST MANAGEMENT

The concept of integrated pest management (IPM) in both agriculture and forestry received considerable attention in the literature during the decade 1970-1980. The following sources deal with specific aspects of the concept and its development: Apple and Smith (1976), Baumgartner (1975), Berryman et al. (1978), Brooks et al. (1978), Coulson (1980, 1981), IPM (1979), NAS (1969, 1972, and 1975), Rabb and Guthrie (1972), Stark (1977), Stark and Gittins (1973), and Waters and Stark (1980).

Most comprehensive definitions of IPM contain four points. First, the foundation of IPM is based on ecological principles. Second, IPM is just one component of forest resource management. Third, the functional goal is to reduce or maintain pest populations at tolerable levels (both economically and socially). Finally, the methodology involves a combination of tactics, i.e., a strategy. The definition of integrated forest pest management (IFPM) provided by Waters (1974) embodies each of the four components identified above: "Simply stated, pest management is the maintenance of destructive agents, including insects, at tolerable levels by the planned use of a variety of preventive, suppressive, or regulatory techniques and strategies that are ecologically and economically efficient. It is implicit that the actions taken must be fully integrated into the total resource management process...in both planning and operation. This means that pest management must be geared to the life span of a tree crop as a minimum and to a longer span where the resource planning horizon so requires."

Many of the definitions of IPM and IFPM have focused in large part on methodology. This circumstance is understandable, as emphasis in developing operational IPM systems has been directed more to methodology than to concept. Formulation of the concepts of IPM preceded implementation by at least a decade.

If the methodological implications of IPM are removed, while retaining economic and social values associated with a forest ecosystem, IPM becomes pest management. Focus is now directed to the "pest" organism rather than to the methodology. From a functional viewpoint pest management is synonymous with population management. Generally, a population becomes a pest problem if it is large enough to cause significant injury. Social and economic values define the meaning of "significant injury" for a particular ecosystem. Viewed in simplest terms, populations become larger by reproduction and immigration and smaller by mortality and emigration. Therefore, pest management (population management) is directed to maintenance of pest populations at tolerable levels by either lowering reproduction or immigration or increasing mortality or emigration. How these ends are accomplished is simply detail (Reece 1979). However, in the next section we will see that manipulating reproduction, mortality, immigration and emigration to predictably manage population size is by no means a simple task in a highly co-evolved system.

In order to understand the relationship of IPM in forestry, we need to consider several additional definitions. Forestry is defined as "the scientific management of forests for the continuous production of goods and services" (Society of American Foresters 1958). A component of forestry practice is forest management which is defined as "the application of business methods and technical forestry principles to the operation of a forest property" (Ibid.). Forests are, or can be, managed for a number of specific purposes including timber production, outdoor recreation, range, watershed, and wildlife and fish. Forest protection is a component of forest management. Traditionally, forest protection has included consideration of insects, disease, and fire. IPM is, therefore, a conceptual methodology for accomplishing forest protection goals in light of multiple forest values. The concept of IPM, as defined above, clearly relates to insects and diseases, but the principles are different for fire. Stark (1979) suggested that the goals and concepts of forest management, forest protection, and IPM could be incorporated using the term "integrated forest protection against pests" (IFPAP).

Multidisciplinary research and applications programs conducted on tree-killing Dendroctonus spp. have been instrumental in defining contemporary concepts of integrated forest pest management. The following references review the development of the concept in considerable detail: Stark (1977), Waters and Stark (1980), Huffaker (1980), and Coulson

(1980). There are five basic components of a forest pest management system: pest population dynamics, forest stand dynamics , evaluation of impact on resource values and management objectives, treatment strategies, and decision analysis or benefit/cost integration (Waters and Stark 1980).

We have labored in some detail on the concept of IFPM, in part, to demonstrate the complex nature of management of bark beetles in forest ecosystems. In succeeding sections, we will focus on the interrelationships between population dynamics of bark beetles and treatment strategies and tactics that have been applied to minimize losses.

POPULATION SURVIVAL MECHANISMS

Knowledge of the structure and function of the population systems of bark beetles is a requisite for a rational approach to development, testing, and implementation of treatment tactics. Survival mechanisms have evolved through time that enhance the perpetuation of bark beetles in the presence of many biotic and abiotic mortality agents. Treatment tactics applied to suppress populations can be viewed simply as additional mortality agents. To be effective, a tactic must disrupt sophisticated survival mechanisms that serve to minimize the effects of mortality agents. Following are examples of survival mechanisms that are characteristic of bark beetle population systems in general. Some of the mechanisms were described in Chapter 8 and will be discussed here only to emphasize their survival value. The degree of development and relative importance of the mechanisms vary with individual bark beetle species. Our approach will be to identify various mechanisms and then describe how they are of benefit to survival. Certainly, there are alternative interpretations of the significance of the survival mechanisms.

Primary Host Selection

Primary host selection is the process whereby a small number of adult beetles, "pioneer beetles", identify a host suitable for colonization.   The exact mechanism for primary host selection has been the subject of debate in the scientific literature.   The process probably involves visual and olfactory cues as well as random searching.   Regardless of the exact mechanism, bark beetles have the ability to distinguish highly susceptible hosts and thereby reduce mortality that would result from unsuccessful attempts to colonize resistant trees.

Secondary Host Selection

Chemical communication systems of bark beetles, which involve production and perception of pheromones and host plant attractants, are highly advanced and have been studied extensively (Borden 1974, Wood and Bedard 1976, Wood 1979, Payne 1979, Chap. 4).   Production and perception of pheromones and plant attractants play critical roles in: (1) identification of suitable hosts, (2) aggregation of sufficient numbers of beetles to overcome tree resistance defenses and insure tree mortality, (3) prevention of "over population", on individual trees and (4) regulation of interspecific and intergeneric competition in the host.

Diel Patterns of Emergence

Many bark beetle species have diel patterns of emergence from hosts, e.g., D. brevicomis and D. frontalis. Generally, the emergence periods are in the morning and afternoon. Fares et al. (1980) have demonstrated that these two periods correspond to times when inversion conditions exist beneath the forest canopy. By contrast, lapse conditions often exist during the interim. Inversion conditions greatly enhance communication by pheromones and plant attractants (Fares et al. 1980), the effect of which is to reduce mortality in dispersing populations by facilitating host identification and location. It is therefore of survival value for emergence patterns to correspond to periods when inversion conditions are most likely to occur.

Patterns of Attack by Adults

Most bark beetle species colonize hosts over a period of time ranging from about ten days to six weeks. This extended period of attack establishes a brood population with an age distribution that corresponds to the length of the colonization period. Developing brood life stages, therefore, have an age class distribution within the host, and age specific mortality agents affect only a portion of the population at any one time. For example, Coulson et al. (1976b, 1980a) demonstrated that the southern pine sawyer (Monochamus titillator) was an interspecific competitor with D. frontalis. However, mortality occurred only to larvae that had not migrated into the outer bark by the time foraging by M. titillator became extensive. A similar phenomenon probability exists for age-specific predation and parasitization.

Patterns of Reemergence and Emergence

Because of the length of time that the process of colonization takes, patterns of reemergence and emergence are extended over periods ranging from ca. 14 days to several weeks. Short term weather-related disasters, therefore, affect only a small segment of the flying population at any one time.

Regulation of Oviposition

For some bark beetle species, e.g., D. frontalis and D. brevicomis, egg population number is a density-dependent function of the population of attacking adults (Coulson et al. 1976a). The number of eggs oviposited by attacking adults decreases as the number of attacking adults increases. Conversely, each female oviposits a larger complement of eggs at low attacking adult density. Density-dependent oviposition is of survival value for several reasons. First, at high population densities, intraspecific competition between late larval stages is minimized. Second, at high population densities, beetles can reemerge after ovipositing only a small portion of their egg complement and deposit additional eggs in other less densely populated hosts. Third, at low attacking adult density, emphasis on oviposition can be concentrated in a small number of trees, minimizing mortality of adults in flights between hosts and in attacks upon resistant hosts.

Allocation of Adults

"Allocation" is a term applied to describe the circumstance where beetles are reemerging and emerging from several hosts within an infestation (Coulson et al. 1979a). Reemergence of attacking adults is a component of the natural history of many bark beetles but it plays a significant role in the population dynamics of $\underline{D}$. brevicomis, $\underline{D}$. frontalis, and $\underline{Ips}$ spp. The significance of the process is manifold. At any one time the attacking population can be comprised of both reemerging and/or emerging beetles. Thus, a continuous supply of adults is available for colonization. A perturbation involving adult populations would have only a short term effect on the development of an infestation (Coulson et al. 1980b). This process of allocation is apparently most developed with $\underline{D}$. frontalis, where continuous growth of infestations over several months is not uncommon.

Migratory Behavior of Larvae

Larval populations of several species of bark beetles, such as $\underline{D}$. frontalis and $\underline{D}$. brevicomis, have been observed to migrate from the phloem into the outer bark. The inner bark of an attacked host deteriorates rapidly after colonization as a result of utilization by bark beetles and other subcortical insects, and as a result of drying and development of microorganisms such as fungi (Wagner et al. 1980). In addition, competitors, such as $\underline{M}$. titillator, and other bark beetles are avoided by moving to the outer bark. For example, $\underline{D}$. frontalis and $\underline{Ips}$ avulsus often occur together in the same host, but since the $\underline{I}$. avulsus larvae do not migrate into the outer bark, it is likely that the two species coexist because larval development is spatially isolated, and competition for space is minimized.

Utilization of a Common Host by Several Species

During periods of low population numbers, bark beetles often exploit extremely susceptible hosts, like lightning-struck trees. During periods where there are not sufficient adults available to successfully attack and kill a host, several bark beetle species can be found utilizing the same host. Apparently, the typical patterns of resource partitioning are relaxed and there is a greater degree of coexistence within a particular host.   For example, it is not uncommon to find relatively equal numbers of D. frontalis, D. terebrans, I. avulsus, I. grandicollis and I. calligraphus in trees struck by lightning. Perhaps the combination of species provides an attacking adult population large enough to kill the host. From a survival standpoint, this behavior results in creation of a habitat suitable for all species to develop, at a time when no one species is abundant enough to kill the host.

For effective management, these nine characteristics of bark beetle populations must be addressed by treatment tactics.   Because the mechanisms are interrelated and often act contemporaneously, it is difficult to predict intuitively the outcome of a perturbation on a population.   This circumstance has necessitated the construction and use of mathematical models of population dynamics.   Models can be used to predict the efficacy of proposed treatment tactics (Coulson et al. 1979b).   Utilization of mathematical models forces the pest manager to clearly define the mode of action of the proposed tactic. Furthermore, a number of different tactics can be investigated individually and later combined into various strategies for subsequent testing.   Use of computer technology in evaluating tactics and strategies can be achieved at a small fraction of the cost of field investigations.   Although testing in the field will always be a requisite for validating simulated results, field testing should be reserved for those tactics and strategies that are demonstrated to show promise (Coulson et al. 1979b).

TACTICS AND STRATEGIES EMPLOYED TO SUPPRESS BARK BEETLE POPULATIONS

Historically, a great deal of research emphasis has been directed towards remedial procedures (treatment tactics) to "control" bark beetle populations in forests. This research was often based on intuition or prejudice that stemmed from the individual researcher's background and perception of a particular bark beetle-host system. Some very imaginative tactics resulted from the approach. Unfortunately, efficacy for most of the tactics could not be critically investigated. Inadequate sampling and estimation procedures precluded field measurements of proposed modes of action and their efficacy. Furthermore, mathematical models of population dynamics were generally not available for testing by computer simulation. When these various tactics were tested in the field, the findings were often inconclusive. It was not possible to separate extraneous sources of variation from the field tests. It is probably also fair to state that the magnitude and complexity of evaluating treatment tactics were underestimated. Early efforts focused on tactics whose sole objective was to kill insects at one place and at one time. These tactics simply caused a temporary perturbation in the population structure of the invertebrate community. Preoccupation with immediate observable results in efforts to control insects precluded basic research on fundamental ecological principles of forest development and bark beetle interactions. As we indicated earlier, this approach is the antithesis of the rapidly maturing concept of IFPM. In this section, we review briefly the types of controls utilized and their respective merits. We are concerned more with the philosophy of control of scolytids rather than specific details of methodology. Current information on methodologies can be obtained from university forestry schools and/or the U.S. Forest Service, and Forest Insect and Disease Management units throughout the country. In our concluding section we do, however, present an example of a detailed opinion of an integrated pest management approach to population regulation of those scolytids which do periodically cause great damage.

Biological Control

Biological control may be achieved by conservation and/or augmentation of the natural enemy complex or by introduction of exotic natural enemies. Introductions of exotics have been categorized as (1) major - programs which included adequate planning, operation, and evaluation of the establishment of introduced agents and their subsequent impact; (2) secondary - opportunistic programs, often utilizing surplus natural enemies from major programs against other target species, usually inadequately evaluated; and (3) futile - programs doomed by inadequate selection of natural enemies and poor techniques (Turnock et al. 1976).

Scolytids, like most long established insect groups, possess a rich complex of natural parasites and predators which are unquestionably important in regulating population densities (Chap. 5). However, in the history of biological control against scolytids in North America, there have been no major programs, one secondary program, and one futile introduction. For example, in 1892-93, A.D. Hopkins arranged

for the introduction of a clerid beetle, a general predator
of scolytids in Europe, for control of the southern pine
beetle; however, the predator did not become established. In
the 1930's, a "futile" attempt was made to import from Europe
parasites and predators of Dendroctonus obesus (McGugan and
Coppel 1962). The reasons for this lack of activity in
biological control of destructive scolytids are unclear.
Logically and biologically, there is as much chance for
success in this complex of insects as against other forest
pests (C.B. Huffaker, pers. comm., Chap. 5). Others (Turnock
et al. 1976) believe that the tendency of coniferous forests
to grow in simple communities, which are relatively unstable
and prone to wide fluctuations in insect abundance, makes
biological control more difficult, and therefore, has
probably discouraged practitioners of the science. Our
increasing use of mathematical models of population dynamics
permits testing of the impact of complexes of natural enemies
and may identify circumstances where it is feasible to
consider implementing biological control programs using
various parasites and predators (F.M. Stephen pers. comm.).

Insectivorous birds are an important component of the
regulatory processes which maintain community homeostasis, or
some degree of numerical balance between all populations of
organisms in forest communities (Bruns 1960, Thomas et al.
1975, Dickson et al. 1979). Many species of birds feed on
insects, but cavity nesting birds, particularly woodpeckers,
appear to have the greatest effect on scolytid populations.
Some 85 species of North American birds excavate holes or
exploit existing cavities in trees, usually in dead or
deteriorating trees called snags. Seventy-four of these 85
species are insectivorous, and 19 of the insectivores are
woodpeckers (Scott et al. 1977). Woodpeckers are capable of

strong numerical response to bark beetle infestations, and they are active throughout the year (Buckner 1966). For example, three species of woodpeckers are important predators of the spruce beetle in Colorado, particularly the northern three-toed woodpecker (Massey and Wygant 1963). Heavy feeding on infested trees can reduce the beetle population by 98%. In California, of the 11 identified avian predators, three species of woodpeckers had the most important impact on a population of bark beetles (Otvos 1970).

The importance of cavity nesting birds has been recognized by the U.S. Forest Service. In 1977 a national snag policy was enacted which required all regions to provide habitats for cavity nesting and snag-dependent wildlife species and guidelines for the desirable attributes and density of snags per unit area are currently being developed (Balda 1975, DeGraaf and Evans 1979, Jackson et al. 1979, McClelland and Frissell 1975, Scott 1978, Scott et al. 1977 and 1978, Thomas et al. 1975).

Major habitat disturbances have a profound effect on the distribution of birds and other animals. The effect of management practices on natural controls such as predation has not been adequately investigated. For example, Hagar (1960) found great differences in the avian communities in mature stands versus the young stands that grew in their place following clearcuts. Logging eliminated some of the species, did not affect some, and even attracted some species. Hooper et al. (1973) found that the cover provided by foliage less than 12 feet high accounted for 56% of the variation in densities of nesting birds, whereas the mixture of coniferous and deciduous foliage greater than 12 feet high accounted for 66% of the diversity of species. They concluded, for recreational areas at least, reasonably dense and diverse bird populations help to achieve major management goals. Nesting boxes have proven effective in Europe (Franz 1961) and California (Dahlsten 1967) for enriching the avian fauna. At this time, however, our knowledge of the effect of forest management practices on avian fauna is rudimentary. We can only suggest that forest managers be aware that their activities are likely to interfere with the activities of these important regulatory agents.

Parasitic nematodes are common in scolytids (Nickle 1962, 1963, Massey 1966, 1974). These organisms live internally in the host beetle, and although they rarely kill the beetles, they depress the viability and fecundity of their hosts. There is disagreement over their effectiveness in regulating beetle populations (Poinar 1970). Reid (1958) found that oviposition of infested beetles was reduced by 30 percent in one generation but oviposition was not affected in subsequent generations (Reid 1962). On the other hand, a nematode was credited with the decline of an outbreak of the fir engraver, Scolytus ventralis, in New Mexico (Massey 1964) and some reduction in initial flight period of the parasitized Douglas-fir beetles, D. pseudotsugae, prompted the speculation that nematode infection might limit dispersal (Atkins 1961). These observations led biological control experts to believe that nematodes could be used for control, but efforts to explore their use against the Douglas-fir beetle have been unsuccessful. Nickle (1971) observed a profound effect by a nematode on the behavior of the shot-hole borer, Scolytus rugulosus, and attempted to manipulate this parasite to attack the vector of Dutch elm disease, S. multistriatus, without success (Nickle, pers. comm.).

Mites of many groups are commonly associated with scolytids (Hunter and Davis 1963, Lindquist 1964, Boss and

Thatcher 1970, Moser and Roton 1971). Some species use the insect simply as a means of dispersal, while others prey upon other mites, and others parasitize the beetles (Kinn 1970). Each of these behaviors has an effect on the population dynamics of the host. Southern pine beetle adults have been so heavily laden with mites they could not fly (Fronk 1947); Atkins (1961) found that mites did not affect the duration of Douglas-fir beetle flight but did affect wing beat frequency when clustered at the tips of the elytra. Mites are among the principal natural enemies of the southern pine beetle (Berisford 1980), and Walters (1956) credited mite populations with destroying over 30 percent of the eggs of the Douglas-fir bark beetle. However, the overall controlling effect of mites is probably of minor importance (Kinn 1970) and fewattempts have been made to utilize them as biological control agents.

In summary, there is a rich and diverse complex of natural enemies interacting with scolytids in natural coniferous forest communities. However, manipulating natural enemy populations in order to regulate bark beetle pests is a difficult undertaking and the potential responses of the organisms are poorly understood.

Behavior-Modifying Chemicals

"Behavior-modifying chemicals" is a term coined to cover a considerable array of compounds produced by the insect of the host plant which influence the insect's behavior (Chap. 4). In general, three approaches to the use of these chemicals are being investigated: (1) the use of aggregating pheromones to direct the insect to traps or baited trees;

(2) the use of repellent chemicals to keep the insects off the trees; and (3) the inundation of an area with aggregating pheromones in order to prevent or reduce mating. Although these approaches show considerable promise for use in pest management systems, only a single example of successful use in forest situations has been developed (Wood 1979).

McLean and Borden (1975) used a pheromone for detecting and estimating the need for control of the ambrosia beetle, Gnathotricus sulcatus, in a mill storage area, and an efficient cost-effective suppression program using pheromone-baited traps was developed (McLean and Borden 1977, 1978). Approximately 65% of a population attacking the storage area was trapped in 1976. Improvements in formulations and trap numbers and locations have since increased trapping efficiency and resulted in significant economic savings (J.H. Borden, pers. comm.). This technique is recommended only for those areas where logs are concentrated after harvesting and before milling, not in the natural forest. This example is the only proven operational use of pheromones for the successful control of a forest insect and it provides a model for similar pests in similar situations.

Despite the general lack of success, due largely to the enormous complexity of the problem (Wood 1979), investigation of the use of behavior modifying chemicals in pest management is intense, and has included those of the mountain pine beetle (Moeck 1980, Pitman et al. 1978), the western pine beetle (Browne 1978, Tilden et al. 1979, Wood 1980), the Douglas-fir beetle (Furniss et al. 1974, Ringold et al. 1975, Rudinsky and Michael 1974), southern pine beetle (Johnson and Coster 1979, Payne et al. 1979), black turpentine beetle and some species of Ips (Birch and Svihra 1979), and probably others.

We believe that, like the use of pesticides, the application of pheromone treatments in the natural forest will be limited to discrete situations similar to the one successful example cited above. Intensive ecological, physiological, and biological research on each insect is necessary before we begin tampering on a large scale with this aspect of insect behavior. For the ubiquitous tree killers, such as Dendroctonus, we question the cost-effectiveness and efficacy of population manipulation by pheromones except in conjunction with sound managerial and silvicultural practices.

Mechanical Control

Here we discuss the various physical attempts at reducing scolytid populations, primarily the tree killers (Dendroctonus). The various techniques used have been reviewed by Klein (1978). In desperation, techniques such as electrocution of beetles in standing trees, the use of explosive cables wrapped around trees, and spraying the bole with fuel oil and igniting a brush pile around the tree have been tried. The least said about these tactics, the better. Trap trees, felling or rendering trees "attractive" by baits or mechanical injury, have been widely used to attract insects to sites where they are destroyed. Various claims of success have been made but they constitute short-term solutions at best and one wonders if the "cure" is worse than the disease. Felling infested trees and destroying beetle broods by incineration, toxic sprays, drowning, exposure to solar heat, peeling and destroying the bark, etc., have also been used, largely ineffectively. In such treatments, no consideration has been given to the fact that beneficial insects are destroyed and cost-effectiveness of the treatment is seldom, if ever, measured. Such methods provide no long-term effect on the population densities of the insects.

The reproductive capacity of the many survivors soon offsets any temporary reduction in population numbers. The point that is overlooked, as Klein (1978) points out, is that it is what we are not doing, rather than what we do, that renders all such individual tree treatments ineffective. We are not altering the condition which is responsible for the problem.

Chemical Control

The tree boring habit of most scolytids makes the use of toxic chemicals largely impractical. Two principal methods, preventative or protective sprays and injections directly into the tree, are used. The least expensive means of dispersal of chemicals, by aircraft, is ineffective; the lethal agent must be applied by highly labor-intensive means. The great expense incurred, therefore, restricts the use of this control tactic to special situations, such as cone and seed orchards, campgrounds, urban street plantings and parks, arboretums, and research plots. For example, chemical means have been tried to control cone beetles (Conopthorus spp.) in seed orchards (Ebel et al. 1975, Furniss and Carolin 1977, Hedlin 1974). The most effective means for reaching the developmental stages within the cones is by the use of systematic poisons injected into the tree, but there is danger of phytotoxicity at the dosage levels required (Annila 1973). Again, it is not practical to consider control of cone insects in extensive natural stands.

    Klein (1978) reviewed the history of control attempts against the mountain pine beetle and we agree with his conclusions that control in individual trees is, at best, no more than a delaying action. If control is terminated before the infestation runs its course, the extent of tree mortality will be essentially the same as if no control was attempted. Successful prevention was achieved only in restricted areas when preventative sprays were thorough and maintained throughout the course of the outbreak--frequently a lengthy and costly process. Individual tree protection is not practical under forest conditions but is feasible in the discrete situations mentioned above. In fact, we believe, with Klein, that chemical control has no place in the forest against tree-killing bark beetles. The problem is an ecological one and can best be handled by the ecological and managerial strategies. On the other hand, chemical control of scolytids and wood borers in harvested logs in concentrated areas, large accumulations of slash, mill yards and stored and "in-use" products is feasible and can be done cost-effectively with minimal insult to the environment.

Other Methods of Control

There are many other tactics for disposing of insects or affecting their population numbers (National Academy of Sciences 1969). Few of these have been attempted in natural forests so we can only speculate on their potential. Although the introduction of disease organisms shows great promise in some forest pest communities, we know of no significant entomopathogenic organisms associated with scolytids (but see Chap. 5). Presumably, the protected environment in which they live and the relative isolation of "families" within the tree limits establishment and spread of disease.

The development of resistant strains of trees has long been considered as a viable option, but the adaptive capacity of scolytids over the long time span necessary to develop and test resistant trees would, in our opinion, make this an unprofitable venture. The encouragement and enhancement of natural resistance, developed through the eons of successful coevolution, is another matter and is treated above and in Chapter 7.

Genetic engineering of forest pests is being considered as a possible futuristic control method (Smith and von Borstel 1972). Although it smacks of science fiction, recent advances in the science of genetics makes such an approach feasible. Many behavioral traits of forest insects are genetically controlled (Chap. 3) and may be able to be manipulated by insect geneticists.

Other futuristic approaches include the use of insect hormones to prevent or inhibit development, and use of antimetabolites, which block metabolic pathways and the use of feeding deterrents. These would have to be delivered in a manner similar to pesticides and so are of limited applicability. The success of the sterilization control program for the screwworm (Knipling 1960) evoked great interest in this technique. Unfortunately, there are probably no scolytids in the coniferous forests of North America which fit the criteria necessary for successful use of this approach (Knipling 1965).

ECOLOGICAL ROLES OF BARK BEETLES RELATIVE TO FOREST MANAGEMENT VALUES

The task of evaluating or interpreting the probable ecological roles of bark beetles in forest ecosystems is substantially more difficult to accomplish than defining social and economic impacts. There are several basic problems that complicate interpretation. First, the evaluation should be directed to conditions that existed prior to the intervention of forest cultivation and management practices. Second, the distribution and abundance of most tree species today may have little resemblance to patterns that existed in primitive forests where the ecological interactions had their origins. Third, forest management goals emphasize management or at least suppression of tree mortality agents such as bark beetles, other insects, disease, and fire. Therefore, an evaluation of the probable roles of bark beetles in forest ecosystems must be based on an interpretation founded on historical evidence and cast into a framework of ecological theory. Fortunately, there has been considerable research conducted in recent years on basic pattern and process of forest ecosystems (Bormann and Likens 1979a).

Interpretative views of the role of Dendroctonus bark beetles in combination with fire have been developed by Amman (1977), Peterman (1978), and Schowalter et al. (1981). All three views consider periodic perturbation as a primary factor influencing evolution of ecosystem structure and function. Many ecosystems have become dependent on periodic perturbation for regeneration and cycling of limited nutrients. Functionally, both bark beetles and fire serve as natural harvesters and, as such, are (or were) responsible for periodic perturbations. Certain disease organisms have a similar function. These "consumers" tailor nutrient turnover rates to fit resource availability and slow the loss of nutrients to the marine ecosystem.

The shifting mosaic of communities within the ecosystem is important for the persistence of the ecosystem. Ecosystem research reported by Bormann and Likens (1979b) supports Loucks' (1970) view of perturbation as a means of truncating community development at a point in time prior to

senescence.   Senescent communities are less able to play a
regulatory   role   in   ecosystem   function   and   have   fewer
r-selected or exploitive species which increase ecosystem
resilience   following   perturbation.   Fire   periodically
rejuvenates   patches   of   the   ecosystem   by   restarting
development   at   an   early   stage.   Bark   beetles   potentially
regulate this process by thinning old or stressed stands,
thereby providing concentrations of fuel to enhance the
effect of subsequent fire.   The resulting dynamic mosaic of
communities,   representing   various   stages   of   succession,
increases the relative stability of the ecosystem by reducing
the impact of perturbation (Bormann and Likens 1979a).

For all intents and purposes, bark beetle outbreaks and
fire   in   mature   coniferous   stands   are   unacceptable
perturbations when viewed from forest management objectives
for   timber   production,   recreation,   wildlife   and   fish,
watershed, and range.   The common occurrence of bark beetle
outbreaks in mature and overmature forests should be the
expected outcome where forest managers have allowed these
conditions to occur.   Suppression of bark beetle populations
in mature and overmature forests is therefore not a realistic
expectation, which is evidenced by the consistent failure of
remedial treatment tactics applied in outbreak conditions.
In the next section, we will describe a strategy to reduce
losses to bark beetles that stresses prevention of the
conditions that lead to outbreaks.

INTEGRATED FOREST AND PEST MANAGEMENT--AN EXAMPLE

The raison d'etre for pest management is the achievement of
perceived goals of forest management.   Without explicit goals
and   objectives   for   forest   management,   "protection"   or

"control" of forest pests is ill-defined, ill-organized, and ineffective. Planning for forest protection against those agents (pests) which constrain or thwart management objectives must be part of the initial planning process. Moreover, the management of any forest unit against a single pest may be counter-productive because that forest has coevolved with many potential "damaging" agents into a naturally operating ecosystem. Protection considerations should include all potential agents which can run counter to anthropomorphic perceptions of forest management.

For an example of forest and pest management, we have chosen a lodgepole pine ecosystem, with the mountain pine beetle as the principal insect pest in that system. Dwarf mistletoe and fire have played a role in the evolution of lodgepole pine forests and have emerged as equally significant factors in imposing constraints on management objectives. In addition, there are at least six species of pathogenic root fungi that attack lodgepole pine; the effects of their interactions on pest management are substantial (Kulhavy et al. 1978). We believe that because there are great similarities between this pine ecosystem and other forest-pest systems, the basic principles presented here should apply to most forests where Dendroctonus species are present.

Lodgepole pine, Pinus contorta, is one of the most widespread pine species in North America (Figure 9.1). Ecologically, it is characterized as follows: (1) it is a seral species with low shade tolerance; (2) it possesses serotinous cones which require high temperatures for seed release; (3) it regenerates rapidly in large numbers, often leading to dense stagnated stands; (4) young trees grow rapidly, older trees slowly; (5) it is highly susceptible to dwarf mistletoe infestation; (6) it is susceptible to attack

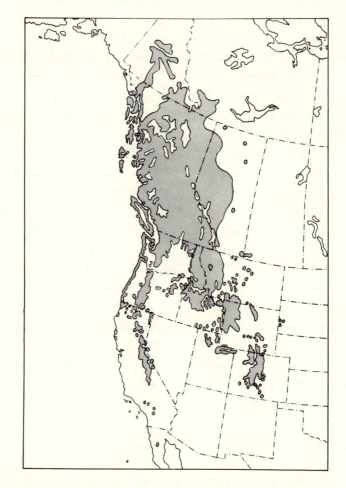

Figure 9.1.   The distribution of lodgepole pine, Pinus
contorta (from Critchfield 1980).

from mountain beetle at an early age (ca. 80 yrs.); (7) stagnation, disease, and insects result in large accumulations of fuel leading to intense fires which, in turn, results in the continued regeneration of this species, and (8) it has the capability to grow on almost any forest site (Pfister and Daubenmire 1975). From viewpoint of the forest and forest pest manager, there are four principal, interacting, damaging agents: the mountain pine beetle, dwarf mistletoe, root diseases, and fire.

Pfister and Daubenmire (1975) recognized four major successional roles for lodgepole pine: (1) minor seral, where lodgepole is a minor component of mixed-species stands and will be replaced by shade tolerant species in 50-200 years; (2) dominant seral, where lodgepole is the dominant cover type in mixed-species stand but will be replaced in 100 - 200 years; (3) persistent, where it is the dominant cover type of even-aged stands that apparently will not be replaced by shade-tolerant species; and (4) climax, where it is the only tree species capable of growing on a particular site.

The problems of management in the minor and dominant seral situations differ only in degree. In both, lodgepole will eventually be replaced by shade-tolerant species. Short-term loss of lodgepole pine productivity will be offset by long-term productivity of other species. Such succession under careful management (or even naturally) will ultimately result in a mosaic of stands of different species that should lessen future problems caused by the mountain pine beetle. The greatest havoc wreaked by the mountain pine beetle is in the persistent and climax groups (Wellner 1978) where lodgepole pine is dominant.

There are two broad categories of forest management objectives, where timber or fiber values are primary, and where other resource values, such as wildlife and recreation, are primary (Amman et al. 1977). Forests that are committed to recreation, wilderness, or other uses may not require action against pests such as the mountain pine beetle. Indeed, if viewed ecologically, "pests" may enhance the values to which such forests are committed. The pest management program outlined below applies to situations where timber or fiber values are primary and some degree of management is necessary. Effective forest protection, particularly against Dendroctonus spp., cannot be achieved by sporadic direct attacks against the pest. It requires sustained, planned efforts in harmony with the ecology and successional growth of the forest stands.

The forest pest management process begins with a clear definition of management objectives. These objectives should consider, as explicitly as possible, how management of the present forest will affect the generations of trees to follow (Peterman 1978). The next step is to determine the current status of forest site and stand conditions and status and history of the principal pests. Where possible, forest stands should be rated as to the probability of its infestation by principal pests. Four such hazard rating systems for the mountain pine beetle (Amman et al. 1977, Mahoney 1978, Safranyik et al. 1975) and dwarf mistletoe (Hawksworth 1978a) are available. All but one of the systems for the pine beetle are based on standard forest mensurational techniques. There has not been sufficient implementation of any one to rank them according to merit. Hazard-rating of stands permits the forest manager to establish priorities so that those areas of greatest risk can be treated first.

Based on the analysis of the current situation, predictions should be made of the natural course of events in the management unit and potential changes as a result of mountain pine beetle or other damaging agents. Estimates of stand growth rates, species composition, tree size composition, and the effects of mountain pine beetle attack (Crookston et al. 1978) and dwarf mistletoe infection (Edminster 1978) are now possible. Output from these models enables the forest manager to examine the possible consequences of various management options, consider management alternatives, and to decide whether pest management is indicated and if so, to select the best probable management approach and treatment strategies.

There are several obvious considerations in pest management which are applicable to any forest situations. These include forest management practices designed to enhance or at least conserve soil moisture, soil organic matter, decomposition and nutrient cycling processes, and soil nutrients (Stoszek 1978). These goals can be accomplished by: (1) using equipment causing minimum soil compaction, displacement of top soil, or erosion; (2) using slash disposal practices that do not further decrease nutrients in the soil; (3) fostering conditions conducive to increasing the rate of decomposition of organic matter and nutrient cycling; (4) reducing intertree competition for moisture and nutrients to maintain vigorous growth; and (5) avoiding physical damage of the residual stand that would increase vulnerability to insect and fungal attacks.

Finally, recommendations can be made to minimize the probability of abortion of management objectives by the mountain pine beetle. For example, the following specific recommendations include some consideration of dwarf mistletoe and root pathogens (Amman 1978, Berryman et al. 1978, Safranyik et al. 1975, Scharpf and Parmeter 1978).

1. Infestations in stands over 80 years of age with an average diameter above 8 inches (20 cm), especially in high

risk areas, are potentially epidemic. Such stands should be harvested immediately. If beetle-infested trees are present, the treatment area should include all stands containing more than one infested tree per acre. If an outbreak is in progress and the area exceeds that which can be treated practically, those areas with the largest diameter trees, at the periphery of the infestation, should receive cutting priority. Whether beetles are present in significant numbers or not, management planners need to remember that the larger the tree diameter, the greater the risk of attack and initiation of an outbreak. Harvesting before trees reach sizes conducive to beetle outbreaks is an effective method of preventing future losses. Further, probability of attack in a high hazard area increases significantly above age 100. Cutting practices in such stands are, however, conducive to the spread of dwarf mistletoe, which spreads more rapidly in open than in dense stands (Hawksworth 1978b). Further, if large diameter, overstory trees are infested with dwarf mistletoe, there is a high likelihood that adjacent overstory and understory trees will be also. Plans should include examination of neighboring residual trees and removal of infested trees.

2. One recommendation for minimizing future mountain pine beetle losses is the breaking up of continuous susceptible lodgepole pine forests by harvesting in small patches. This results in a mosaic of different age and size classes which reduces the area likely to be infested at any one time and reduces the potential for widespread outbreaks of the mountain pine beetle. Dwarf mistletoe, however, builds up rapidly in open, multistoried stands. Experience tells us that management practices such as selective cutting and preferential removal of certain species greatly exacerbates the dwarf mistletoe problem in lodgepole pine stands. Again,

plans should include an assessment of the mistletoe problem and additional removal of infested trees in accordance with dwarf mistletoe control guidelines.

3. Thinning to a basal area (stand density) suitable to the site has shown promise in reducing mountain pine beetle in ponderosa pine stands (Sartwell and Stevens 1975) and should be effective in lodgepole pine forests as well, since vigor of remaining individuals will generally be improved. However, such a practice in mistletoe-infected stands may be incompatible with mistletoe control recommendations because the rate of mistletoe infection increases with host vigor and its spread is more rapid in open than in dense stands. This practice should be used only in mistletoe-free stands or thinning criteria should include removal of mistletoe-infested trees of all sizes.

4. Mixed stands are a priori less susceptible to both build-up of large mountain pine beetle populations and spread of dwarf mistletoe since both organisms are generally host-specific. Maintenance of mixed stands, however, may be difficult to achieve and impractical in persistent and climax stands of lodgepole pine.

5. Sanitation thinnings and cuttings may be a more appealing option in the near future. This practice is seldom used against the mountain pine beetle due to its excessive costs. When both dwarf mistletoe and mountain pine beetle erupt, however, thinning may be an economical alternative. For dwarf mistletoe, sanitation cuttings (where all visibly infected trees are cut) are recommended only in stands where less than about 40% of the trees are infected and stands are less than 30 years old. Sanitation thinnings (where the emphasis is on spacing and cutting trees that are most heavily infected) are usually applied to stands where the trees are older than 30 years but are not recommended in

heavily infested stands because the remaining infected trees show little response to thinning. In both of these cases, mountain pine beetle infested trees could be removed for little additional expense.

6.   Clearcutting in persistent and climax stands of lodgepole pine appears to be the most satisfactory management practice to address the two principal pests of lodgepole pine.   Social and regulatory constraints are the principal barriers to its use. Clearcutting in large patches and with due regard to mistletoe guidelines can create a mosaic of age classes which reduce the potential of large scale outbreaks of the mountain pine beetle. Eradication of dwarf mistletoe is possible using clearcutting, if it is done properly. The top priority for Forest Service dwarf mistletoe control funds is for residual removal--cleancut rather than clearcut. If ecologically acceptable, broadcast burning assists in eradication of dwarf mistletoe and in rapid regeneration of lodgepole pine.   In persistent and climax stands, this practice is compatible with natural succession.

7.   In all silvicultural practices, stumps should be treated to prevent innoculation by airborne spores of pathogenic fungi. This could well be the best investment of the treatments outlined above. Where severe pockets of root diseases spread as a result of root contact between trees in the stand, species conversion should be considered.

8. Protection of individual trees of high value, such as in picnic areas, campgrounds, visitor centers and summer and permanent homesites, can be achieved by chemical sprays, deep watering during droughty periods, fertilization, avoidance of mechanical injury to the trees, and avoidance of excessive compaction around the root collar.

9.   We believe that integrated management of any forest pest is possible only in intensively managed forests. There

is little justification for any treatment in unmanaged forests. <u>Prior</u> to implementation of any treatment strategy, the monetary, social, and environmental risks associated with trees being killed by the mountin pine beetle and associated pests must be weighed against the costs of preventing or reducing these losses. <u>Subsequent</u> to their implementation, the success of the treatment must be monitored.

10. Evolution of Bark Beetle Communities

K. B. STURGEON and J. B. MITTON

The association between bark beetles and their host trees is an ancient one. Although there is a paucity of evidence linking insects and plants during their early evolution in the Paleozoic, fossil evidence shows clearly that both insects and land plants evolved rapidly and simultaneously as soon as ozone accumulated in the upper atmosphere to levels that substantially reduced the damaging effects of ultraviolet radiation (Smart and Hughes 1973). In the Permian, fossils which can be assigned to the insect order Coleoptera appear coincident with the appearance of fossils that characterize the order Coniferales, the plant taxon that gave rise to all modern coniferous families (Foster and Gifford 1974). By the Triassic Period, which began 280 million years ago, we can find fossils of petrified wood showing galleries characteristic of Scolytid feeding habits (Ross 1965). This ancient coexistence has provided ample opportunity for reciprocal evolution to produce numerous intricate interactions involving large numbers of species. For example, Dendroctonus frontalis alone is associated with over 90 species of insects and an additional 90 species of mites (Moser and Roton 1971, Moser et al. 1971b, Moser 1975, 1976). Largely because bark beetle communities often have substantial impact on agricultural species or valuable natural resources, there has been a large amount of research into their basic biology. These large and intricate communities, by virtue of the great amount of basic knowledge accumulated over the last century, constitute a valuable resource for the study of evolutionary mechanisms. This is not to say that more information is not needed; few of the speculations offered here can be tested rigorously with available data. But the data do provide a substantial base from which testable hypotheses may be generated.

SPECIATION

Scolytids have radiated to include over 6000 species. They can be found in both the Old and New Worlds where they exhibit a broad diversity of adaptations for colonization of almost every form of plant life and nearly every part of those plants. To what can we attribute their success?

Extensive adaptive radiations in parasitic insects are typically associated with four conditions (Price 1980) all of which are easily met by Scolytids: 1) there is a large diversity of hosts available to be exploited; 2) the host target is large either in body size, in population size, or in its geographic range; 3) sufficient time for colonization has occurred; and 4) there have been sufficient pressures to select for specialized adaptations.   We have seen that virtually all higher plants have been exploited by these insects and that time periods involved for colonization can be measured in geologic epochs (conditions 1 and 3). If we consider the host targets of the tree-killing <u>Dendroctonus</u> species, it is apparent that forest trees are not only large in size, allowing for ample opportunity for resource partitioning with a tree, but also they grow in extensive stands over wide geographic areas, permitting the evolution of host or geographic races which vary in morphology, behavior and/or physiology (condition 2). Finally, the vast majority of Scolytids are monophagous or oligophagous, attacking only one or a few species within a plant genus (Table 10.2) testifying to the highly specialized nature of their adaptive responses (condition 4)--a matter to be discussed in more depth in the section to follow on coevolution.

Although we can demonstrate that bark beetles have developed under conditions associated with extensive adaptive radiations in parasitic insects, we have not explained the evolutionary processes by which the radiations occur. If we are to more fully understand this group of insects, we must search for models of speciation that fit their life histories and habits and attempt to generate some predictions that may be tested. Modes of speciation have been considered traditionally in a geographic context without regard for the genetic mechanisms that underlie the process (White 1978). Here we wish to present two models of speciation that consider both geographic and genetic components of the process and which we feel best incorporate what we know about the habits and the habitat of bark beetles.

One model of speciation that fits the life cycle of bark beetles and, in addition, has considerable predictive power is the founder-flush principle first proposed by Carson (1968, 1971, 1975) and recently elaborated by Powell (1978) and Templeton (1980a,b). This model applies to species that occasionally undergo the founding of a new site by a restricted number of individuals, which is followed by an increase or flush of the population during a period of relaxed selection and eventually by a crash in the population during a period of stringent selection. The founding of the population by a restricted group of individuals undoubtedly results in a sample of genotypes not perfectly representative of the original populations. If the founding group is small enough to deviate significantly in genetic composition from the original population, yet is endowed with sufficient genetic variation to respond to selection, the opportunity for a genetic transilience (Templeton 1980a,b) is maximal. Some of the attributes of the founding population that increase its probability for genetic transilience are listed in Table 10.1. Reduced intraspecific competition associated with the founding event may allow genotypes to survive that would not ordinarily have survived in the original population and may produce an array of unique genotypes upon which selection can act during their development in the new environment. Should these separate groups of founders come into contact once again, they may have differentiated sufficiently to bring about some developmental genetic incompatibilities and/or they may not recognize each other as the same species (Figure 10.1). The unique features of this mode of speciation are that speciation may occur very quickly, that it need not involve large portions of the genome, and that the resulting species need not be specifically adapted to their different environments. The species may, in time, develop substantial genetic differences so that morphological and ecological distinctions become evident, but these developments would follow speciation, not

Table 10.1. Attributes of the founder population that either
increase or decrease the chance for genetic transilience
(from Templeton 1980).

Attributes that increase the chance	Attributes that decrease the chance
Average number of offspring large	Average number of offspring small
Reproductive value of founders high	Reproductive value of founders low
Open niche allowing population flush	Population flush not possible
Initial density low	Initial density high
Initial subdivided population structure	Initial panmictic population structure
Overlapping generations	Discrete generations
Assortative mating	Disassortative mating
Sexual selection on the mate recognition system	Rare male or similar sexual selection
Imprinting, partially learned sexual behavior	Sexual behavior totally genetic
Chromosome number, large	Chromosome number, small
Total genomic map, length large	Total genomic map, length small
Crossover suppressors few or easily lost	Crossover suppressors many and not easily lost

precede or cause it. Powell (1978) has modeled this scenario
with Drosophila in the laboratory. Although he was able to
detect the evolution of premating isolation mechanisms in
only a minority of cases, when it did occur, it occurred
rapidly, within fifteen generations after the original
founding event.

How do bark beetles fit this scheme? Each generation, beetles leave their host trees in large numbers to colonize new hosts. These new hosts are spatially separated from one

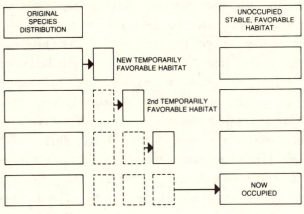

A. Reproductive isolation evolves rapdily while old and new species are allopatric.

ORIGINAL
SPECIES
DISTRIBUTION

UNOCCUPIED
STABLE, FAVORABLE
HABITAT

NEW TEMPORARILY
FAVORABLE HABITAT

2nd TEMPORARILY
FAVORABLE HABITAT

NOW
OCCUPIED

B. Adaptive divergence occurs later while allopatric.

PROGENITOR
SPECIES
DISTRIBUTION

NEW
SPECIES
DISTRIBUTION

C. Sympatry may reinforce isolation already partially formed while allopatric.

PROGENITOR   NEW

Figure 10.1 Possible scheme for multiple founder-flush cycles. Each arrow represents a founder event followed by a flush. Reproductive isolation is postulated to evolve rapidly independent of and preceding adaptive divergence (modified from Powell 1978).

another, often widely scattered throughout the forest community. While the total population of beetles in the forest may be large, the actual numbers colonizing an individual tree are a relatively small sample of that population. Each founding female is capable of laying hundreds of eggs of which only a few are likely to survive to emerge from that host the following year. Thus, each generation, from a few original founders, hundreds of new genotypes are produced that are subsequently subjected to rigorous within-tree selection pressures resulting from brood mortality as high as 99% (Coulson 1979). Bark beetles also experience longer term fluctuations in population size. During periods of endemism characterized by low level infestation, beetles are rare, inhabiting only a few isolated trees. During periods following severe droughts, or as a result of forest fires or pollution damage, these low level endemic populations explode into massive epidemics in which beetles colonize vast tracts of the forest. During these episodic flushes, as well as during the fluctuations in population size that occur each generation, the stage is set where new combinations of genes can be generated and tested by natural selection.

A second model of speciation that may be relevant to adaptive radiations in bark beetles involves speciation events that are associated with a shift in host plant (Bush 1969, 1974, 1975a, b). This model applies to insect species that require a host plant to complete at least some phase of their life cycle. Fruit flies that had been obligately associated with wild hawthorn in New England were observed to be living on apples in 1864 and on cherries in the 1960's. A similar host shift has been documented for another fruit fly from Oregon which shifted from wild to domesticated cherries

in the early 1900's. Mating and deposition of eggs occurs on
the fruit of the host species so that individuals utilizing
different hosts are effectively reproductively isolated.
Thus, a host shift is synonymous with speciation, and because
the newly colonized host occurs in close proximity to the
original host, speciation may occur even though the
genetically diverging fly populations are not separted by
great distances (but see Futuyma and Mayer 1980). Bush
believes that genetic determination of host preference may
involve only one or a few genes concerned with selection of
the host so that the shift need not initially respesent a
repatterning of the genome or extensive genetic shake-up. In
this way, this model is similar to the founder-flush genetic
transilience model which affects a few major genes with
strongly epistatic effects. They differ in that, in the host
shift model, reproductive isolation is selected for directly
while the two host races are sympatric; that is, reproductive
isolation is ecological, involving a preference for mating on
an alternate host. In the founder-flush model, reproductive
isolation evolves fortuitously when the progenitor and the
new species are allopatric (Figure 10.1).

Although these models of speciation have independent
origins and have several points about which they differ, they
are not mutually exclusive. In fact, the life cycle of bark
beetles includes many of the criteria of both models. We
have previously discussed the ways in which the bark beetle
life cycle meets the criteria of the founder-flush cycle. In
what ways do the habits of bark beetles approximate the
criteria of the host shift model and how can the two models
be integrated?

Bark beetles, like fruit flies, select a particular host
tree upon which mating and oviposition occur and in which

their brood will develop. If, during this time of brood development, larvae either become conditioned to prefer the host in which they were reared (Baker et al. 1971) and/or host-adapted genes are accumulated by natural selection via differential mortality within the tree, then those insects emerging from different hosts will be ecologically isolated from one another. Furthermore, if their pheromones are synthesized from or synergized by compounds specific to different species of host tree, as they are in some Scolytids (McKnight 1979), isolation between the incipient host races will be reinforced. Allochronic separation of brood emergence from different hosts may further restrict gene flow between the host races and, in time, speciation may result. Several other mechanisms likely to affect reproductive isolation in bark beetles are listed in Table 10.2.

Whether accumulation of host-adapted genes by these processes will ultimately lead to speciation is an important question of current interest to evolutionary biologists (Stebbins and Ayala 1981). It is precisely this issue that may serve to integrate the two models of speciation. Although evidence suggests that the slow accumulation of host-adapted genes, by itself, need not necessarily lead to speciation, the consolidation of gene complexes adapted to different host plants in combination with a genetic transilience may increase the probability that the genetic changes would result in the formation of a species both reproductively isolated from the progenitor species and already adapted to its new habitat. Host shifts, in conjunction with the genetic events of the founder-flush cycle, may provide a major speciation event for bark beetles. If this is true, then speciation may occur quickly and the resulting species may be morphologically indistinguishable.

Table 10.2.   Mechanisms that affect reproductive isolation among sympatric species of <u>Dendroctonus</u> (modified from Lanier and Burkholder 1974).

A. Pre-mating isolating mechanisms

1. Ecological stratification of the habitat and of the host tree
2. Allochronic differences in flight period or diel activity
3. Pheromone differences
4. Synergist differences
5. Pheromone inhibitor differences
6. Host preferences
7. Host toxicity
8. Courtship behavior specificity; e.g., stridulation
9. Genitalia mismatch

B. Post-mating isolating mechanisms

1. Spermatophore rejection
2. Egg viability
3. Hybrid sterility

With these mechanisms and predictions in mind, we may examine some of the patterns of genetic variation that have been documented for bark beetles. Gel electrophoresis is a technique that was developed relatively recently (Harris 1966, Lewontin and Hubby 1966) and has been used for sampling the genome in order to estimate the amount of genetic variation in natural populations. This method can also be used to provide genetic markers to estimate the extent or manner in which populations, races or species have differentiated (Selander and Johnson 1973, Avise 1974, Ayala et al. 1974, Gottlieb 1977). We used electrophoresis to study the population structure of mountain pine beetles, Dendroctonus ponderosae, from several sites where they were inhabiting three different host trees growing adjacent to one another (Sturgeon 1980). We found that beetles collected from the three hosts were differentiated morphologically as well as genetically (Table 10.3). This finding suggests either that genetically different mountain pine beetles are selecting different host trees and/or that selection due to host-mediated mortality is occurring within the tree. In either case, we have detected genetic differentiation associated with host species in these populations. We found an additional enticing piece of information. Deviations in genotypic proportions from those expected on the basis of the Hardy-Weinberg equilibrium law were detected at one locus in all of these populations. Although several factors can account for the deficiencies in heterozygotes that we found, one explanation is that we actually sampled from what we thought was one population of mountain pine beetles when, in reality, it was several populations between which mating was restricted. Perhaps we actually detected the presence of incipient host races or sibling species in these five populations.

On the other hand, as we mentioned earlier, much evidence suggests that the genes we sample using electrophoresis (that is, structural genes) are not those that are important in speciation events (Wilson 1975). Bush and his associates (Bush 1975b, Boller and Bush 1973) were unable to detect extensive differentiation among any of the host races of fruit flies in spite of the fact that they were apparently isolated in the field. But the allozymes of the mountain pine beetle reveal another facet to the story which may be

Table 10.3. Allele frequencies at two esterase loci of mountain pine beetles from three host trees

Locus	Allele	Host tree			d.f.	X^2	P
		ponderosa	lodgepole	limber			
		N=173	N=112	N=205			
EST-1	2	0.130	0.040	0.083			
					2	13.9	<0.001
	3	0.870	0.960	0.917			
EST-2	2	0.232	0.232	0.234			
					4	9.8	<0.05
	3	0.335	0.379	0.271			
	4	0.433	0.388	0.495			

significant. Populations of mountain pine beetles located less than a mile from one another along one stretch of road in Colorado were significantly differentiated at three of five polymorphic loci surveyed electrophoretically. In another population where four individuals of one host tree were present, beetles collected from the four trees differed at one of the five polymorphic loci. These data suggest that, in fact, bark beetle populations are subdivided into smaller units. Beetles from different trees of the same host, located right next to one another, are differentiated genetically (Sturgeon 1980). This demic substructure, in which founder events and inbreeding may have strong impacts on the genetic characteristics of bark beetle populations, is a condition most conducive to the occurrence of genetic events likely to be involved in speciation (Wright 1940, White 1978, Wilson 1975, Bush et al. 1977) and those most likely to result in a genetic transilience (Templeton 1980a, b). Should a genetic transilience occur in a population subdivided along host lines, a host race may rapidly become a new species that is pre-adapted for life in its new habitat. Thus, ecological properties of bark beetle populations influence their genetic characteristics and the genetic characteristics, in turn, influence their ability to utilize new ecological niches.

The evolutionary significance of this demic sub-structure imposed by the tree-boring habitat may extend to much more than just speciation in bark beetles. Hamilton (1978) has gone so far as to speculate that the breeding structure imposed upon organisms by this habitat is so conducive to evolutionary change that many of the major groups of insects may have diverged there; that is, major patterns of diversity seen in the Class Insecta may have had their origins in the tree-boring habitat. For example, he suggests that this habitat may have been the selective force for the evolution of holometaboly. Clearly, the presence of wings is a hindrance to the effective exploitation of a burrowing habitat, yet it is a distinct advantage for dispersal to locate new hosts. Whether similar evolutionary pressures have contributed to speciation in bark beetles is not known but the ideas are intriguing and the abundance of bark beetles that inhabit the coniferous forests of the west and southeast offer ample opportunities for scientists to pursue answers.

COEVOLUTION

Accompanying the increase in species diversity among bark beetles was a manifold increase in the numbers of species associated with them. Evolution in bark beetles influenced evolution in other species groups. The process by which interacting species affect each other's evolution was named "coevolution" by Ehrlich and Raven (1965). While Darwin (1859) was perhaps the first to recognize the importance of these interactions in affecting evolutionary change, Ehrlich and Raven were the ones that really emphasized the importance of reciprocal selective pressures in generating organic diversity and they specifically called attention to the importance of plant-herbivore interactions. They coined the word coevolution after summarizing the utilization of different plant groups by different families of butterflies with particular attention focused upon chemical characteristics of the plants. For example, they observed that plants produced a series of chemical compounds whose original anti-herbivore function probably occurred purely by chance. However, once protected by these compounds, the plants diversified in the relative absence of insect predation and the chemicals came to characterize entire groups of plants. Eventually, insects evolved the ability to feed on the plants in spite of the toxicity of the compounds. Subsequently, these insects radiated in the absence of competition from other phytophagous insects. For many insects, the chemical "defenses" of the plant became cues which enabled them to locate the plant and to initiate feeding or mating. In some cases, such as in the case of the monarch butterfly that feeds on milkweeds, the chemical was sequestered by the insect and used in its own defense against its predators. In this manner, a diversity of organisms involving several trophic levels became associated into clearly defined communities.

This same sort of evidence can be used to speculate on coevolution in bark beetle communities. Monoterpenes were probably originally formed in gymnosperms as waste or by-products of terpene metabolism. Terpene metabolism serves several vital functions in plants such as the production of abscisic and gibberellic acids and of carotenoid pigments (Harborne 1973). By chance, these low molecular weight terpenoids, the monoterpenes, served to deter several groups of phytophagous insects from attacking the trees. In part because of the absence of predation by insects, the plants radiated to produce what we now refer to as the Order Coniferales and which includes all of our familiar species of gymnosperms such as the pines and the true firs. But today, these trees are all attacked by several species of bark beetles and almost every species of tree is attacked by a Dendroctonus. Presumably, at some time in their evolutionary history, a Dendroctonus progenitor evolved the ability to detoxify one or more monoterpenes and found itself in a new adaptive zone, free to radiate in the absence of competition from other phytophages utilizing gymnospermous bark. As the genus Dendroctonus evolved, the selective pressures exerted on the trees probably led to the development of more elaborate chemical defenses by the trees. Evolutionarily, the more probable route was not to produce additional novel compounds but rather to vary the proportions of those compounds already present so that, today, all of these trees are characterized by the presence of many terpenoids that vary both within and between species (Squillace 1976). Further evolutionary refinements have resulted in the insects use of these chemicals to locate their hosts, to initiate feeding and to synthesize and synergize their population aggregating pheromones. The chemicals have come a long way from their apparent original role as anti-herbivore defenses.

Although the evidence for coevolution from these examples is persuasive, these ideas about coevolution were generated from patterns resulting from interactions that occurred in the past. Evidence that coevolution is occurring currently must be sought in interactions among extant populations. We must demonstrate that changes in the abundance and distribution of organisms caused by their interactions with other organisms have genetic, and therefore, evolutionary consequences. When we look at bark beetle-host tree coevolution from this perspective we can understand how the chemicals today are serving the trees as anti-herbivore defenses as well as serving the beetles as aids in colonization of those trees.

One of us has investigated coevolutionary interactions occurring between two species of bark beetles and their major host tree, ponderosa pine. We believe that an analysis of these interactions offers a partial explanation of the dual roles played by these chemicals. The western pine beetle (WPB), Dendroctonus brevicomis, is a "specialist" on ponderosa pine; that is, it attacks (with minor exception) only ponderosa pine. The part of its range where it is most abundant and it has been most destructive is in California, although it has also been recorded outside of the state. The mountain pine beetle (MPB), Dendroctonus ponderosae, on the other hand, attacks several host trees throughout its range, only one of which is ponderosa pine. Where its range overlaps with the WPB it is found only rarely on ponderosa pine but instead switches to sugar and lodgepole pines. So, although the two species of insect occur sympatrically, they rarely occupy the same host when they are sympatric. Their host in allopatry, ponderosa pine, is chemically polymorphic. Individual trees contain five major monoterpenes, alpha-pinene, beta-pinene, 3-carene, myrcene, and limonene along with, at least, six additional monoterpenes that occur in lower concentration. Throughout its range, trees with a wide diversity of chemical composition have been identified and there is considerable intrapopulation variation (Smith 1977). Knowing that monoterpenes are important to the beetles in selecting and colonizing their hosts, it was reasonable to predict that the variation among individual trees in their chemical composition might influence their resistance or susceptibility to attack by beetles. If one or more combinations of monoterpenes were toxic or repellant to the beetles, these trees should remain untouched during an infestation. Other combinations, containing large quantities of pheromone precursors and synergists ought to be attacked in greater proportion. The hypothesis was tested in two geographic localities, one in California where ponderosa pine is attacked by the WPB (Sturgeon 1979) and one in Colorado where the pine is attacked by the MPB (Sturgeon 1980).

In California, 386 trees from populations with and without history of WPB infestation were analyzed biochemically. The data were subjected to a principal component analysis which is presented in Figure 10.2. The common chemical composition of trees from populations with no history of WPB infestation (closed circles) shows quite a range, but falls within a large grouping to the left in the diagram. In addition, there are a few chemically unique trees that appear on the figure in the upper right. In those populations in which WPB infestations have occurred in the past (open triangles), these chemically unique trees are greatly increased in number. It appears then that chemically unique trees survived infestations in greater proportions than did the

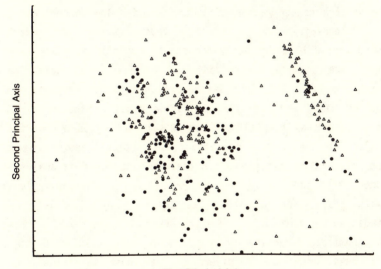

First Principal Axis

Figure 10.2   Two dimensional representation of principal components analysis of individuals from populations with a history of bark beetle infestation (△) and those from populations without known history of predation (●) (from Sturgeon 1979).

chemically common ones.   When we analyze these rare trees
even closer we find that they are characterized by low levels
of alpha-pinene, a pheromone precursor of the WPB, and high
levels of limonene, a known toxin or repellant of the WPB.
The presence of limonene alone is not sufficient to deter the
WPB.   Other trees with high levels of limonene are tolerated
as long as they also have sufficiently high levels of the
necessary   pheromone   precursor,   alpha-pinene.   However,
chemically rare trees without high levels of limonene do not
appear to survive in proportions higher than the common
chemical phenotype.   So, the important combination appears to
be a high concentration of limonene and a low concentration
of alpha-pinene.   The WPB thus appears to exert
simultaneously a frequency-dependent selection pressure (for
the rare chemical phenotype) and a directional selection
pressure (for the right limonene/alpha-pinene combination).
By selecting and rejecting ponderosa pines on the basis of
their chemical composition, the WPB is acting as a powerful
selective agent and is affecting chemical evolution in
populations of the pines.   A discriminant function analysis
shows that the mean chemical composition of the populations
with history of predation has shifted significantly from
those which have never been attacked (Figure 10.3).   The
biochemical diversity of the stands, as measured by the
evenness component of diversity, has been maintained, or
perhaps increased, by this selective predation.

     What does this mean for the WPB?   The next time it
reinfests these forests, it will find fewer of its preferred
trees.   Eventually the beetles will either have to avoid
those populations or evolve the ability to utilize different
chemical phenotypes.   Evidence from insect evolution of
resistance to insecticides suggests that evolution would be

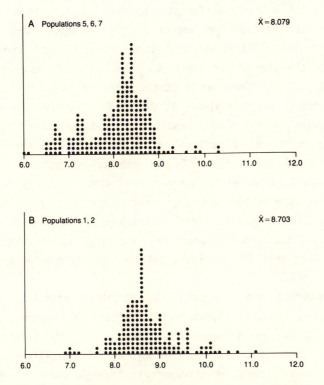

Figure 10.3   Frequency distributions of P. ponderosa from populations with a history of bark beetle predation (A) and from those without a history of predation (B) upon a discriminant function constructed from variance-covariance matrices of five monoterpenes. $F_{5,386}$; P < .001 (from Sturgeon 1979).

an easy task for any insect. However, when we consider the numerous ways in which bark beetles are tied to their host tree, avoidance of the resistant stand may be the more likely alternative. The WPB would need to evolve not only an ability to detoxify different compounds but also to utilize different pheromone precursors and synergists. More than likely, WPB infestations will occur in different locations until the day comes that it runs out of populations containing its preferred trees and it has no other alternative but to evolve. At that time, selective pressures to evolve would be sufficiently intense to greatly increase the probability of survival of individuals with genotypes capable of utilizing different chemical phenotypes. In California, then, we can conclude that monoterpenes serve the beetles as pheromone precursors and synergists and the trees as an anti-herbivore defense. The key to their dual roles lies in the great between-tree diversity in chemistry--a condition that is, itself, maintained by the feeding habits of the WPB.

Ponderosa pines attacked by the MPB present a very different picture. Biochemical diversity of ponderosa pine populations in Colorado is lower than in California populations (Figure 10.4) which may, in part, explain why the MPB can utilize three different host trees there (Sturgeon 1980). The alternative hosts of the MPB in Colorado are lodgepole and limber pine, both of which are also biochemically less variable than California ponderosa pines. It is conceivable that, in Colorado, a lodgepole pine may be biochemically no more different from a ponderosa pine than one biochemical variant of ponderosa pine in California is from another variant of the pine. That is, the breadth of diet available to the MPB in three host trees may be no

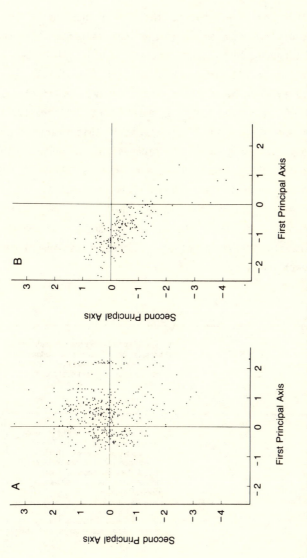

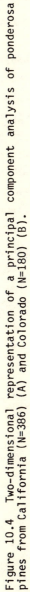

Figure 10.4 Two-dimensional representation of a principal component analysis of ponderosa pines from California (N=386) (A) and Colorado (N=180) (B).

larger than that available to the WPB in one host. The MPB does not discriminate among biochemical variants of ponderosa pine (Sturgeon 1980) but rather it discriminates most probably among its three hosts. Of the total chemical diversity present in a stand, any one MPB is likely to be capable of utilizing only a small part. Therefore, any chemical phenotype that lies outside of this range would be selected for. The greater the biochemical distance among the three hosts, the greater the chance that a MPB will not be able to incorporate all three into its diet. So, by discriminating among the three host species, the MPB simultaneously increases the diversity among the three hosts (by decreasing their overlap) and restricts the diversity within each one (Figure 10.5a,b).

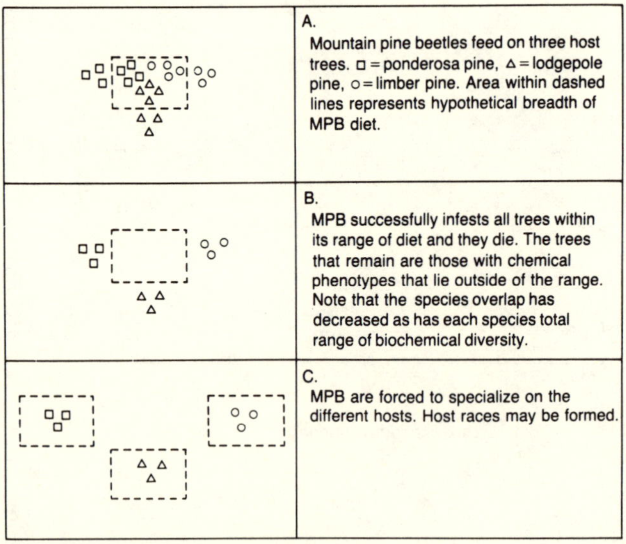

	A. Mountain pine beetles feed on three host trees. □ = ponderosa pine, △ = lodgepole pine, ○ = limber pine. Area within dashed lines represents hypothetical breadth of MPB diet.
	B. MPB successfully infests all trees within its range of diet and they die. The trees that remain are those with chemical phenotypes that lie outside of the range. Note that the species overlap has decreased as has each species total range of biochemical diversity.
	C. MPB are forced to specialize on the different hosts. Host races may be formed.

Figure 10.5   Reciprocal selection pressures exerted between mountain pine beetles and three host trees.

Eventually, beetles using different hosts may be forced to use different pheromone precursors and synergists. In fact, one would predict that MPB populations would rapidly become subdivided into races adapted to each of the three different hosts (Figure 10.5c). In populations where MPB's occur in all three trees sympatrically, beetles collected from the different hosts are differentiated genetically and morphologically (Sturgeon 1980). The host tree is a selective environment for the beetle in several ways. Each host tree presents a formidable barrier to the beetle in terms of its chemical defenses and there is likely to be variation within any beetle population for the ability to detoxify these compounds. The MPB's that we collected from the three hosts are genetically differentiated at two esterase loci (Table 10.3) and esterases have repeatedly been implicated in insect detoxification systems (Krisch 1971). Esterases are among the most polymorphic enzymes known and it is believed that their high levels of polymorphism probably reflect variation in food substrates available to the organisms (Gillespie and Kojima 1968). During their development within the tree, an average of 85-99% of the MPB brood succumbs for various reasons. Perhaps a portion of that mortality can be accounted for by differential survival of genotypes capable of tolerating the different toxins present in the three host trees.

Host trees differ in other ways. The bark of the host tree is the single most important factor buffering the phloem, where the beetle brood develops, from the physical environment (Stark and Dahlsten 1970). Trees with thin bark offer less protection to the developing brood from cold, for example. Limber pine has thin bark relative to either ponderosa or lodgepole pines and we found that beetles collected from limber pine were significantly more homozygous than were beetles collected from either of the two other hosts (Figure 10.6). This suggests that the thin bark of limber pine may have selected for alleles adapted to colder environments. Beetles collected from limber pine are, in fact, more cold hardy than are beetles from the other hosts (Morrow 1972). Furthermore, beetles collected from limber pine are morphologically less variable. Both of these findings indicate that limber pine imposes a more restricted environment upon MPB's than do the other hosts. The thin bark of limber pine appears to have exerted directional selection pressures for particular alleles resulting in reduced genetic variation accompanied by a reduction in morphological variation.

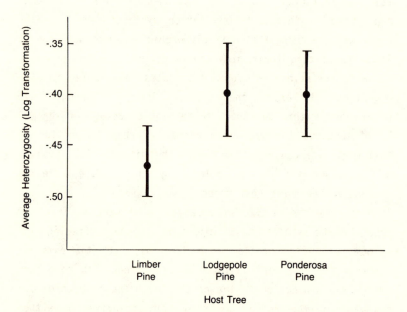

Figure 10.6   Average heterozygosity and 95% confidence intervals for mountain pine beetles in three host trees.

In summary, by looking at coevolutionary interactions occurring in extant populations, we can see that the evolutionary responses are, indeed, reciprocal. The beetles influence the chemical composition of populations of their host trees and the trees affect the genetic composition and morphology of beetle populations. Therefore, we have good reason to assume that bark beetle/host tree coevoluton has been ongoing for millions of years and that it continues today.

An interesting perspective on bark beetle/host tree coevolution is seen in the outbreak of the Dutch elm disease in the United States. In Europe, where the fungus, tree, and beetle have had the opportunity to coevolve, Dutch elm disease is a minor problem. The fungus, Ceratocystis ulmi, and its vector, the lesser European elm bark beetle, Scolytus multistriatus, were accidentally introduced into the U.S. prior to 1909. Records of this disease document its rapid migration east and west and its slower migration to the north where it is probably limited by cold winter temperatures (Strobel and Lanier 1981).

In natural environments characterized by endemic level populations of beetles, the idealized age structure of trees postulated below may be found (Figure 10.7). Under endemic conditions, the beetle only takes older trees; these trees often have had ample time to reproduce, and may now be either injured or senescent. The elm trees in America have not had the opportunity to adjust to the selective pressures exerted by this fungus. We witness the wholesale mortality of trees in their prime. Evolutionary theory, however, allows us to make an optimistic prediction. The elms being struck with the disease are either in the early or middle part of their reproductive period. Therefore, there is a high probability

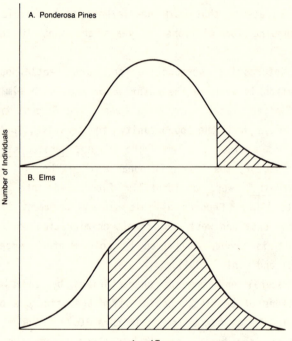

Figure 10.7   Age classes of ponderosa pine killed by mountain pine beetles and of elms killed by Dutch elm disease.

that resistance to the fungus will evolve. Certainly some of the trees now challenged are more resistant than others, and they should continue to reproduce longer than more susceptible trees around them. We hope that, in time, the elms in America will evolve so that their susceptibility to this disease is pushed later and later into the life span typical for this species. Ultimately, age class of infected trees in Figure 10.7B should evolve to look like Fig. 10.7A.

COMMUNITY EVOLUTION

Our understanding of the evolution of community structure is still woefully inadequate even though considerable attention has been focused on it.   Some of the earliest attempts to understand community structure began with descriptive comparisons of patterns seen in tropical and temperate communities. Largely under the leadership of the late Robert H. MacArthur, descriptive ecology became a predictive science that integrated observed patterns with quantitative theory (Cody and Diamond 1975).   He wrote that "to do science is to search for repeated patterns".   Pattern implies repetition and repetition implies that prediction is possible (MacArthur 1972).   Here we are concerned with patterns observed in plant communities and the herbivores associated with them.   Are there patterns to be found?   If so, do they help us understand how these communities evolved?   There are a relatively limited number of studies that have addressed the issue of plant/herbivore community structure.   Striking patterns have been observed but, to-date, interpretation of the patterns remains embroiled in controversy.

Biologists have always been fascinated by islands because their patterns of community structure in many ways differ markedly from the adjacent mainland communities from which they were obviously derived.   The "theory of island biogeography" (MacArthur and Wilson 1967) was originally developed to describe these actual island patterns but the authors were aware that it could easily be extended to include "islands" such as patches of forest or species of trees isolated from other such patches or species. In fact, the theory of island biogeography has revealed some interesting generalizations about plant/herbivore communities (Opler 1974, Seifert 1975, Strong 1974a, 1974b, Strong and Levin 1979, Strong et al. 1977, Tepedino and Stanton 1976, Cornell and Washburn 1979). The number of herbivore species that any one host "island" can support (be that host island one plant or a group of plants) is proportional to the area of that host island, and an upper limit to the number of species is established rather quickly. Host plants and their parasites regularly conform to this species-area relationship (Figure 10.8A) which when plotted on a double logarithmic scale produces a linear relationship (Figure 10.8B). The upper limit to the number of species found in any area is established for several reasons: 1) the host island is a finite resource; that is, its resource base (e.g. its size or distribution) cannot continue to expand indefinitely; 2) there is a limit to the amount of resource partitioning that can occur before population sizes of the herbivores become so small as to greatly increase the probability of their extinction; or 3) there is a limit to the amount of overlap allowed among the herbivores before competitive exclusion of one or the other results (MacArthur 1972). An alternative theory, developed by T.R.E. Southwood (1960, 1961, 1973) suggests that the evolutionary age of the host plant, rather than its area, is a better indicator of the number of parasitic species that it can support. Those plants that have been around longer will support more parasites. A third hypothesis, the "insect-plant coevolution theory", was suggested by Claridge and Wilson (1978). This theory recognizes the special role that plant chemical defenses may have played in determining the feeding habits of insects. This theory may be interpreted to mean that the resource base of the host, primarily through its elaboration of complex chemical defenses, is able to expand essentially

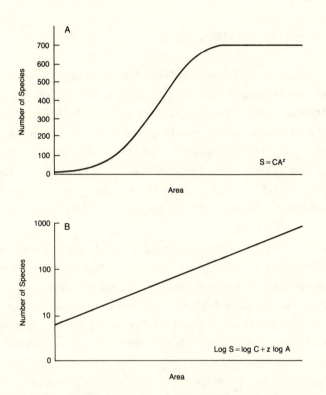

Figure 10.8   Species-area relationship.   S = number of species, C = constant which varies between taxa and from place to place, A = area of island, Z = slope or rate at which new species are added.

indefinitely. This view is perhaps most vividly expressed by Gillett (1962),

> Pest pressure is the inevitable, ubiquitous factor in evolution which makes for an apparently pointless multiplicity of species in all areas in which it has had time to operate

and by Whittaker (1969),

> I conclude that species diversity of plant communities is an evolutionary product, subject to self-augmentation through time. There is no ceiling or saturation level for the diversity which results; the chemical differentiation of the higher plants implies virtually unlimited potentialities for the addition of different species...; plant and insect communities may show indefinite, indeterminate evolutionary increase in diversity.

Most recently, Lawton and Strong (1981) suggest that other factors, such as host plant phenology, may be more important determinants on species-area relationships than are either competition or coevolution. Unfortunately, we do not have enough information of qualitative biochemical differences among plants to assess their contribution to patterns of species richness.

Species-area relationships may also be found in bark beetle communities. If we choose one taxon, the genus Pinus, we can minimize the effects of evolutionary age of the host plant as a confounding variable and focus our attention on two other theories, the island biogeographic theory and the insect-plant coevolution theory. What we find is that host plants with larger geographic areas support larger numbers of Scolytids (r = 0.72, P< 0.001) (Figure 10.9). Our limited knowledge of the biochemistry of these trees, does not allow us to determine whether biochemically complex hosts support more Scolytids than do biochemically restricted ones. However, some interesting observations can be made from Fig. 10.9. More than likely, geographic area is positively correlated with habitat diversity so that trees which are

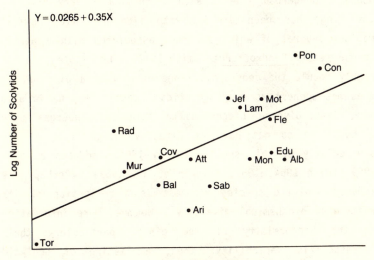

Figure 10.9 The number of Scolytid species on _Pinus_ species of the western United States (r = 0.72, P < 0.001). TOR = P. torreyana, ARI = P. aristata, BAL = P. balforiana, MUR = P. muricata, RAD = P. radiata, COU = P. coulteri, ATT = P. attenuata, SAB = P. sabiniana, MON = P. monophylla, ALB = P. albicaulis, EDU = P. edulis, FLE = P. flexilis, LAM = P. lambertiana, JEF = P. jeffreyi, MOT = P. monticola, CON = P. contorta, PON = P. ponderosa.

found in many different environments are likely to be attacked by a correspondingly greater number of beetles. For example, ponderosa pine occupies thousands of square miles throughout the western United States, and taxonomists have recognized three morphological and physiological races or varieties. Throughout its range, it is attacked by no less than 36 different species of Scolytids (Furniss and Carolin 1977). On the other hand, only a few Scolytids have been recorded on bristlecone pine which grows in relatively small, mountain-top stands in the southwestern United States. However, ponderosa pine is also biochemically very diverse. Its range has been divided into five distinct chemical regions several of which are each associated with a nearly unique set of Dendroctonus (Smith 1977). Lodgepole pine has nearly double the geographic range of ponderosa pine but it does not support as many Scolytids. Lodgepole pine is also much less diverse biochemically than is ponderosa pine although we do not have as extensive information on the biochemistry of this species (Mirov 1961, Anderson et al., 1969, Smith 1964, 1967, Zavarin et al. 1969). Perhaps its reduced Scolytid species richness is a reflection of its restricted biochemical diversity. We need more information on the biochemistry of the pines, particulary their intraspecific variability; then, a listing of Scolytid species by chemical races rather than by taxonomic variety could be generated. We know that California ponderosa pines are far more variable than Colorado pines. Do they support more Scolytids? We cannot now distinguish between these hypotheses but this avenue of investigation holds promise to help us understand how bark beetle community structure has evolved.

One of the more interesting complexities of bark beetle community structure involves the common occurrence of mutualism. The conditions under which coevolutionary relationships evolve towards mutual benefit is one of the many issues that theoreticians continue to address (Roughgarden 1975, Slatkin and Smith 1979, Wilson 1980). A theory that appears to fit the ecology and genetic structure of bark beetles is the theory of structured demes presented by Wilson (1980). This theory attempts to explain evolution of weak altruism and mutual benefit by modifying contemporary selection models that implicitly or explicitly rely upon homogeneous gene frequencies within natural populations. Genetic variation from place to place within demes (these areas within demes are trait groups in the terminology of Wilson) makes modeling of evolutionary dynamics more difficult, but if Wilson is correct, it also allows the evolution of adaptations, such as mutualism, not possible in more conventional models.

For example, one of the more complex relationships that exists in the bark beetle community is that involving bark beetles and their symbiotic microorganisms. Rigorous thinking about evolution within populations caused by fitness variation between individuals has discouraged phrases about the "ecological role," or "function," or "purpose" of a species in a community. We have been taught to think of each individual as acting selfishly, or at most looking out for closely related kin (Hamilton 1964), but we are uncertain about the mechanisms by which highly organized, mutualistic relationships evolve. It is interesting enough to consider that tree-killing fungi are vectored from tree to tree by bark beetles and that the fungi, in turn, benefit the beetle vector by modifying conditions in the tree such that they are conducive to beetle survival and reproduction. It is difficult to grasp the evolution of a mutualistic association that includes beetle consumption of the fungi during brood development. Has the fungus evolved to enhance the survival of the beetle, which is both vector and predator? By what mechanism did this evolve? Individual selection models would not predict this outcome (Wilson 1980) because, while the fungus does benefit itself in that it ensures itself a vector, it also benefits all other fungi and any other organism with for which it must compete for space during the journey from tree to tree. Any fungal "cheater" genotype--one which does not allow itself to be used as beetle food--would be selected for. It could, for example, allocate more biomass to a more rapid or more extensive colonization of the tree, thereby increasing its chances of locating many different beetles to serve as vectors to different host trees. Yet, it appears that beetles and fungi have evolved for their mutual benefit. Beetles vector fungi in special chambers called mycangia (Chap. 3), and the fungi aid the beetle in overcoming the defenses of the trees (Chap. 6) and by providing food for the developing larvae. The fungi may even regulate water conditions in expiring trees. The fungi rely solely on beetles for transport--many are not found in natural populations apart from the beetle. Wilson (1980) attempts to explain some of these complex phenomena using his theory of structured demes while simultaneously reconciling the different approaches of the evolutionary ecologists (who conceive of all adaptations in terms of individual fitness) and the ecosystems ecologists (who speak in terms of species "professions" or "roles" in community "function"). Much of this theory is untested but once again, bark beetle communities offer an opportunity in which to seek answers.

APPENDIX

Host Utilization of Scolytidae

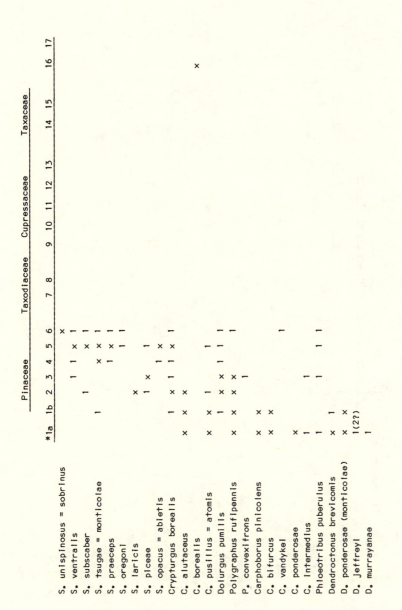

Appendix: Host utilization of Scolytidae

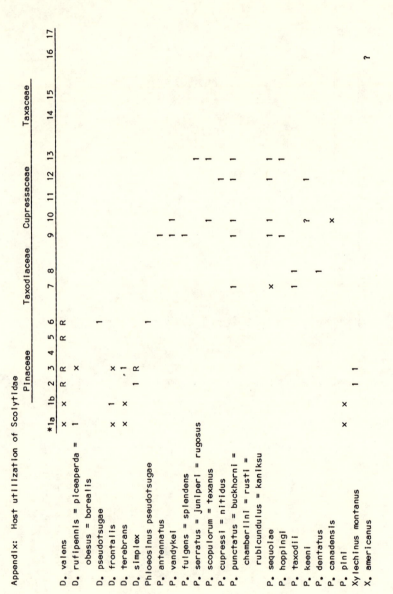

Appendix: Host utilization of Scolytidae

	Pinaceae							Taxodiaceae		Cupressaceae					Taxaceae			
	*1a	1b	2	3	4	5	6	7	8	9	10	11	12	13	14	15	16	17
Sclerus annectans	1			x		1												
S. pubescens				x		1												
Pseudohylesinus nobilis = furnissi					1	x												
P. sericeus				1	1	x	1											
P. pini	x																	
P. nebulosus		?		x	1	x	1				x							
P. dispar	1				1	x	1											
P. granulatus					1	x	1											
P. tsugae = keeni = obesus = similis					1	x												
P. pullatus						1												
P. sitchensis				1(2)														
Hylastes ruber						1	1											
H. nigrinus = yukonis		1	1		1	1	1											
H. tenuis = pusillus = parvus	x	x																
H. macer	x	x	1															
H. gracilis	x			1	1													
H. longicollis	1				1													
H. porculus	x	x																
H. salebrosus	x	x	x															
Hylurgops subcostulatus	x	x		x		x												
H. rugipennis	x					1	1											
H. reticulatus	x																	
H. porosus = lecontei	x	x	x															
H. pinifex	x	x	x															

Appendix: Host utilization of Scolytidae

	Pinaceae							Taxodiaceae		Cupressaceae					Taxaceae			
	*1a	1b	2	3	4	5	6	7	8	9	10	11	12	13	14	15	16	17
Trypodendron lineatum = bivittatum	x	x	x	x	x	x	x										x	
T. rufitarsus	x	?	?	x	?	?	1											
T. scabricollis	x	x	x	x	x													
Cryphalus pubescens	1	1						1										
C. fraseri						x	x											
C. rubentis			1															
C. ruficollis			x		x	1	1											
Gnathotricus aciculatus	1	?	1	1	x	x	1	1										
G. retusus		?		1	x	1	1	1		1								
G. sulcatus	x	x	x		x												x	
G. materiarus	1	1	x	x		x					x						x	
Conophthorus ponderosae = flexilis = monticolae = contortae	1																	
C. monophyllae	1																	
C. lambertianae	1	1																
C. radiatae	1																	
C. coniperda = taedae	1	1																
C. resinosa	x																	
Pityophthorus keeni	1																	
P. boycei = catulus	x	x																
P. pinguis	x																	
P. ramiperda = fivazi	1	1																
P. scalptor	x																	
P. toralis	x																	

Appendix: Host utilization of Scolytidae

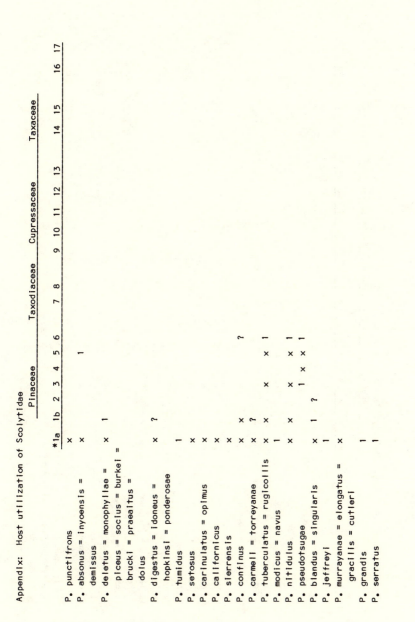

	Pinaceae									Taxodiaceae		Cupressaceae					Taxaceae		
	*1a	1b	2	3	4	5	6	7	8	9	10	11	12	13	14	15	16	17
P. punctifrons	x																	
P. absonus = inyoensis = demissus	x					1												
P. deletus = monophyllae = piceus = socius = burkei = brucki = praealtus = dolus	x		1															
P. digestus = idoneus = hopkinsi = ponderosae	x	?					?											
P. tumidus	1																	
P. setosus	x																	
P. carinulatus = opimus	x																	
P. californicus	x																	
P. sierrensis	x																	
P. confinus	x	x																
P. carmelii = torreyanae	x	?																
P. tuberculatus = rugicollis	x	x		x	x													
P. modicus = navus	1																	
P. nitidulus	x	x		x	1	x	1											
P. pseudotsugae					1	x	1											
P. blandus = singularis	1			1	x		1											
P. jeffreyi	1		?															
P. murrayanae = elongatus = gracilis = cutleri	x																	
P. grandis	1																	
P. serratus	1																	

Appendix: Host utilization of Scolytidae

	Pinaceae							Taxodiaceae		Cupressaceae					Taxaceae			
	*1a	1b	2	3	4	5	6	7	8	9	10	11	12	13	14	15	16	17
P. confertus	x	x				x												
P. electus	1																	
P. fuscus = smithi	1																	
P. venustus = artifex	x	x				1												
P. bassetti				1		1												
P. intextus = shepardi			x	x														
P. pullcarius	x	x																
P. opaculus = abietus	x	x	x	x		x	1											
P. biovalis	1			1		1												
P. balsameus = patchi	1			1		1												
P. dentifrons				1														
P. cariniceps	x	x																
P. annectens	x	x															x	
P. puberlus																		
Ips spinifer = sabinianae	x	x																
I. latidens = longidens = guildi	x	x		1	x													
I. concinnus = chamberlini	x			1			?											
I. mexicanus = radiatae	x	x																
I. emarginatus	x	x																
I. pini = oregoni = laticollis	x	x		x														
Ips integer	x																	
I. plastographus plastographus	1																	
I. p. maritimus	x																	
I. calligraphus = ponderosae	x	x																

Appendix: Host utilization of Scolytidae

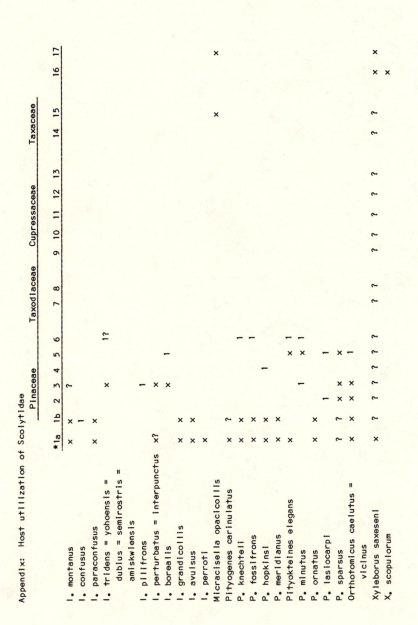

	Pinaceae						Taxodiaceae		Cupressaceae					Taxaceae				
	*1a	1b	2	3	4	5	6	7	8	9	10	11	12	13	14	15	16	17
I. montanus	x	x																
I. confusus		1																
I. paraconfusus	x	x																
I. tridens = yohoensis =				x			1?											
dubius = semirostris =																		
amiskwiensis					1													
I. pilifrons				1														
I. perturbatus = interpunctus	x?			x														
I. borealis				x		1												
I. grandicollis	x	x																
I. avulsus	x	x																
I. perroti	x																	
Micracisella opacicollis	x															x		x
Pityogenes carinulatus	x	?																
P. knechteli	x	x																
P. fossifrons	x	x					1											
P. hopkinsi	x	x			1		1											
P. meridianus	x	x																
Pityokteines elegans	x					x	1											
P. minutus				1		x	1											
P. ornatus	x																	
P. lasiocarpi	?	?	1			1												
P. sparsus	?	x	x	x	x	x	1											
Orthotomicus caelutus =	x	?	x	x	?	?	?											
vicinus																		
Xyleborus saxeseni	x	?	?				?	?	?	?	?	?	?	?	?	?	x	x
X. scopulorum																	x	

Glossary

Adaptive radiation. Evolutionary divergence of a group of organisms into a series of different niches or adaptive zones.

Aggregation pheromone. A pheromone which attracts conspecific individuals, usually of both sexes, to a given area.

Alkaloid. Nitrogenous compounds (apx. 5,500 forms) distributed widely in the roots, leaves, and fruits of angiosperms. Many of these are physiologically active, and they may serve as plant defenses.

Allele. One of two or more forms of a gene.

Allochronic. Pertaining to a temporal separation between two or more groups of organisms in the timing or development of their growth and development.

Allomone. A chemical emitted by an organism that induces in an organism of another species a behavioral or physiological response of adaptive benefit to the emitter, but not to the perceiver.

Allopatric. Having separate geographical ranges.

Allopatric speciation. The evolution of reproductive isolation (barriers to hybridization) in allopatry.

Altruism. Risky or potentially self-destructive behavior performed for the benefit of others.

Ambrosia beetles. Small compact beetles of the family Scolytidae. Ambrosia beetles, sometimes called timber beetles, bore into the sapwood of trees and feed on fungi (see ambrosial) that they inoculate into their galleries. Ambrosial beetles generally attack dead trees and fallen logs.

Ambrosial. Pertaining to a growth form of fungi associated with bark and ambrosia beetles. Usually catenulate cells forming a compact palisade lining on the walls of galleries and pupal cells.

Anti-aggregation pheromone. An epideictic pheromone which acts in opposition to an aggregation pheromone.

Antibody. A protein, produced by the immune system, that attacks specific antigens by agglutinating them or precipitating them from solutions.

Antigen. A protein or polysaccharide on the surface of cells or compounds that can be recognized by the immune system.

Aposematic. Serving as a warning such as the brilliant coloration associated with distasteful or poisonous individuals.

Apostatic selection. Selective predation on the most abundant forms in a population. This frequency-dependent selection may maintain several forms within a population.

Aposymbiotic. Separated from its symbiotes.

Axenic. Free from associated individuals.

Bark beetles. Small compact, cylindrical beetles of the family Scolytidae. Bark beetles, sometimes called engraver beetles, live in and beneath the bark and mine the surface of the sapwood but do not enter it. Many species in this group must kill a tree to successfully raise their brood.

Biological species concept. A definition of species that focuses primarily upon patterns of reproductive isolation.

Callow adult. A recently metamorphosed adult bark beetle, easily identified by its light coloration.

Cambium. The soft formative tissue in stems and roots of dicotyledonous and gymnospermous trees and shrubs that gives rise to xylem (wood) and phloem (inner bark).

Cardiac glycoside. Plant defenses, such as calotropin and oleandrin, which are bitter tasting, toxic, and emetic to higher organisms.

Chitin. A polysaccharide forming the hard exoskeleton of insects.

Chlorinated hydrocarbon. An organic molecule with a backbone of carbon and one or more attached chlorine atoms. Some of the better known examples of this class of substances are dieldrin and DDT.

Chromosome. A threadlike structure in the nucleus of eukaryotic cells that contains genes arranged in a linear sequence.

Classification. The ordering of organisms into groups on the basis of their relationships.

Cleptoparasitoid. A parasitoid that preferentially attacks a host already parasitized by another species rather than an unparasitized host.

Climatic release. Favorable weather (or abiotic) conditions that allow a build up of populations to abnormally high levels.

Climax. The final, stable community in an ecological succession, able to reproduce itself indefinitely under existing conditions.

Climax species. One of the species characteristic of the terminal stage of succession, the climax.

Coevolution. The process of evolutionary change in response to reciprocal pressures among sympatric species. The reciprocal pressures may be among competing species, between predator and its prey, or between a host and its parasites.

Community. A sympatric assemblage of species.

Competition. Negative interactions between species utilizing limited resources such as nesting sites or common food resources.

Condensed tannin. Compounds with tannin properties structurally formed by the condensation of flavan-3-ols (cathechin) or flavan-3, 4-diols.

Corpus allatum. The secretory organ which produces juvenile hormone in insects.

Corpus cardiacum. The neurohemal organ which stores and releases insect brain hormones, and also secretes other hormones.

Cortical blister resin. Cyst-like resin ducts or cavities formed at the cambium-sapwood interface in response to wounding. Typical of true firs.

Deme. Local population.

Diapause. A temporary arrest of growth in the larvae of insects.

Digestibility reducing compound. Any of a large array of plant secondary compounds, such as phenols and tannins, that precipitate soluble proteins when cells are disrupted. These compounds render the leaves and stems of plants inedible.

Directional selection. Selection upon a continuous character that progressively moves the population mean in one direction.

Disruptive selection. Selection upon a continuous distribution that favors two or more phenotypes. This type of selection, if unchecked by other forces, will diversify a single population into a population marked by distinct forms, or possibly into separate populations with dissimilar phenotypes.

DNA. Deoxyribonucleic acid, the genetic material.

Ectodermal. Pertaining to or derived from the outer layer of a multicellular animal.

Egg gallery. Cells, tunnels and vestibules excavated in bark or phloem to receive eggs.

Electroantennogram. Recording via oscilloscope of the summed depolarization of sensory cells on an insect antenna.

Electrophoresis. A method of separating proteins which employs both buffered solutions and an electric field. Proteins migrate through a supporting medium such as acrylamide or starch, and are separated on the basis of the electric charge of the protein, its size, and its three dimensional shape.

Elytra. The anterior, leathery or chitinous wings of beetles which cover the hind wings. Elytra are not used in flight.

Endemic population. Infestation characterized by low numbers of organisms in a restricted locality and by relatively constant population sizes.

Endopterygote. Having the wings developing internally; as in insects with complete metamorphosis.

Endosymbiosis. Symbiotes living inside cells or tissues.

Entomopathogen. An agent, usually a microorganism, that causes a disease of an insect.

Epideictic pheromone. A pheromone which regulates the density of a population, e.g. to minimize competition for food resources.

Epidemic population. Infestation characterized by large numbers of organisms of widespread occurrence and by rapidly increasing population sizes.

Epistasis. Interaction of genes at different loci.

Equilibrium. A stable state maintained by opposing forces.

Esterase. A class of enzymes whose functions are believed to include detoxification of substances foreign to the body.

Evolution. A change in the genetic constitution of a population.

Exopterygote. With the wings developing on the outside of the body, ás in insects with simple metamorphosis.

Founder flush mode of speciation. Speciation as a result of genetic events associated with the founding of a population by a few individuals followed by a rapid increase and a subsequent decrease in their numbers.

Frequency-dependent selection. A form of natural selection in which the fitness of a genotype depends upon its frequency in the population.

Fungi imperfecti. Fungi that are not known to reproduce sexually or the asexual reproducing state of a fungus known to reproduce sexually.

Gene. The unit of inheritance; also, the segment of DNA coding for a functional product, such as messenger RNA or a protein.

Generalist. A species that utilizes a relatively broad range of resources. For example, a phytophagous insect feeding on plants from several or many families.

Generalized protein complexing agent. Digestibility-reducing substances that reduce the availability of substrate peptide groups to digestive enzymes by complexing with the peptide groups. May complex with digestive enzymes, nucleic acids, polysaccharides, proteins, and other substances.

Genetic transilience. Evolution as a result of genetic changes associated with the founding of a population by a few individuals and by subsequent inbreeding without a severe reduction in genetic variability.

Genome. The genetic contents of all of the chromosomes; the total genotype.

Genotype. The genetic composition of an individual, usually referring to a single locus.

Hardy-Weinberg equilibrium. Equilibrium state of allelic and genotypic frequencies that occurs in a large population in which individuals are mating at random and in which mutation, migration, and selection are absent. Frequencies will remain constant.

Heartwood. The non-functional wood core of a tree trunk.

Herbivore. An animal that eats plants.

Hemimetabolous. Pertaining to species which undergo incomplete metamorphosis.

Heredity. The transmission of genetic material or genetically determined traits from one generation to another.

Heterozygous. Having dissimilar alleles at a particular gene.

Histocompatibility system. A gene or set of tightly linked genes that comprise an antigen-antibody system.

Holometabolous. Pertaining to species which undergo complete metamorphosis.

Homologous. Having the same embryological origin. For example, the wing of a bird and the arm of a human are homologous structures.

Homozygous. Having identical alleles at a particular gene.

Hormone. A chemical secreted in one part of the body that is transported to another part of the body where it exerts its effect.

Host races. Populations that have begun to differentiate morphologically, behaviorally and/or physiologically and between which mating is restricted.

Host selection principle. Feeding preferences of herbivorous individuals reflect, in part, larval feeding experiences. This principle is attributed to A.D. Hopkins.

Host shift. Change in preference by a parasitic organism for its species of host.

Hydrolyzable tannin. Compounds with tannin properties consisting of esters of glucose, or sometimes other polyols, with gallic acid, m-digallic acid, hexahydroxydiphenic acid or their congeners. Divided into gallotannins or ellagitannins depending on the acids obtained during hydrolysis.

Hyperparasitoid. An organism that parasitizes another parasitoid.

Identification. The assignment of unidentified individuals into an established classification.

Induced resin system. A resin system that produces resin in response to a wound, attack, or infection.

Inquiline. An organism that lives in the nest or abode of another species.

Instar. An insect larva between molts. Four or five instars are common in scolytids.

Inversion. A modification in the genetic material in which the order of genes is inverted.

Island biography. Theory that predicts the numbers of species found on an island based on its size and its distance from the mainland.

Juvenile hormone. Hormone produced by the corpora allata which regulates metamorphosis in immature insects, and reproduction-related functions, in adults.

Kairomone. A chemical emitted by an organism that induces in an organism of another species a behavioral or physiological response of adaptive benefit to the perceiver but not the emitter.

Klinotaxis. Orientation toward or away from a stimulus in which an organism moves alternately right and left, sequentially comparing the intensity of a stimulus on each side of the body.

Koch's rules of proof. A set of rules intended to insure rigor in the search for the source of a disease. The rules are, briefly: the suspected agent must be present in every case of the disease; it must be isolated into pure culture; a pure culture of it must, when inoculated into a susceptible host, give rise to the disease; it must be present in and recoverable from the experimentally diseased host.

Larva. The immature, feeding, wingless stage of a holometabolous insect.

Leucoanthocyanidins. Condensed tannins.

Lignin. An organic substance that, with cellulose, forms the chief part of woody tissue.

Locus. A position or place on a chromosome; a gene.

Macrosymbiotic organism. Macrosymbiote; a symbiote that is multicellular and composed of differentiated tissues.

Metamorphosis. A radical change in the morphology of insects from an actively feeding larva to an immobile pupa, and then to the adult form, the imago.

Microsomal detoxification system. A system capable of altering toxins to an innocuous form. The system is associated with microsomes, cellular inclusions packed with ribosomes and segments of endoplasmic reticulum. Composed primarily of non-specific, mixed-function oxidase enzymes that catalyze many reactions resulting in more polar, excretable products.

Monogamy. A mating system in which reproductive efforts involve one male and one female.

Monophagous. Pertaining to a species that preys upon or eats only one species.

Monoterpene. Low molecular weight, volatile hydrocarbons produced by plants and known to affect the physiology or behavior of animals, e.g. alpha-pinene, limonene.

Morphological species concept. A definition of biological
species that relies primarily upon morphological
similarity.

Multiple parasitoidism. The situation when more than one
parasitoid species occurs simultaneously in or on the body
of the host.

Mutation. A change in the genetic material.

Mutualism. A symbiotic relationship in which both species
benefit by the interaction.

Mycetophagous. Fungus eating.

Natural selection. A mechanism for genetic change that
relies upon differential reproduction of different
genotypes or phenotypes.

Oleoresin. See resin.

Oleoresin pressure. The pressure within the preformed resin
system of gymnospermous trees.

Oligogenic. Pertaining to a character controlled by two or
perhaps a few genes.

Oligophagous. Pertaining to a species which preys upon or
eats just a few species.

Oviposit. To lay or deposit eggs.

Parasite. An organism that lives and feeds in or on another
species; the host is injured by the parasite but is not
usually killed.

Parasitoid. An organism that feeds in or on another organism
for a relatively long time, consuming all or most of its
tissues, and eventually killing it. The immature
parasitoid usually feeds on one host while the adult is
free living, oftentimes feeding on flower nectaries or
honeydew produced by aphids.

Pesticide. A chemical used to kill pests.

Phenol. A class of secondary plant substances having six to ten carbons arranged in one or two benzene rings and believed to play role in plant defense. Examples of phenols are hydroquinone, pyrogallol, and juglone.

Phenotype. The physical manifestation of the genotype. We often think of the phenotype as the physiognomy of an individual, but phenotype may also refer to the physiological performance of an individual or some other attribute not immediately obvious to an observer.

Phenotypic resistance. Resistance to a parasite or parasitoid through some physical attribute such as a thick integument or a covering of dense hair.

Pheromone. A chemical emitted by an organism which induces a behavioral or physiological response in another organism of the same species.

Phoretic. Pertaining to phoresy; a type of symbiosis where the symbiote obtains transportation from the host.

Photic reaction. A behavioral response to light.

Pioneer beetles. The first few scolytid beetles which make the initial attack on a new host.

Pioneer species. A species associated with the earliest stage of succession.

Polygamy. A mating system in which one individual mates with several or many of its opposite sex.

Polygenic. Determined by two or more genes.

Polymorphism. The existence within a population of two or more genotypes for a given trait the rarest of which exceeds an arbitrarily assigned low frequency.

Polyphagous. Preying upon or eating many species.

Predation. Capture, killing and consumption of prey.

Predator. An animal that attacks and feeds on another organism (its prey).

Preformed resin system. A system of canals or ducts containing resin that are found in several tissues of conifers.

Primary attraction. Kairomonal attraction emanating from the hosts, usually trees or logs, of scolytid beetles.

Primary production. The accumulation of biomass by photosynthesis.

Proteinase inhibitors. Proteins or polypeptides that bind specifically to proteolytic enzymes.

Protein complexing agent. See generalized protein complexing agent and specific protein complexing agent.

Pseudo-resistance. Ability to repel the attack of a parasite or parsitoid due to circumstances provided by environmental circumstances rather than a genetically determined resistance.

Qualitative defense. A strategy of plant defense against herbivores that relies primarily upon toxins.

Quantitative defense. A strategy of plant defense against herbivores that relies primarily upon digestibility-reducing compounds.

Recombination. The exchange of segments of chromosome between homologous chromosomes.

Reproductive isolating mechanism. Properties of organisms that prevent their mating with individuals of other populations.

Reproductive value. The likely contribution of an individual at a particular age to the growth of a population.

Resin. A complex mixture of terpenoid compounds found in coniferous trees and believed to play a role in their defense.

Resin acids. Organic compounds having the empirical formula $C_{20}H_{30}O_2$, and possessing a decahydrophenantrene ring structure.

Resin duct. A canal that stores and transports resin. Resin ducts are found throughout the tissues of conifers, including leaves, wood, bark, and cones.

Resistance. A property of not being susceptible to a specific disease, parasite, or parasitoid.

Saprophage. An animal that feeds on dead organic matter.

Sapwood. The functional part of the wood between the inner bark and the heartwood.

Sclerotin. An insoluble protein that serves to stiffen the chitin of the insect cuticle.

Secondary attraction. Attraction to hosts attacked by scolytids, usually due to host-produced kairomones and beetle-produced aggregation pheromones.

Secondary plant substance. A wide variety of chemical compounds synthesized by plants whose functions are largely unknown. Many of these were first thought to be waste products of primary metabolism, but are now believed to serve as defenses against herbivores.

Segregation. The regular separation of homologous chromosomes to different gametes, primarily in the first division of meiosis.

Semiochemical. A chemical involved in communication between organisms.

Sibling species. Species with a high degree of morphological similarity, but which differ in other biological characteristics and do not interbreed.

Soluble nitrogen. Nitrogen in chemical configurations that will dissolve in water.

Specialist. A species using a relatively narrow range of resources. For example, a phytophagous insect feeding only on one species, or a few very closely related species.

Species/area relationship. The existence of an orderly relation between the size of a sample area and the number of species found in that area.

Specific protein complexing agent. Digestibility-reducing substances that are specific inhibitors of proteolytic enzymes.

Stridulation. To produce a grating sound by rubbing together parts of the body.

Structural genes. Genes that code for RNA or proteins.

Succession. The progressive change in animal and plant communities towards a characteristic final state, or climax community.

Superparasitoidism. The situation when more individuals of a parasitoid species occur in a host than can survive.

Symbiology. The study of symbiosis.

Symbiosis. The living together of two species.

Sympatric. Having overlapping geographic ranges.

Sympatric speciation. Differentiation of populations without geographic isolation so that the potential for interbreeding is present.

Systematics. The study of the kinds and diversity of organisms and of the relationships among them.

Tannin. Any of a group of phenolic compounds that form complexes with proteins through hydrogen bonds formed between phenolic hydroxyls of the tannin and peptide groups of the protein. May also bind with nucleic acids, polysaccharides and other compounds.

Taxonomy. The theoretical study of classification, including its bases, principles, procedures, and rules.

Tibia. In insects, the fourth segment of the leg, between the femur and tarsus.

Toxin. Usually small molecular weight molecules that disrupt physiological systems or metabolic processes. Usually produced in small amounts; poison.

Trachea. One of the air transporting tubes of the respiratory system of insects.

Trait groups. Subunits of a population or deme that are homogeneous with respect to ecological interactions and between which gene frequencies may differ.

Translocation. A type of mutation in which a segment of chromosome has broken off and become attached to another, non-homologous chromosome.

Typological thinking. An earlier perspective of systematists and taxonomists that disregarded variation within species.

Vector. An agent, usually an animal, that transports a pathogen to its host.

Xylomycetophagous. Pertaining to an animal that eats wood-inhabiting fungi.

References

Agarwal, R. and N. Krishananda. 1976. Preference to oviposition and antibiosis mechanism to jassids (Amrasia devastans Dist.) in cotton (Glossypium sp.). pp. 13-22 In T. Jermy (ed.). The Host-Plant in Relation to Insect Behavior and Reproduction. Plenum, New York.

Allen, H.J.W. 1976. Aulonicum trisulcum Fourc. (Coleoptera: Colydiidae) in Gloucestershire. Entomologist's Monthly Magazine 111:39.

Amman, G.D. 1972. Mountain pine beetle brood production in relation to thickness of lodgepole pine phloem. J. Econ. Entomol. 65:138-140.

Amman, G.D. 1973. Population changes of mountain pine beetle in relation to elevation. Environ. Entomol. 2:541-547.

Amman, G.D. 1975. Abandoned mountain pine beetle galleries in lodgepole pine. U.S.D.A. For. Serv. Res. Paper INT-197.

Amman, G.D. 1977. Role of the mountain pine beetle in lodgepole pine ecosystems: Impact on succession. pp. 3-18 In W.J. Mattson (ed.). The Role of Arthropods in Forest Ecosystems. Springer-Verlag, New York.

Amman, G.D. 1978. Biology, ecology and causes of outbreaks of the mountain pine beetle in lodgepole pine forests. Pp. 39-53 In A.A. Berryman, G.D. Amman, R.W. Stark and D.L. Kibbee (eds.). Theory and practice of mountain pine beetle management in lodgepole pine forests--a symposium. April 25-27. College of Forest Resources, University of Idaho, Moscow.

Amman, G.D. 1980. Incidence of mountain pine beetle abandoned galleries in lodgepole pine. U.S.D.A. Res. Note. INT-284.

Amman, G.D. and B.H. Baker. 1972. Mountain pine beetle influence on lodgepole pine stand structure. J. For. 70:204-209.

Amman, G.D. and V.E. Pace. 1976. Optimum egg gallery densities for the mountain pine beetle in relation to lodgepole pine phloem thickness. U.S.D.A. For. Serv. Res. Pap. INT-209.

Amman, G.D., M.D. McGregor, D.B. Cahill and W.H. Klein. 1977. Guidelines for reducing losses of lodgepole pine to the mountain pine beetle in unmanaged stands in the Rocky Mountains. U.S.D.A. For. Serv. Gen. Tech. Rept. INT-36. 19 p.

Anderson, A.B., R. Riffer, and A. Wong. 1969. Monoterpenes, fatty and resin acids of Pinus contorta and Pinus attenuata. Phytochem. 8:2401-2403.

Anderson, R.F. 1948. Host selection by the pine engraver. J. Econ. Entomol. 41:596-602.

Anderson, R.F. 1960. Forest and Shade Tree Entomology. John Wiley and Sons. New York. 428 pp.

Anderson, R.F. 1977. Dispersal and attack behavior of the southern pine engraver, Ips grandicollis Eich. (Coleoptera: Scolytidae). pp. 17-26 In H.M. Kuhlman and H.C. Chiang (eds.). Insect Ecology. Univ. of Minn. Agric. Res. Stn. Tech. Bull. 310.

Anderson, W.W., C.W. Berisford, and R.H. Kimmich. 1979. Genetic differences among five populations of the southern pine beetle. Ann. Entomol. Soc. Amer. 72:323-327.

Andrewartha, H., and L. Birch. 1954. The distribution and abundance of animals. Univ. of Chicago Press, Chicago. 782 pp.

Angst, M.E. and G.N. Lanier. 1979. Electroantennogram responses of two populations of Ips pini (Coleoptera: Scolytidae) to insect-produced and host tree compounds. J. Chem. Ecol. 5:131-140.

Annila, E. 1973. Chemical control of spruce cone insects in seed orchards. Metsantut Kimuslaitoksen Julkaisuja 78(8): 5-25.

Apple, J.D. and R.F. Smith (eds.). 1976. Integrated Pest Management. Plenum Press, New York. 200 pp.

Ashraf, M. and A.A. Berryman. 1969. Biology of Scolytus ventralis (Coleoptera: Scolytidae) attacking Abies grandis in Northern Idaho. Melanderia 2:1-23.

Ashraf, M. and A.A. Berryman. 1970. Biology of Sulphuretylenchus elongatus (Nematoda: Sphaerulariidae), and its effect on its host, Scolytus ventralis (Coleoptera: Scolytidae). Can Entomol. 102(2): 197-213.

Atkins, M.D. 1959. A study of the flight of the Douglas-fir beetle, Dendroctonus pseudotsugae Hopk. (Coleoptera: Scolytidae) I. Flight preparation and response. Can. Entomol. 91:283-291.

Atkins, M.D. 1961. A study of the flight of the Douglas-fir beetle, Dendroctonus pseudotsugae Hopk. (Coleoptera: Scolytidae) III. Flight capacity. Can. Entomol. 93:467-474.

Atkins, M.D. 1966a. Behavioral variation among scolytids in relation to their habitat. Can. Entomol. 98:285-288.

Atkins, M.D. 1966b. Laboratory studies on the behavior of the Douglas-fir beetle, Dendroctonus pseudotsugae. Can. Entomol. 98:953-991.

Atkins, M.D. 1969. Lipid loss with flight in the Douglas-fir beetle. Can. Entomol. 101:164-165.

Atkins, M.D. and S.H. Farris. 1962. A contribution to the knowledge of flight muscle changes in the Scolytidae (Coleoptera). Can. Entomol. 94:25-32.

Averhoff, W.W. and R.H. Richardson. 1974. Pheromonal control of mating patterns in Drosophila melanogaster. Behav. Genet. 4:207-225.

Averhoff, W.W. and R.H. Richardson. 1976. Multiple pheromone system controlling mating in <u>Drosophila</u> melanogaster. Proc. Nat. Acad. Sci. USA 73:71-72.

Avise, J.C. 1974. Systematic value of electrophoretic data. System. Zool. 23:465-481.

Ayala, F.J. 1975. Genetic differentiation during the speciation process. Evol. Biol. 8:1-78.

Ayala, F.J., and J.R. Powell. 1972. Allozymes as diagnostic characters of sibling species of <u>Drosophila</u>. Proc. Nat. Acad. Sci. USA 69:1094-1096.

Ayala, F.J., M.L. Tracey, L.G. Barr, J.F. McDonald, and S. Perez-Salas. 1974. Genetic variation in natural populations of five <u>Drosophila</u> species and the hypothesis of selective neutrality of protein polymorphisms. Genetics 77:343-384.

Baird, A.B. 1938. The Canadian insect pest review. Supplement to No. 1. Dept. Agric. Canada, Div. Ent. 16:77-154.

Baker, B.H., G.D. Amman, and G.C. Trostle. 1971. Does the mountain pine beetle change hosts in mixed lodgepole and whitebark pine stands? USDA Forest Service Res. Note INT-151.

Baker, B.H. and J.A. Kemperman. 1974. Spruce beetle effects on a white spruce stand in Alaska. J. For. 72:423-425.

Baker, J.E. and D.M. Norris. 1967. A feeding stimulant for <u>Scolytus</u> <u>multistriatis</u> (Coleoptera: Scolytidae) isolated from the bark of <u>Ulmus</u> <u>americana</u>. Ann. Entomol. Soc. Amer. 60:1213-1215.

Baker, W.L. 1972. Eastern forest insects. USDA Forest Serv. Misc. Publ. 1175. 642 pp.

Bakke, A. 1975. Aggregation pheromone in the bark beetle <u>Ips</u> <u>duplicatus</u> (Sahlberg). Norw. J. Entomol. 22:67-69.

Bakke, A. 1976. Spruce bark beetle <u>Ips</u> <u>typographus</u>: pheromone production and field response to synthetic pheromones. Naturwiss. 63:92.

Bakke, A. 1978. Aggregation pheromone components of the bark beetle Ips acuminatus. Oikos 31:184-188.

Bakke, A. and T. Kvamme. 1978. Kairomone response by the predators Thanasimus formicarius and Thanasimus rufipes to the synthetic pheromone of Ips typographus. Norw. J. Entomol. 25:41-43.

Bakke, A. and T. Saether. 1978. Granbarkbillen kan fanges i rorfeller. Skogeieren 65:10.

Bakke, A., P. Froyen, and L. Skattebol. 1977. Field response to a new pheromonal compound isolated from Ips typographus. Naturwiss. 64:98-99.

Balda, R.P. 1975. The relationship of secondary cavity nesters to snag densities in western coniferous forests. USDA Forest Serv. Wildld. Habitat Tech. Bull. 7. Albuquerque, NM. 37 pp.

Baldwin, P.H. 1960. Overwintering of woodpeckers in bark beetle-infested spruce-fir forests of Colorado. Proc. 12th Int. Ornith. Congr., Helsinki (1958), pp. 71-84.

Ball, J.C. and D.L. Dahlsten. 1973. Hymenopterous parasites of Ips paraconfusus (Coleoptera: Scolytidae) larvae and their contribution to mortality. I. Influence of host tree and tree diameter on parasitization. Can. Entomol. 105:1453-1464.

Bannon, M. 1936. Vertical resin ducts in the secondary wood of the Abietinae. New Phytol. 35:11-47.

Barr, B.A. 1969. Sound production in Scolytidae (Coleoptera) with emphasis on the Genus Ips. Can. Entomol. 101:636-672.

Barras, S.J. 1967. Thoracic mycangium of Dendroctonus frontalis (Coleoptera: Scolytidae) is synonymous with a secondary female character. Ann. Entomol. Soc. Amer. 60:486-487.

Barras, S.J. 1969. Penicillium implicatum antagonistic to Ceratocystis minor and C. ips. Phytopath. 59:520.

Barras, S.J. 1970. Antagonism between Dendroctonus frontalis and the fungus Ceratocystis minor. Can. Entomol. Soc. Amer. 63:1187-1189.

Barras, S.J. 1972. Improved White's for surface sterilization of pupae of Dendroctonus frontalis. Jour. of Econ. Entomol. 65:1504.

Barras, S.J. 1973. Reduction of progeny and development in the southern pine beetle following removal of symbiotic fungi. Can. Entomol. 105:1295-1299.

Barras, S.J. 1975. Release of fungi from mycangia of southern pine beetles observed under a scanning electron microscope. Z. Ang. Entomol. 79:173-176.

Barras, S.J. 1979. Forest ecosystem approach to tree-pest interaction. Proc. Western Forest Insect Work Conference, Boise, Idaho, March 7-9, 1979.

Barras, S.J. and J.D. Hodges. 1969. Carbohydrates of inner bark of Pinus taeda as affected by Dendroctonus frontalis and associated microorganisms. Can. Entomol. 101:489-493.

Barras, S.J. and J.E. Marler. 1974. Identification of bacterial flora in the digestive tract of the southern pine beetle Dendroctonus frontalis (Zimm.). Final Rept. USFS, FS-SO-2203-1.22, 10 pp.

Barras, S.J. and T. Perry. 1971a. Gland cells and fungi associated with prothoracic mycangium of D. adjunctus (Coleoptera: Scolytidae). Ann. Entomol. Soc. Amer. 64:123-6.

Barras, S.J. and T. Perry. 1971b. Leptographium terebrantis sp. nov. associated with Dendroctonus terebrans in loblolly pine. Mycopath. Mycol. Appl. 43:1-10.

Barras, S.J. and T. Perry. 1972. Fungal symbionts in the prothoracic mycangium of Dendroctonus frontalis (Coleoptera: Scolytidae) Z. Ang. Entomol. 71:95-104.

Barras, S.J. and T.J. Perry. 1975. Interrelationships among microorganisms, bark or ambrosia beetles, and woody host tissue: an annotated bibliography, 1965-1974. U.S.D.A. For. Serv. Gen. Tech. Rept. SO-10.

Barras, S.J. and J.J. Taylor. 1973. Varietal Ceratocystis minor identified from mycangium of Dendroctonus frontalis Mycopath. ep. Mycol. Applicata 50:293-305.

Barson, G. 1976. Laboratory studies on the fungus Verticillium lecanii, a larval pathogen of the large elm bark beetle (Scolytus scolytus). Annals of Applied Biology 83:207-214.

Barson, G. 1977. Laboratory evaluation of Beauvaria bassiana as a pathogen of the larval stage of the large elm bark beetle, Scolytus scolytus. Jour. Invert. Path. 29:361-366.

Batra, L.R. 1963. Ecology of ambrosia fungi and their dissemination by beetles. Trans. Kansas Acad. of Science. Vol. 66:213-236.

Batra, L.R. (ed.). 1979. Insect-fungus symbiosis. Allanheld, Osmun and Co. New Jersey, p. 276.

Batzli, G., and F. Pitelka. 1975. Vole cycles: test of another hypothesis. Am Nat. 109:482-487.

Bauer, J. and J.P. Vite. 1975. Host selection by Trypodendron lineatum. Naturwiss. 62:539.

Baumgartner, D.M. (ed.). 1975. Management of Lodgepole Pine Ecosystems. Wash. State Univ., Coop. Ext. Serv. 1:405 pp.

Beaver, R.A. 1966. The biology and immature stages ofEntedon leucogramma (Ratzburg) (Hymenoptera: Eulophidae) a parasite of bark beetles. Proc. Roy. Ent. Soc. London (A) 41:37-41.

Beaver, R.A. 1967. The regulation of population density in the bark beetle Scolytus scolytus (F.). J. Anim. Ecol. 36:435-451.

Beaver, R.A. 1974. Intraspecific competition among bark
beetle larvae (Coleoptera: Scolytidae). J. Anim. Ecol.
43:455-467.

Beaver, R.A. 1977. Bark and ambrosia beetles in tropical
forests. Proc. Symp. on For. Pests and Dis. in S.E. Asia.
Bongor, Indonesia (1976). Biotrop. Spec. Publ. No.
2:133-149.

Beck, S.D. 1965. Resistance of plants to insects. Ann. Rev.
Entomol. 10:207-232.

Beck, S.D. and J. Reese. 1976. Insect-plant interactions:
Nutrition and metabolism. pp. 41-92 In J. Wallace and R.
Mansell (eds.). Biochemical Interaction between Plants and
Insects. Plenum, New York.

Bedard, W.D. 1933. The relation of parasites to mountain pine
beetle control in western white pine. Unpublished report,
Forest Insect Field Station, Coeur d'Alene, Idaho. Pacific
Southwest Forest and Range Exp. Station files, U.S.D.A.
Forest Service, 4 pp.

Bedard, W.D. 1937. Biology and control of the Douglas-fir
beetle   Dendroctonus   pseudotsugae   Hopkins
(Coleoptera-Scolytidae) with notes on associated insects.
Ph.D. thesis, Washington State College, 73 pp.

Bedard, W.D. 1961. Media for the rearing of immature bark
beetles (Scolytidae). Ph.D. Thesis, Univ. of California,
Berkeley, 54 p.

Bedard, W.D. 1965. The biology of Tomicobia tibialis
(Hymenoptera: Pteromalidae) parasitizing Ips confusus
(Coleoptera: Scolytidae) in California. Contrib. Boyce
Thompson Inst. 23:77-82.

Bedard, W.D. 1966. A ground phloem medium for rearing
immature bark beetles (Scolytidae). Ann. Entomol. Soc.
Amer. 59:931-938.

Bedard, W.D. and D.L. Wood. 1974. Programs utilizing pheromones in survey or control. Bark beetles--the western pine beetle. pp. 441-449 In M.C. Birch (ed.). Pheromones. North-Holland Pub. Co., Amsterdam.

Bedard, W.D., P.E. Tilden, D.L. Wood, R.M. Silverstein, R.G. Brownlee, J.O. Rodin. 1969. Western pine beetle: field response to its sex pheromone and a synergistic host terpene, myrcene. Science 164:1284-1285.

Bedard, W.D., P.E. Tilden, D.L. Wood, K.Q. Lindahl, Jr. and P.A. Rauch. 1980b. Effects of verbenone and trans-verbenol on the response of Dendroctonus brevicomis to natural and synthetic attractant in the field. J. Chem. Ecol. 6:997-1013.

Bedard, W.D., P.E. Tilden, D.L. Wood, K.Q. Lindahl, Jr., R.M. Silverstein, and J.O. Rodin. 1980a. Field responses of the western pine beetle and one of its predators to host- and beetle-produced compounds. J. Chem. Ecol. 6:625-641

Bennett, R.B. and J.H. Borden. 1971. Flight arrestment of tethered Dendroctonus pseudotsugae and Trypodendron lineatum (Coleoptera: Scolytidae) in response to olfactory stimuli. Ann. Entomol. Soc. Amer. 64:1273-1286.

Bennett, W. 1954. The effect of needle structure upon the susceptibility of hosts of the pine needle miner (Exotleia pinifoliella Chamb.). Can. Entomol. 86:49-54.

Bennett, W.H. and W.M. Ciesla. 1971. Southern pine beetle Dendroctonus frontalis Zimm. U.S.D.A. For. Serv. For. Pest Leaflet 49.

Berisford, C.W. 1974. Hymenopterous parasitoids of the eastern juniper bark beetle, Phloesinus dentatus (Coleoptera: Scolytidae). Can. Entomol. 106:869-872.

Berisford, C.W. 1980. The southern pine beetle. pp. 31-52 in R.C. Thatcher, J.L. Searcy, J.E. Cosler and G.D. Hertell (eds.). Expanded southern pine beetle research and application program. U.S.D.A. For. Serv. Tech. Bull. 16-31.

Berisford, C.W., H.M. Kulman and R.L. Pienkowski. 1970. Notes on the biologies of Hymenopterous parasites of Ips spp. bark beetles in Virginia. Can. Entomol. 102:484-490.

Berisford, C.W., H.M. Kulman, R.L. Pienkowski, and H.J. Keikkenen. 1971. Factors affecting distribution and abundance of Hymenopterous parasites of Ips spp. bark beetles in Virginia (Coleoptera: Scolytidae) Can. Entomol. 103:235-239.

Berlocher, S.H. 1979. Biochemical approaches to strain, race, and species discriminations. pp. 137-144 In M.A. Hoy and J.J. McKelvey, Jr. (eds.). Genetics in Relation to Insect Management. Rockefeller Foundation.

Berlocher, S.H. 1980. An electrophoretic key for distinguishing species of the genus Rhagoletis (Diptera:Tephritidae) as larvae, pupae, or adults. Ann. Entomol. Soc. Amer. 73:131-137.

Berryman, A.A. 1966. Studies on the behavior and development of Enoclerus lecontei (Wolcott), a predator of the western pine beetle. Can. Entomol. 98:519-526.

Berryman, A.A. 1967. Preservation and augmentation of insect predators of the western pine beetle. J. For. 65:260-262.

Berryman, A.A. 1968a. Distributions of Scolytus ventralis attacks, emergence, and parasites in grand fir. Can. Entomol. 100:57-68.

Berryman, A.A. 1968b. Estimation of oviposition by the fir engraver, Scolytus ventralis (Coleoptera: Scolytidae). Ann. Entomol. Soc. Amer. 61:227-228.

Berryman, A.A. 1969. Responses of Abies grandis to attack by Scolytus ventralis (Coleoptera: Scolytidae). Can. Entomol. 101:1033-1041.

Berryman, A.A. 1970. Overwintering populations of Scolytus ventralis (Coleoptera: Scolytidae) reduced by extreme cold temperatures. Ann. Entomol. Soc. Amer. 63:1194-1196.

Berryman, A.A. 1972. Resistance of conifers to invasion by bark beetle-fungal associations. BioScience 22:598-602.

Berryman, A.A. 1973a. Management of mountain pine beetle populations in lodgepole pine ecosystems. Pp. 627-650 In D. Baumgartner (ed.). Management of Lodgepole Pine Ecosystems--A symposium. Washington State University, Cooperative Extension Service, October 9-11.

Berryman, A.A. 1973b. Population dynamics of the fir engraver Scolytus ventralis (Coleoptera: Scolytidae). I. Analysis of population behavior and survival from 1964-1971. Can. Entomol. 105:1465-1488.

Berryman, A.A. 1974. Dynamics of bark beetle populations: towards a general productivity model. Environ. Entomol. 3:579-585.

Berryman, A.A. 1976. Theoretical explanation of mountain pine beetle dynamics in lodgepole pine forests. Environ. Entomol. 5:1225-1233.

Berryman, A.A. 1978. A synoptic model of the lodgepole pine/mountain pine beetle interaction and its potential application in forest management. pp. 98-105 In A.A. Berryman, G.D. Amman, R.W. Stark and D.L. Kibbee (eds.). Theory and practice of mountain pine beetle management in lodgepole pine forests. College of Forest Resources, Univ. of Idaho, Moscow.

Berryman, A.A. 1979. Dynamics of bark beetle populations: analysis of dispersal and redistribution. Mitt. Schweiz. Ent. Gesel. 52:227-234.

Berryman, A.A. 1980. Threshold theory and its application in population management. (In press) In G.R. Conway, (ed.). The Management of Pest and Disease Systems. Pergamon Press, IIASA Series.

Berryman, A.A. 1981. Population Systems--A General Introduction. Plenum Press, N.Y. 222 pp.

Berryman, A.A. 1982. Biological control, threshholds, and pest outbreaks. Environ. Entomol. (In press).

Berryman, A.A. and M. Ashraf. 1970. Effects of _Abies grandis_ resin on the attack behavior and brood survival of _Scolytus_ _ventralis_ (Coleoptera: Scolytidae). Can. Entomol. 102:1229-1236.

Berryman, A.A. and L.V. Pienaar. 1973. Simulation of interspecific competition and survival of _Scolytus_ _ventralis_ broods (Coleoptera: Scolytidae). Environ. Entomol. 2:447-459.

Berryman, A.A. and L.C. Wright. 1978. Defoliation, tree condition, and bark beetles. pp. 81-87 In M.H. Brookes, R.W. Stark, and R.W. Campbell (eds.). The Douglas-fir tussock moth: A synthesis. U.S.D.A. Tech. Bull. 1585. 331 pp.

Berryman, A.A., G.D. Amman, R.W. Stark, and D.L. Kibbee (eds.). 1978. Theory and practice of mountain pine beetle management in lodgepole pine forest--a symposium. April 25-27, 1978. College of Forest Resources, University of Idaho, Moscow.

Bethlahmy, N. 1974. More streamflow after a bark beetle epidemic. J. of Hydrology. 23:185-189.

Bhakthan, N.M.G., J.H. Borden and K.K. Nair. 1970. Fine structure of degenerating and regenerating flight muscles in a bark beetle, _Ips confusus_. I. Degeneration. J. Cell. Sci. 6:807-820.

Bhakthan, N.M.G. K.K. Nair and J.H. Borden. 1971. Fine structure of degenerating and regenerating flight muscles in a bark beetle, Ips confusus. II. Regeneration. Can. J. Zool. 49:85-89.

Billings, R.F. and R.I. Gara. 1975. Rhythmic emergence of Dendroctonus ponderosae (Coleoptera: Scolytidae) from two host species. Ann. Entomol. Soc. Amer. 68:1033-1036.

Billings, R.F., R.I. Gara and B.F. Hrutfiord. 1976. Influence of ponderosa pine resin volatiles on the response of Dendroctonus ponderosae to synthetic trans-verbenol. Environ. Entomol. 5:171-179.

Bingham, R.T. 1966. Breeding blister rust resistant western white pine, III. Comparative performance of clonal and seedling lines from rust-free selections. Silvae Genetica 15:160-164.

Birch, M.C. 1978. Chemical communication in pine bark beetles. Amer. Sci. 66:409-419.

Birch, M.C. 1980. The evolution of chemosensory specificity and diversity in Ips. Paper presented at 16th Int. Congr. Entomol., Kyoto, Japan.

Birch, M.C. and D.L. Wood. 1975. Mutual inhibition of the attractant pheromone response by two species of Ips (Coleoptera: Scolytidae). J. Chem. Ecol. 1:101-113.

Birch, M.C. and D.M. Light. 1977. Inhibition of the attractant pheromone response in Ips pini and I. paraconfusus (Coleoptera: Scolytidae): field attraction of ipsenol and linalool. J. Chem. Ecol. 3:257-267.

Birch, M.C. and P. Svihra. 1979. Exploiting olfactory interactions between species of Scolytidae. pp. 135-138 In W.E. Waters (ed.). Current Topics in Forest Entomology. U.S.D.A. For. Serv. Gen. Tech. Rep. WO-8.

Birch, M.C., D.M. Light and K. Mori. 1977. Selective inhibition of response of Ips pini to its pheromone by the S-(-)-enantiomer of ipsenol. Nature 270:738-739.

Birch, M.C., P. Svihra, T.D. Paine and J.C. Miller. 1980. Influence of chemically mediated behavior on host tree colonization by four cohabiting species of bark beetles. J. Chem. Ecol. 6:395-414.

Birch, M.C., P.E. Tilden, D.L. Wood, L.E. Browne, J.C. Young and R.M. Silverstein. 1977. Biological activity of compounds isolated from air condensates and frass of the bark beetle, Ips confusus. J. Insect Physiol. 23:1373-1376.

Birch, M.C., D.M. Light, D.L. Wood, L.E. Browne, R.M. Silverstein, B.J. Bergot, G. Ohloff, J.R. West and J.C. Young. 1980. Pheromonal attraction and allomonal interruption of Ips pini in California by the two enantiomers of ipsdienol. J. Chem. Ecol. 6:703-717.

Blackman, M.M. 1931. The Black Hills beetle (Dendroctonus ponderosae Hopk.) N.Y. State College Forest. Tech. Pub. 36: 97 pp.

Blandford, W.F.H. 1895-1905. Family Scolytidae. Biologia Centrali-Americana, Insects, Coleoptera 4:97-298.

Blight, M.M., L.J. Wadhams and M.J. Wenham. 1978. Volatiles associated with unmated Scolytus scolytus beetles on English elm: differential production of alpha-multistriatin and 4-methyl-3-heptanol, and their activities in a laboratory bioassay. Insect Biochem. 8:135-142.

Blight, M.M., L.J. Wadhams and M.J. Wenham. 1979a. The stereoisomeric composition of the 4-methyl-3-heptanol produced by Scolytus scolytus and the preparation and biological activity of the four synthetic stereoisomers. Insect Biochem. 9:525-533.

Blight, M.M., L.J. Wadhams and M.J. Wenham. 1979b. Chemically mediated behavior in the large elm bark beetle, Scolytus scolytus. Bull. Entomol. Soc. Amer. 25:122-124.

Blight, M.M., F.A. Mellon, L.J. Wadhams and M.J. Wenham. 1977. Volatiles associated with *Scolytus scolytus* beetles on Engligh elm. Experienta 33:845-846.

Blight, M.M., C.J. King, L.J. Wadhams and M.J. Wenham. 1978. Attraction of *Scolytus scolytus* (F.) to the components of Multilure, the aggregation pheromone of *S. multistriatus* (Marsham) (Coleoptera: Scolytidae). Experientia 34:1119-1120.

Blight, M.M., L.J. Wadhams, M.J. Wenham and C.J. King. 1979. Field attraction of *Scolytus scolytus* (F.) to the enantiomers of 4-methyl-3-heptanol, the major component of the aggregation pheromone. J. For. 52:83-90.

Blum, M.S. 1970. The chemical basis of insect sociality. pp. 61-94 In M. Beroza (ed.). Chemicals Controlling Insect Behavior. Academic Press, New York.

Boller, E.G., and G.L. Bush. 1973. Evidence for genetic variation in populations of the European cherry fruit fly, *Rhagoletis cerasi* (Diptera: Tephritidae) based on physiological parameters and hybridization experiments. Entomol. Exp. Appl. 17:279-293.

Bordasch, R.P. and A.A. Berryman. 1977. Host resistance to the fir engraver beetle, *Scolytus ventralis* (Coleoptera: Scolytidae) 2. Repellency of *Abies grandis* resins and some monoterpenes. Can. Entomol. 109:95-100.

Borden, J.H. 1967. Factors influencing the response of *Ips confusus* (Coleoptera: Scolytidae) to male attractant. Can. Entomol. 99:1164-1193.

Borden, J.H. 1974. Aggregation pheromones in the Scolytidae. pp. 135-160 In M.C. Birch (ed.). Pheromones. North Holland, Amsterdam.

Borden, J.H. 1977. Behavioral responses of coleoptera to pheromones, allomones and kairomones. pp. 169-198 In H.H. Shorey and J.J. McKelvey, Jr. (eds.). Chemical Control of Insect Behavior: Theory and Application. John Wiley and Sons, New York.

Borden, J.H. and R.B. Bennett. 1969. A continuously recording flight mill for investigating the effect of volatile substances on the flight of tethered insects. J. Econ. Entomol. 62:782-785.

Borden, J.H., R.G. Brownlee and R.M. Silverstein. 1968. Sex pheromone of Trypodendron lineatum (Coleoptera: Scolytidae): production, bio-assay, and partial isolation. Can. Entomol. 100:629-636.

Borden, J.H., L. Chong, J.A. McLean, K.N. Slessor and K. Mori. 1976. Gnathotrichus sulcatus: synergistic response to enantiomers of the aggregation pheromone sulcatol. Science 192:894-896.

Borden, J.H., L. Chong, K.N. Slessor, A.C. Oehlschlager, H.D. Pierce, Jr., and B.S. Lindgren. 1981. Allelochemic activity of aggregation pheromones between three sympatric species of ambrosia beetles (Coleoptera: Scolytidae). Can. Entomol. 113:557-563.

Borden, J.H., J.R. Handley, B.D. Johnston, J.G. MacConnell, R.M. Silverstein, K.N. Slessor, A.A. Swigar and D.T.W. Wong. 1979. Synthesis and field testing of 4,6,6-lineatin, the aggregation pheromone of Trypodendron lineatum (Coleoptera: Scolytidae). J. Chem. Ecol. 5:681-689.

Borden, J.H., J.R. Handley, J.A. McLean, R.M. Silverstein, L. Chong, K.N. Slessor, B.D. Johnston and H.R. Schuler. 1980b. Enantiomer-based specificity in pheromone communication by two sympatric Gnathotrichus species (Coleoptera: Scolytidae). J. Chem. Ecol. 6:445-456.

Borden, J.H. and C.J. King. 1977. Population aggregation pheromone produced by male Scolytus scolytus (F.) (Coleoptera: Scolytidae). For. Comm. Res. and Dev. Pap. No. 118.

Borden, J.H., C.J. King, B.S. Lindgren, L. Chong, D.R. Gray, A.C. Ochlschlager, K.N. Slessor and H.D. Pierce, Jr. 1982. Variations in response of Trypodendron lineatum from two continents to semiochemicals and trap form. Environ. Entomol. (in press).

Borden, J.H., B.S. Lindgren and L. Chong. 1980a. Ethanol and alpha-pinene as synergists for the aggregation pheromones of two Gnathotrichus species. Can. J. For. Res. 10:303-305.

Borden, J.H. and J.A. McLean. 1980. Pheromone-based suppression of ambrosia beetles in industrial timber processing areas. In E.R. Mitchell (ed.). Management of Insect Pests with Semiochemicals. Plenum Pub. Co., New York.

Borden, J.H. and M. McClaren. 1970. Biology of Cryptoporus volvatus (Peck) Shear (Agaricales: Polyporaceae) in southwestern B.C.: distribution, host species, and relationship with subcortical insects. Syesis 3:145-154.

Borden, J.H., K.K. Nair and C.E. Slater. 1969. Synthetic juvenile hormone: induction of sex pheromone production in Ips confusus. Science 166:1626-1627.

Borden, J.H., A.C. Oehlschlager, K.N. Slessor, L. Chong and H.D. Pierce, Jr. 1980c. Field tests of isomers of lineatin, the aggregation pheromone of Trypodendron lineatum (Coleoptera: Scolytidae). Can. Entomol. 112:107-109.

Borden, J.H. and C.E. Slater. 1968. Induction of flight muscle degeneration by synthetic juvenile hormone in Ips confusus (Coleoptera: Scolytidae). Z. Vergl. Physiol. 61:366-368.

Borden, J.H. and C.E. Slater. 1969. Flight muscle volume change in Ips confusus (Coleoptera: Scolytidae). Can. J. Zool. 47:29-32.

Borden, J.H., T.J. VanderSar and E. Stokkink. 1975. Secondary attraction in the Scolytidae: an annotated bibliography. Simon Fraser University Pest Mgt. Pap. No. 4. 97 pp.

Borden, J.H. and D.L. Wood. 1966. The antennal receptors and olfactory response of Ips confusus (Coleoptera: Scolytidae) to male sex attractant in the laboratory. Ann. Entomol. Soc. Amer. 59:253-261.

Bormann, F.H. and G.E. Likens. 1979a. Catastrophic disturbance and the steady state in northern hardwood forests. Amer. Scientist. 67:660-669.

Bormann, F.H., and G.E. Likens. 1979b. Patterns and Process in a Forested Ecosystem. Springer-Verlag. New York. 253 pp.

Borror, D.J., D.W. DeLong, and C.A. Triplehorn. 1976. An Introduction to the Study of Insects. 4th ed., Holt, Reinhart, and Winston, New York. 852 pp.

Boss, G.D. and T.O. Thatcher. 1970. Mites associated with Ips and Dendroctonus in southern Rocky Mountains with special reference to Iponemus truncatus (Acarina: Tarsonemidae). U.S.D.A. For. Serv. Res. Note RM-171.

Brand, J.M. and S.J. Barras. 1977. The major volatile constituents of a basidiomycete associated with the southern pine beetle. Lloydia 40:318-399.

Brand, J.M., J.W. Bracke, A.J. Markovetz, D.L. Wood and L.E. Browne. 1975. Production of verbenol pheromone by a bacterium isolated from bark beetles. Nature 254:136-7.

Brand, J.M., J.W. Bracke, L.N. Britton, A.J. Morkovetz and S.J. Barras. 1976. Bark beetle pheromones: production of verbenone by a mycangial fungus of D. frontalis. J. Chem. Ecol. 2:195-199.

Brand, J.M., J. Schultz, S.J. Barras, L.J. Edson, T.L. Payne and R.L. Hedden. 1977. Bark beetle pheromones: enhancement of Dendroctonus frontalis (Coleoptera: Scolytidae.) aggregation pheromone by yeast metabolites in laboratory bioassays. J. Chem. Ecol. 3:657-666.

Brattsten, L. 1979. Biochemical defense mechanisms in herbivores against plant allelochemics. pp. 199-270 In G. Rosenthal and D. Janzen (eds.). Herbivores: Their Interaction with Secondary Plant Metabolites. Academic Press, New York.

Brian, P.W. 1957. The ecological significance of antibiotic production. pp. 168-188 In Society for General Microbiology, Microbial Ecology. Cambridge Univ. Press. London.

Bright, D.E. 1963. Bark beetles of the genus Dryocoetes (Coleoptera:Scolytidae) in North America. Ann. Entomol. Soc. Amer. 56:103-115.

Bright, D.E. 1968. Review of the tribe Xyleborini in America north of Mexico (Coleoptera:Scolytidae). Can. Entomol. 100:1288-1323.

Bright, D.E. 1969. Biology and taxonomy of bark beetles in the genus Pseudohylesinus Swaine (Coleoptera:Scolytidae). Univ. California Publ. Entomol. 54. 49 pp.

Bright, D.E. 1981. Taxonomic monograph of the genus Pityophthorus Eichhoff in North and Central America. Mem. Ent. Soc. Canada 118:1-378.

Bright, D.E., Jr. and R.W. Stark. 1973. The bark and ambrosia beetles of California: Scolytidae and Platypodidae. Bull. Calif. Insect Survey 16. Univ. of Calif. Press, Berkeley. 169 pp.

Brongniart, C. 1877. Note sur des perforations observees dans deux morceaux de bois fossile. Ann. Soc. Entomol. France 7:215-220.

Brooks, M.H., R.W. Stark, and R.W. Campbell (eds.). The Douglas-fir Tussock Moth: A Synthesis. USDA For. Serv. and SEA. Tech. Bull. 1585.

Brower, L.P. and J.V.Z. Brower. 1964. Birds, butterflies, and plant poisons: a study in ecological chemistry. Zoologica 49:137-59.

Brown, W.L., Jr. 1968. An hypothesis concerning the function of the metaplural glands in ants. Amer. Natur. 102:188-191.

Brown, W.L., Jr., T.E. Eisner and R.H. Whittaker. 1970. Allomones and kairomones: transpecific chemical messengers. Bioscience 20:21-22.

Browne, L.E. 1978. A trapping system for the western pine beetle using attractive pheromones. J. Chem. Ecol. 4:261-275.

Bruns, H. 1960. The economic importance of birds in forests. Bird Study 7(4):193-208.

Bucher, G.E. 1960. Potential bacterial pathogens of insects and their characteristics. J. Insect. Path. 2:172-195.

Bucher, G.E. 1973. Definition and identification of insect pathogens. Annals New York Acad. Sci. 217:8-17.

Buchner, P. 1965. Endosymbiosis of animals with plant microorganisms. Interscience Publishers, J. Wiley and Sons, Inc., New York.

Buckner, C.H. 1966. Vertebrate predators in forest insect control. Ann. Rev. Entomol. 11:449-470.

Bush, G.L. 1969. Sympatric host race formation and speciation in frugivorous flies of the genus Rhagoletis (Diptera: Tephritidae). Evolution 23:237-251.

Bush, G.L. 1974. The mechanism of sympatric host race formation in the true fruit flies (Tephritidae). In M.J.D. White, (ed.). Genetic Mechanisms of Speciation in Insects. Australian and New Book Co., Sydney.

Bush, G.L. 1975a. Modes of animal speciation. Ann. Rev. Ecol. Syst. 6:339-364.

Bush, G.L. 1975b. Sympatric speciation in phytophagous parasitic insects, In P.W. Price (ed.). Evolutionary Strategies of Parasitic Insects and Mites. Plenum Publ. Corp., New York.

Bush, G.L., S.M. Case, A.C. Wilson, and J.L. Patton. 1977. Rapid speciation and chromosomal evolution in mammals. Proc. Nat. Acad. Sci. 74:3942-3946.

Bushing, R.W. 1965. A synoptic list of the parasites of Scolytidae (Coleoptera) in North America north of Mexico. Can. Entomol. 97:449-492.

Byers, J.A. and D.L. Wood. 1980a. Interspecific inhibition of the response of the bark beetles, Dendroctonus brevicomis LeConte and Ips paraconfusus Lanier, to their pheromones in the field. J. Chem. Ecol. 6:149-164.

Byers, J.A. and D.L. Wood. 1981. Interspecific effects of pheromones on the attraction of the bark beetles, Dendroctonus brevicomis and Ips paraconfusus in the laboratory. J. Chem. Ecol. J. Chem. Ecol. 7:9-18.

Byers, J.A., D.L. Wood, L.E. Browne, R.A. Fish, B. Piatek and L.B. Hendry. 1979. Relationship between a host plant compound, myrcene, and pheromone production in the bark beetle Ips paraconfusus. J. Insect Physiol. 25:477-482.

Byrne, K.J., A.A. Swigar, R.M. Silverstein, J.H. Borden and E. Stokkink. 1974. Sulcatol: population aggregation pheromone in the scolytid beetle, Gnathotrichus sulcatus. J. Insect Physiol. 20:1895-1900.

Cabrera, H. 1978. Phloem structure and development in lodgepole pine. pp. 54-63 In A.A. Berryman, G.D. Amman, R.W. Stark, D. Kibbee (eds.). Theory and Practice of Mountain Pine Beetle Management in Lodgepole Pine Forests. Moscow, Idaho.

Cade, S.C., B.F. Hrutfiord, and R.I. Gara. 1970. Identification of a primary attractant for Gnathotrichus sulcatus isolated from western hemlock logs. J. Econ. Entomol. 63:1014-1015.

Caird, R.W. 1935. Physiology of pines infested with bark beetles. Bot. Gaz. 96:709-733.

Callaham, R.Z. 1966a. Needs in developing forest trees resistant to insects. pp. 469-473 In H.D. Gerhold, R. McDermott, E. Schreiner, and J. Winieski (eds.). Breeding Pest-Resistant Trees. Pergamon Press, Oxford.

Callaham, R.Z. 1966b. Nature of resistance of pines to bark beetles. pp. 197-201 In H.D. Gerhold, R. McDermott, E. Schreiner, and J. Winieski (eds.). Breeding Pest-Resistant Trees. Pergamon Press, Oxford.

Callaham, R.Z. and M. Shifrine. 1960. The yeasts associated with bark beetles. For. Sci. 6:146-154.

Calvert, W.H., L.E. Hedrick, and L.P. Brower. 1979. Mortality of the monarch butterfly (Danaus Plexippus L.): Avian predation at five overwintering sites in Mexico. Science 204:847-851.

Cameron, E.A. and J.H. Borden. 1967. Emergence patterns of Ips confusus (Coleoptera: Scolytidae) from ponderosa pine. Can. Entomol. 99:236-244.

Camors, F.B., Jr. and T.L. Payne. 1973. Sequence of arrival of entomophagous insects to trees infested with the southern pine beetle. Env. Entomol. 2:267-270.

Carle, P. 1974. Mise en evidence d'une attraction secondaire d'origine sexuelle chez Blastophagus destruens Woll. (Coleoptera: Scolytidae). Ann. Zool. - Ecol. Anim. 6:539-550.

Carle, P. 1975. Dendroctonus micans Kug. (Coleoptera: Scolytidae), the giant bark-beetle of European spruce beetle (bibliographical note). Revue Forestiere Francaise 27:115-128.

Carle, P. 1978. Essais d'attraction en laboratoire et en foret de Blastophagus (piniperda L. et destruens Woll.). pp. 92-101 In Les Pheromone Sexuelles des Insects. I.N.R.A. Centre de Recherches d'Avignon Station de Zoologie, Montfavet, France.

Carle, P., C. Descoins and M. Gallois. 1978. Pheromones des Blastophagus (piniperda L. et destruens Woll). pp. 87-91 In Les Pheromone Sexuelles des Insectes. I.N.R.A. Centre de Recherches d'Avignon Station de Zoologie, Montfavet, France.

Carson, H.L. 1959. Genetic conditions which promote or retard the formation of species. Cold Springs Harbor Symp. Biol. 24:87-105.

Carson, H.L. 1968. The population flush and its genetic consequences. In R.C. Lewontin (eds.). Population Biology and Evolution. Syracuse Univ. Press, New York.

Carson, H.L. 1971. Speciation and the founder principle. Stadler Symp. 3:51-70.

Carson, H.L. 1975. The genetics of speciation at the diploid level. Am. Nat. 109:83-92.

Castello, J.D., C.G. Shaw and M.N. Furniss. 1976. Isolation of Cryptoporus volvatus and Fomes pinicola from Dendroctonus pseudotsugae. Phytopath. 66:1431-1434.

Cates, R.G. 1980. Feeding patterns of monophagous, oligophagous, and polyphagous insect herbivores: The effect of resource abundance and plant chemistry. Oecologia 46:22-31.

Cates, R.G. and D.F. Rhoades. 1977. Patterns in the production of antiherbivore chemical defenses in plant communities. Biochem. Syst. and Ecol. 5:185-193.

Chamberlin, W.J. 1939. The bark and timber beetles of North America north of Mexico. Oregon State College Cooperative Association, Corvallis. 513 pp.

Chamberlin, W.J. 1958. The Scolytidae of the Northwest--Oregon, Washington, Idaho and British Columbia. Oregon State Monographs. Studies in Entomology No. 2. Corvallis. 208 pp.

Chansler, J.F. 1967. Biology and life history of Dendroctonus adjunctus (Coleoptera: Scolytidae) Ann. Entomol. Soc. Amer. 60:760-767.

Chapman, J.A. 1957. Flight muscle change during adult life in the Scolytidae. Canada Dept. For., Bi-Mon. Progr. Rept. 13:3-4.

Chapman, J.A. and E.D.A. Dyer. 1969. Cross attraction between the Douglas-fir beetle (Dendroctonus pseudotsugae Hopk.) and the spruce beetle [D. obesus (Mann.)]. Canada Dept. Fish. and For., Bi-Mon. Res. Notes 25:31.

Charles, V.K. 1941. A preliminary check list of the entomophagus fungi of North America. U.S.D.A. Bur. Plant Ind., Insect Pest Survey. Bull. 21:707-85.

Chitty, D. 1960. Population processes in the vole and their relevance to general theory. Can. J. Zool. 38:99-113.

Chitty, D. 1971. The natural selection of self-regulatory behavior in animal populations. 95 pp. In I.A. McLaren (ed.). Natural Regulation of Animal Populations. Atherton Press, New York.

Choudhury, J.H. and J.S. Kennedy. 1980. Light versus pheromone-bearing wind in the control of flight direction by bark beetles, Scolytus multistriatus. Physiol. Entomol. 5:207-214.

Christian, J.J. and D.E. Davis. 1971. Endocrines, behavior and population. 195 pp. In I.A. McLaren (ed.). Natural Regulation of Animal Populations. Atherton Press, New York.

Claridge, M.F., and M.R. Wilson. 1978. British insects and trees: a study in island biogeography or insect/plant coevolution? Am. Nat. 112:451-456.

Clark, L.R., P.W. Geier, R.D. Hughes and R.F. Morris. 1967. The ecology of insect populations in theory and practice. Methuen Co., London, 232 pp.

Claussen, C.P. 1940. Entomophagous Insects. McGraw-Hill, New York, 688 pp.

Cobb, F.W. and M. Krstic, E. Zavarin and H.W. Barber. 1968a. Inhibitory effects of volatile oleoresin components on Fomes annosus and four Ceratocystis species. Phytopath. 58:1327-1335.

Cobb, F.W., D.L. Wood, R.W. Stark and P.R. Miller. 1968b. Photochemical oxidant injury and bark beetle (Coleoptera: Scolytidae) infestation of ponderosa pine. II. Effect of injury upon physical properties of oleoresin, moisture, content, and phloem thickness. Hilgardia 39:127-134.

Cobb, F.W., D.L. Wood, R.W. Stark, and J.R. Parmeter. 1968c. Photochemical oxidant injury and bark beetle (Coleoptera: Scolytidae) infestation of ponderosa pine. IV. Theory on the relationship between oxidant injury and bark beetle infestation. Hilgardia 39:141-152.

Cody, M.L. and J.M. Diamond. 1975. Ecology and Evolution of Communities. The Belknap Press of Harvard Univ. Press, Cambridge. 545 pp.

Cole, D. 1973. Estimation of phloem thickness in lodgepole pines. U.S.D.A. For. Serv. Res. Pap. INT-148. 10 pp.

Cole, W.E. 1962. The effects of intraspecific competition within mountain pine beetle broods under laboratory conditions. U.S.D.A. For. Serv. Res. Note. INT-97. 4 pp.

Cole, W.E. 1973. Interaction between mountain pine beetle and dynamics of lodgepole pine stands. U.S.D.A. For. Serv. Res. Pap. INT-197. 6 p.

Cole, W.E. 1975. Interpreting some mortality factor interactions within mountain pine beetle broods. Env. Entomol. 4:97-102.

Cole, W.E. and G.D. Amman. 1969. Mountain pine beetle infestations in relation to lodgepole pine diameters. U.S.D.A. For. Serv. Res. Note INT-95. 7 pp.

Cole, W.E. and D.B. Cahill. 1976. Cutting strategies can reduce probabilities of mountain pine beetle epidemics in lodgepole pine. J. For. 294-297.

Conn, J. 1981. Pheromone production and control mechanisms in Dendroctonus ponderosae Hopkins. Misc. Thesis. Simon Fraser University, Burnaby, B.C., Canada.

Cooke, R. 1977. The Biology of Symbiotic Fungi. John Wiley and Sons. 282 pp.

Cornelius, R.O. 1955. How forest pests upset management plans in the Douglas-fir region. J. For. 53:711-713.

Cornell, H.V., and J.O. Washburn. 1979. Evolution of richness-area correlation for cynipid gall wasps on oak trees: a comparison of two geographic areas. Evolution 33:257-274.

Coster, J.E. 1970. Production of aggregating pheromones in re-emerged parent females of the southern pine beetle. Ann. Entomol. Soc. Amer. 63:1186-1187.

Coster, J.E., T.L. Payne, L.J. Edson and E.R. Hart 1978. Influence of weather on mass aggregation of southern pine beetles at attractive host trees. Southwest Entomol. 3:14-20.

Coulson, R.N. 1979. Population dynamics of bark beetles. Ann. Rev. Entomol. 24:417-447.

Coulson, R.N. 1980. Evolutions of concepts of integrated pest management in forest. J. Georgia Entomol. Soc. 16:301-315.

Coulson, R.N., W.S. Fargo, P.E. Pulley, D.N. Pope, and A.M. Bunting. 1979a. Spatial and temporal patterns of emergence for within-tree populations of Dendroctonus frontalis (Coleoptera: Scolytidae). Can. Entomol. 111:273-287.

Coulson, R.N., R.M. Feldman, W.S. Fargo, P.H. Sharpe, G.L. Curry, and P.E. Pulley. 1979b. Evaluating suppression tactics of Dendroctonus frontalis in infestations. pp. 27-44 In J.E. Coster and J.L. Searcy (eds.). Evaluating control tactics for the Southern pine beetle--a symposium. U.S.D.A. Forest Serv. Tech. Bull. 1613.

Coulson, R.N., A.M. Mayyasi, J.L. Foltz, F.P. Hain, and W.C. Martin. 1976a. Resource utilization by the southern pine beetle, Dendroctonus frontalis Zimm. Can. Entomol. 108:353-362.

Coulson, R.N., A.M. Mayyasi, J.L. Foltz, F.P. Hain, and W.C. Martin. 1976b. Interspecific competition between Monochamus titillator and Dendroctonus frontalis. Environ. Entomol. 5:235-247.

Coulson, R.N., P.L. Payne, J.E. Coster, and M.W. Housewearp. 1972. The southern pine beetle Dendroctonus frontalis Zimm. (Coleoptera: Scolytidae) with blue stain fungi and yeasts during brook development in lodgepole pine. Can. Entomol. 103:1495-1503.

Coulson, R.N., D.N. Pope, J.A. Gagne, W.S. Fargo, P.E. Pulley, L.J. Edson, and T.L. Wagner. 1980a. Impact of foraging by Monochamus titillator on within-tree populations of D. frontalis. Entomophaga 25:155-170.

Coulson, R.N., P.E. Pulley, D.N. Pope, W.S. Fargo, and L.J. Edson. 1980b. Continuous population estimates for Dendroctonus frontalis occurring in infestations. Res. Popul. Ecol. 22:117-135.

Craighead, F.C. 1921. Hopkins host-selection principle as related to certain cerambycid beetles. J. Agric. Res. XXII(4):189-220.

Craighead, F.C. 1928. Interrelation of tree killing bark beetles (Dendroctonus) and blue stains. J. Forestry 26:886-887.

Critchfield, W.B. 1980. Genetics of lodgepole pine. U.S.D.A. For. Serv. Res. Pap. WO-37.

Crookston, N.L., R.C. Roelke, D.G. Burnell, and A.R. Stage. 1978. Evaluation of management alternatives for lodgepole pine stands using a stand projection model. In A.A. Berryman, G.D. Amman, R.W. Stark, and D.L. Kibbee (eds.). Theory and practice of mountain pine beetle management in lodgepole pine forests. College of Forest Resources, University of Idaho, Moscow.

Crowson, R.A. 1960. The phylogeny of Coleoptera. Ann. Rev. Entomol. 5:111-134.

Crowson, R.A. 1968. A Natural Classification of the Families of Coleoptera. Nathaniel Lloyd, London. 195 pp.

Dahlsten, D.L. 1967. Nesting boxes for the encouragement of insectivorous hole-nesting birds in California. Calif. Christmas Tree Growers Bull. 60:14-15.

Dahlsten, D.L. 1970. Parasites, predators, and associated organisms reared from western pine beetle infested bark samples. pp. 75-79 In R.W. Stark and D.L. Dahlsten (eds.). Studies on the population dynamics of the western pine beetle, Dendroctonus brevicomis Le Conte (Coleoptera: Scolytidae). University of California, Div. of Agric. Sci.

Dahlsten, D.L. and R.W. Bushing. 1970. Insect parasites of the western pine beetle. pp. 113-118. In R.W. Stark and D.L. Dahlsten, (eds.). Studies on the population dynamics of the western pine beetle, Dendroctonus brevicomis Le Conte (Coleoptera: Scolytidae). University of California, Div. of Agric. Sci.

Dahlsten, D.L., W.A. Copper, and K.A. Sheehan. Cold mortality of the western pine beetle at McCloud Flat in California during the 1970-1971 cold snap. Unpublished manuscript.

Dahlsten, D.L. and F.M. Stephen. 1974. Natural enemies and insect associates of the mountain pine beetle, Dendroctonus ponderosae (Coleoptera: Scolytidae) in sugar pine. Can. Entomol. 106:1211-1217.

Darwin, C. 1859. On the Origin of Species by Means of Natural Selection, or the Preservation of Favored Races in the Struggle for Life. John Murray, London. 502 pp.

Daterman, G.S., J.A. Rudinsky, and W.P. Nagel. 1965. Flight patterns of bark and timber beetles associated with coniferous forests of western Oregon. Oreg. State Univ. Tech. Bull. No. 87.

Davidson, A.G. and R.M. Prentice (eds.). 1967. Important forest insects and diseases of mutual concern to Canada, United States, and Mexico. North American Forestry Comm. FAO Publ. by Dept. Forest. and Rural Development. Ottawa, Canada. 248 pp.

Davidson, R.W. 1955. Wood staining fungi associated with bark beetles in Engelmann spruce in Colorado. Mycologia, 47:58-67.

Davidson, R.W. 1958. Additional species of Ophiostomataceae from Colorado. Mycologia 50:661-670.

Davidson, R.W. 1966. New species of Ceratocystis from conifers. Mycopathol. Mycol. Appl. 28:273-286.

Davidson, R.W. 1971. New species of Ceratocystis. Mycologia 1:5-15.

Davidson, R.W. and R.C. Robinson-Jeffrey. 1965. New records of Ceratocystis europhioides and C. huntii with Verticicladiella imperfect stages from conifers. Mycologia 57:488-490.

Day, B.W. 1976. The axiology of pest control. Agrichemical Age 19:5-6.

Day, P.H. 1974. Genetics of Host-Parasite Interaction. W.H. Freeman and Company, San Francisco.

Day, P.H. 1977. The Genetic Basis of Epidemics in Agriculture. Ann. New York Acad. Sci. Vol. 287, 400 pp.

deBary. A. 1879. Die Erscheinung der Symbiose. Trubner, Strassburg.

DeGraaf, R.M. and K.E. Evans. 1979. Management of north central and northeastern forests for nongame birds. USDA For. Serv. Gen. Tech. Rep. NC-51. 268 pp.

DeLeon, D. 1934. An annotated list of the parasites, predators, and other associated fauna of the mountain pine beetle in western white pine and lodgepole pine. Can. Entomol. 66:51-61.

DeLeon, D. 1935. The biology of Coeloides dendroctoni Cushman (Hymenoptera-Braconidae) an important parasite of the mountain pine beetle (Dendroctonus monticolae Hopk). Ann. Entomol. Soc. Amer. 28:411-424.

DeMars, C.J., Jr., A.A. Berryman, D.L. Dahlsten, I.S. Otvos, and R.W. Stark. 1970. Spatial and temporal variations in the distribution of the western pine beetle, its predators and parasites and woodpecker activity in infested trees. pp. 80-101 In R.W. Stark and D.L. Dahlsten (eds.). Studies on the population dynamics of the western pine beetle, Dendroctonus brevicomis Le Conte (Coleoptera: Scolytidae). University of California, Div. of Agric. Sci.

DeMars, C.J., Jr., D.L. Dahlsten, and R.W. Stark. 1970. Survivorship curves for eight generations of the western pine beetle in California, 1962-1965, and a preliminary life table. In R.W. Stark and D.L. Dahlsten (eds.). Studies of the population dynamics of the western pine beetle, Dendroctonus brevicomis Le Conte (Coleoptera: Scolytidae). Univ. of Calif., Div. Agric. Sci. Berkeley.

Dement, W. and H. Mooney. 1974. Seasonal variation in the production of tannins and cyanogenic glucosides in the chaparral shrub, Heteromeles arbutifolia. Oecologia 15:65-76.

Dewey, J., W. Ciesla, and H. Meyer. 1974. Insect defoliation as a predisposing agent to a bark beetle outbreak in eastern Montana. Environ. Entomol. 3:722.

Dickens, J.C. and T.L. Payne. 1977. Bark beetle olfaction: pheromone receptor system in Dendroctonus frontalis. J. Insect Physiol 23:481-489.

Dickens, J.C. and T.L. Payne. 1978. Olfactory-induced muscle potentials in Dendroctonus frontalis: effects of trans-verbenol and verbenone. Experientia 34:463-464.

Dickson, J.G., R.R. Connor, R. R. Fleet, J.A. Jackson, and J.C. Kroll (eds.). 1979. The Role of Insectivorous Birds in Forest Ecosystems. Academic Press, New York, 381 pp.

Dixon, J.C. and E.A. Osgood. 1961. Southern pine beetle: A review of present knowledge. USDA Forest Serv. Res. Pap. SE-128. 34 pp.

Dixon, P. 1973. Biology of Aphids. Edward Arnold, New York. 58 pp.

Dixon, W.N. and T.L. Payne. 1979. Aggregation of Thanasimus dubius on trees under mass-attack by the southern pine beetle. Environ. Entomol. 8:178-181.

Dobzhansky, Th. 1937. Genetics and the Origin of Species. Columbia University Press, New York.

Dolinger, P.M., P.R. Ehrlich, W.L. Fitch and D.E. Breedlove. 1973. Alkaloid and predation patterns in Colorado lupine populations. Oecologia 13:191-204.

Dolph, R.E., Jr. 1966. Mountain pine beetle damage in the Pacific Northwest 1955-1966. U.S.D.A. Forest Serv. Pacific Northwest Region, Portland, Oregon. 36 pp.

Dowden, P.B. 1962. Parasites and predators of forest insects liberated in the United States through 1960. U.S.D.A. For. Serv. Agr. Hdbk. 226. 70 pp.

Dyer, E.D.A. 1973. Spruce beetle aggregated by the synthetic pheromone frontalin. Can. J. For. Res. 3:486-494.

Dyer, E.D.A. 1975. Frontalin attractant in stands infested by the spruce beetle, Dendroctonus rufipennis (Coleoptera: Scolytidae). Can. Entomol. 107:979-988.

Dyer, E.D.A. and C.M. Lawko. 1978. Effect of seudenol on spruce beetle and Douglas-fir beetle aggregation. Environ. Canada, Bi-Mon. Res. Notes 34:30-32.

Eaton, C.B. 1956. Jeffrey pine beetle. USDA Forest Service Forest Pest Leaflet No. 11, 7 pp.

Ebel, B.H., T.H. Flavell, L.E. Drake, H.O. Yates, III, and G.D. DeBarr. 1975. Southern pine cone and seed insects. U.S.D.A. For. Serv. Gen. Tech. Rep. SE-8. 40 pp.

Edminster, C.B. 1978. Development of growth and yield models for dwarf-mistletoe infested stands. pp. 172-177 In Proceedings Symposium on Dwarf Mistletoe Control through Forest Management. U.S.D.A. For. Serv. Gen. Tech. Rep. PSW-31.

Edmunds, G.F. and D.N. Alstad. 1978. Coevolution in insect herbivores and conifers. Science 199:941-945.

Edson, L.J. 1978. Host colonization and the arrival sequence of the mountain pine beetle and its insectan associates. Ph.D. Thesis, Univ. of California, Berkeley. 196 pp.

Egger, A. 1974. The biology of Pityophthorus micrographus and some of its natural enemies as factors in natural control. Centralblatt fur das Gesamte Forstwesen 91:158-165 (In German).

Ehrlich, P., A. Ehrlich, J. Holdren. 1977. Ecoscience: Population, Resources, and Environment. W.H. Freeman, San Francisco, Calif. 1051 pp.

Ehrlich, P.R. and P.H. Raven. 1965. Butterflies and plants: a study in coevolution. Evolution 18:586-608.

Ehrman, L. and P. Parsons. 1976. The Genetics of Behavior. Sinauer Associates, Inc. Sunderland.

Eichhorn, O. and P. Graf. 1974. On some timber bark beetles and their enemies (in German). Anzeiger fur Schadlingskunde, Pflanzenschutz Unweltzschutz. 47:129-135.

Elliott, E.W., G.N. Lanier and J.B. Simeone. 1975. Termination of aggregation by the European elm bark beetle, Scolytus multistriatus. J. Chem. Ecol. 1:283-289.

Elliott, W.J., G. Hromnak, J. Fried and G.N. Lanier. 1979. Synthesis of multistriatin enantiomers and their action on Scolytus multistriatus (Coleoptera: Scolytidae). J. Chem. Ecol. 5:279-287.

Ellis, D.E. 1939. Ceratostomella ips associated with Ips lecontei in Arizona. Phytopathology 29:556-57.

Erickson, J.M., and P.P. Feeny. 1974. Sinigrin: A chemical barrier to larvae of the black swallowtail butterfly, Papilio polyxenes. Ecol. 55:103-111.

Fares, Y., P.J.H. Sharpe, and C.E. Magnuson. 1980. Pheromone dispersion in forests. J. Theor. Biol. 84:335-359.

Farmer, L.S. 1965. The phloem-yeast complex during infestations of the mountain pine beetle in lodgepole pine. Ph.D. thesis, Univ. of Utah. 124 pp.

Farris, S.H. 1965. A preliminary study of the mycangia in the bark beetles Dendroctonus ponderosae Hopk., Dendroctonus obesus Mann., and Dendroctonus pseudotsugae Hopk. Can Dept. For. Bi-monthly Progr. Rept. 21:3-4.

Farris, S.H. 1969. Occurrence of mycangia in the bark beetle Dryocoetes confusus (Coleoptera:Scolytidae). Can Entomol. 101:527-532.

Feeny, P. 1970. Seasonal changes in oak leaf tannins and nutrients as a cause of spring feeding by winter moth caterpillars. Ecology 51:565-581.

Feeny, P. 1976. Plant apparency and chemical defense in biochemical interaction between plants and insects. pp. 1-40 In J. Wallace and R. Mansell (eds.). Biochemical Interaction Between Plants and Insects. Plenum Press, New York.

Fenner, F. 1968. The Biology of Animal Viruses. II. The Pathogenesis and Ecology of Viral Infestions. Academic Press, New York.

Ferrell, G.T. 1971. Host selection by the fir engraver, Scolytus ventralis (Coleoptera: Scolytidae): preliminary field studies. Can. Entomol. 103:1717-1725.

Ferron, P. 1977. Influence of relative humidity on the development of fungal infection caused by Beauveria bassiana (Fungi Imperfecti: Moniliales) in imagines of Acathoscelides obtectus (Coleoptera: Bruchidae). Entomophaga 22:393-396.

Finney, J.R. and C. Walker. 1979. Assessment of a field trial using the DD-136 strain of Neoaplectana sp. for the control of Scolytus scolytus. J. Invert. Path. 33:239-241.

Fish, R.H., L.E. Browne, D.L. Wood and L.B. Hendry. 1979. Pheromone biosynthetic pathways: conversions of deuterium labelled ipsdienol with sexual and enantioselectivity in Ips paraconfusus Lanier. Tetrahedron Let. 17:1465-1468.

Fisher, R.A. 1930. The Genetical Theory of Natural Selection. Clarendon Press, Oxford.

Fiske, W.F. 1908. Notes on insect enemies of wood boring Coleoptera. Proc. Ent. Soc. Wash. 9:23-27.

Fitzgerald, T.D. and W.P. Nagel. 1972. Oviposition and larval bark-surface orientation of Medetera aldrichii (Diptera: Dolichopodidae): response to a prey-liberated plant terpene. Ann. Entomol. Soc. Amer. 65:328-330.

Flangas, A.L. and J.G. Dickson. 1961. The genetic control of pathogenicity, serotypes and variability in Puccinia sorghi. Amer. J. Bot. 48:275-285.

Flor, H.H. 1956. The complementary genetic systems in flax and rust. Adv. Genet. 8:29-54.

Fockler, C.E. and J.H. Borden. 1972. Sexual behavior and seasonal mating activity of Trypodendron lineatum (Coleoptera: Scolytidae). Can. Entomol. 104:1841-1853.

Fockler, C.E. and J.H. Borden. 1973. Mating activity and ovariole development of Trypodendron lineatum: effect of a juvenile hormone analogue. Ann. Entomol. Soc. Amer. 66:509-512.

Forde, M. 1964. Inheritance of turpentine composition in Pinus attentuata x radiata hybrids. New Zeal. J. Bot. 2:53-59.

Foster, A.S., and E.M. Gifford, Jr. 1974. Comparative Morphology of Vascular Plants. W.H. Freeman and Co., San Francisco.

Fraenkel, G. 1959. The raison d'etre of secondary plant substances. Science 129:1466-1470.

Francke, W. and V. Heeman. 1974. Lockversuche bei Xyloterus domesticus L. und X. lineatus Oliv. (Coleoptera: Scolytidae) mit 3-Hydroxy-3-methylbutan-2-on. Z. Angew. Entomol. 75:67-72.

Francke, V.W. and V. Heeman. 1976. Das Duftstoff-bouquet des grossen Waldgartners Blastophagus piniperda L. (Coleoptera: Scolytidae). Z. Angew. Entomol. 82:117-119.

Francke, W., V. Heeman, B. Gerken, J.A.A. Renwick and J.P. Vite. 1977. 2-Ethyl-1,6-dioxaspiro [4.4]nonane, principal aggregation pheromone of Pityogenes chalcographus (L.). Naturwiss. 64:590-591.

Francke, W., V. Heeman and K. Heyns. 1974. Fluchtige inhaltstoffe von Ambrosiakafern (Coleoptera: Scolytidae), I.Z. Naturforsch. 29C:243-245.

Francke-Grosmann, H. 1963. Some new aspects in forest entomology. Ann. Rev. Entomol. 8:415-438.

Francke-Grosmann, H. 1965a. Ein symbioseorgan bie dem borkenkafer Dendroctonus frontalis Zimm. Naturwiss. 52:143.

Francke-Grosmann, H. 1965b. On the symbiosis of xylo-mycetophagous and phloeophagous Scolytoidea with wood-inhabiting fungi. Holz and Organismen Internationales. Symposium Berlin-Dahlem. pp. 503-522 (English Trans.). Duncker and Humbolt. Issue 1. 1966.

Francke-Grosmann, H. 1967. Ectosymbiosis in wood-inhabiting insects. pp. 141-205 In S.M. Henry (ed.). Symbiosis Vol. II. Academic Press, New York and London.

Franklin, R.T. 1970. Observations on the blue stain-southern pine beetle relationship. J. Georg. Entoml. Soc. 5:53-57.

Franz, J.M. 1961. Biological control of pest insects in Europe. Ann. Rev. Entomol. 6:183-200.

Freeland, W.J. 1974. Vole cycles: another hypothesis. Amer. Natur. 108:238-245.

Fronk, W.D. 1947. The southern pine beetle--its life history. Va. Agric. Exp. Stn. Tech. Bull. 108, 12 pp.

Furniss, R.L. and V.M. Carolin. 1977. Western forest insects. U.S.D.A. For. Serv. Misc. Publ. No. 1339. 654 pp.

Furniss, M.M., B.H. Baker and B.B. Hostetler. 1976. Aggregation of spruce beetles (Coleoptera) to seudenol and repression of attraction by methylcyclohexenone in Alaska. Can. Entomol. 108:1297-1302.

Furniss, M.M., G.E. Daterman, L.N. Kline, M.D. McGregor, G.C. Trostle, L.F. Pettinger, and J.A. Rudinsky. 1974. Effectiveness of the Douglas-fir beetle anti-aggregative pheromone methylcyclohexenone at three concentrations and spacings around felled host trees. Can. Entomol. 106:381-392.

Furniss, M.M., L.N. Kline, R.F. Schmitz and J.A. Rudinsky. 1972. Tests of three pheromones to induce or disrupt aggregation of Douglas-fir beetles (Coleoptera: Scolytidae) on live trees. Ann. Entomol. Soc. Amer. 65:1227-1232.

Furniss, M.M. and R.L. Livingston. 1979. Inhibition by ipsenol of pine engraver attraction in northern Idaho. Environ. Entomol. 8:369-372.

Furniss, M.M. and J.A. Schenk. 1969. Sustained natural infestation by the mountain pine beetle in seven new Pinus and Picea hosts. J. Econ. Entomol. 62:518-519.

Furniss, M.M. and R.F. Schmitz. 1971. Comparative attraction of Douglas-fir beetles to frontalin and tree volatiles. U.S.D.A. For. Serv. Res. Pap. INT-96.

Futuyma, D.J. and G.E. Mayer. 1980. Non-allopatric speciation in animals. Syst. Zool. 29:254-271.

Futuyma, D.J. and M. Slatkin. 1982. Coevolution. Sinauer Press, Sunderland, MA.

Gara, R.I. 1963. Studies on the flight behavior of Ips confusus (LeC.) (Coleoptera: Scolytidae) in response to attractive material. Contrib. Boyce Thompson Inst. 22:51-66.

Gara, R.I. and J.E. Coster. 1968. Studies on the attack behavior of the southern pine beetle. III. Sequence of tree infestation within stands. Contrib. Boyce Thompson Inst. 24:77-85.

Gara, R.I. and J.P. Vite. 1962. Studies on the flight patterns of bark beetles (Coleoptera: Scolytidae) in second growth ponderosa pine forests. Contrib. Boyce Thompson Inst. 21:275-289.

Gara, R.I., J.P. Vite and H.H. Cramer. 1965. Manipulation of Dendroctonus frontalis by use of population aggregating pheromones. Contrib. Boyce Thompson Inst. 23:55-56.

Geiszler, D.R. and R.I. Gara. 1978. Mountain pine beetle attack dynamics in lodgepole pine. pp. 182-187 In A.A. Berryman, G.D. Amman, R.W. Stark and D.L. Kibbee (eds.). Theory and practice of mountain pine beetle management in lodgepole pine forests--a symposium. College of Forest Resources, Univ. Idaho, Moscow.

Georghiou, G.P. 1972. The evolution of resistance to pesticides. Ann. Rev. Ecol. Syst. 3:133-168.

Georghiou, G.P. and C.E. Taylor. 1977. Genetic and biological influences in the evolution of insecticide resistance. J. Econ. Entomol. 70:319-323.

Gerken, B., S. Grune, J.P. Vite and K. Mori. 1978. Response of European populations of Scolytus multistriatus to isomers of multistriatin. Naturwiss. 65:110-111.

Gillespie, J.H., and K.I. Kojima. 1968. The degree of polymorphisms in enzymes involved in energy production compared to that in nonspecific enzymes in two Drosophila ananassae populations. P.N.A.S. 61:582-585.

Gillett, J.B. 1962. Pest pressure, and underestimated pressure in evolution. Syst. Assoc. Publ. 4:37-46.

Goeden, R. and D. Norris, Jr. 1964. Attraction of Scolytus quadrispinosus (Coleoptera: Scolytidae) to Carya spp. for oviposition. Ann. Entomol. Soc. Amer. 57:141-146.

Goheen, D.J., and F.W. Cobb, Jr. 1978. Occurrence of Verticicladiella wagenerii and its perfect state, Ceratocystis wageneri sp. nov., in insect galleries. Phytopath. 68:1192-1195.

Goheen, D.J. and F.W. Cobb, Jr. 1978. Occurrence of Verticicladiella wagerseii sp. nov., in insect galleries. Phytopathology 68:1192-1195.

Gore, W.E., G.T. Pearce, G.N. Lanier, J.B. Simeone, R.M. Silverstein, J.W. Peacock and R.A. Cuthbert. 1977. Aggregation attractant of the European elm bark beetle, Scolytus multistriatus. Production of individual components and related aggregation behavior. J. Chem. Ecol. 3:429-446.

Gottlieb, L.D. 1977. Electrophoretic evidence and plant systematics. Ann. Missouri Bot. Gard. 64:161-180.

Gouger, R.J. 1972. Interrelationships of Ips avulsus (Eichh.) and associated fungi. Dist. Abst. 32:6453-B.

Gouger, R.J., W.C. Yearian, R.C. Wilkinson. 1975. Feeding and reproductive behavior of Ips avulsus. Florida Entomologist 58:221-229.

Graham, K. 1959. Release by flight exercise of a chemotropic response from photopositive domination in a scolytid beetle. Nature 184:283-284.

Graham, K. 1961. Air swallowing: a mechanism in photic reversal of the beetle Trypodendron. Nature (Lond.) 191:519-520.

Graham, K. 1962. Photic behavior in the ecology of the ambrosia beetle Trypodendron lineatum. Proc. 11th Int. Congr. Entomol. (1960) 2:226.

Graham, K. 1963. Concepts of Forest Entomology. Reinhold. New York. 388 pp.

Graham, K. 1967. Fungal-insect mutualism in trees and timber. Ann. Rev. Entomol. 12:105-126.

Graham, K. 1968. Anaerobic induction of primary chemical attractancy for ambrosia beetles. Can. J. Zool. 46:905-908.

Graham, S.A. and F.B. Knight. 1965. Principles of Forest Entomology. McGraw-Hill, New York. 417 pp.

Gurando, O.V. and D.B. Tsarichkova. 1974. Change in the sex apparatus of females of the small pine bark beetle caused by parasitic nematodes (In Russian) Zakhist Roslin 20:35-40.

Guttman, S.I., T.K. Wood, and A.A. Karlin. 1981. Genetic differentiation along host plant lines in the sympatric Enchenopa binotata Say complex (Homoptera: Membracidae). Evolution 35:205-217.

Gyorfi, J. 1952. Orszagos Termeszettudomany; Muzeum Eukonyue N.S. 2, 113. Budapest.

Hackwell, G.A. 1973. Biology of Lasconotus subcostulatus (Coleoptera: Colydiidae) with special reference to feeding behavior. Ann. Entomol. Soc. Amer. 66:62-65.

Hadfield, T.S. and K.W. Russell. 1978. Dwarf mistletoe management in the Pacific Northwest. In Dwarf mistletoe control through forest management--a symposium. U.S.D.A. For. Serv. Gen. Tech. Rep. PSW-31.

Hadwiger, L., and M. Schwochan. 1969. Host resistance responses--an induction hypothesis. Phytopathol. 7:13-22.

Hagar, D.C. 1960. Interrelationships of logging, birds, and timber regeneration in the Douglas-fir region of Northwestern California. Ecology 41:116-125.

Hamilton, W.D. 1964a. The genetical theory of social behavior, I. J. Theoret. Biol. 7:1-16.

Hamilton, W.D. 1964b. The genetical theory of social behavior, II. J. Theoret. Biol. 7:17-52.

Hamilton, W.D. 1978. Evolution and diversity under bark. In L.A. Mound and N. Woloff (eds.). Diversity of Insect Faunas. Blackwell Scientific Publ., Oxford.

Hamrick, J.L., Y.B. Linhart and J.B. Mitton. 1979. Relationships between life history characteristics and electrophoretically-detectable genetic variation in plants. Ann. Rev. Sys. Ecol. 10:173-200.

Handlirsch, A. 1906-1908. Die fossilen Insekten und die Phylogenie der rezenten Formen. Leipsig, 1433 pp.

Hanover, J.W. 1966a. Inheritance of 3-carene concentration in Pinus monticola. For. Sci. 12:447-451.

Hanover, J.W. 1966b. Environmental variation in the monoterpenes of Pinus monticola Dougl. Phytochem. 5:713-717.

Hanover, J.W. 1966c. Genetics of terpenes. I. Gene control of monoterpene levels in Pinus monticola Dougl. Heredity 21:73-84.

Hanover, J.W. 1975. Physiology of the resistance to insects. Ann. Rev. Ent. 20:75-95.

Hanover, J. 1980. Breeding forest trees resistant to insects. pp. 487-511 In F.G. Maxwell and P.R. Jennings (eds.). Breeding Plants Resistant to Insects. John Wiley and Sons, New York.

Happ, G.M. 1968. Quinone and hydrocarbon production in the defensive glands of Eleodes longicollis and Tribolium castaneum (Coleoptera: Tenebrionidae). J. Insect Physiol. 14:1821-1837.

Happ, G.M., C.M. Happ and S.J. Barras. 1971. Fine structure of the prothoracic mycangium, a chamber for the culture of symbiotic fungi, in the southern pine beetle, D. frontalis. Tissue and Cell 3:295-308.

Happ, G.M., C.M. Happ and S.J. Barras. 1975. Bark beetle-fungal symbiosis. III. Ultrastructure of conidiogenesis in a Sporothrix ectosymbiont of the southern pine beetle. Can. J. Bot. 53:2702.

Happ, G.M., C.M. Happ, and S.J. Barras. 1976. Bark beetle fungal symbiosis. II. Fine structure of a basidiomycetous ectosymbiont of the southern pine beetle. Can. J. Bot. 54:1049-1062.

Harborne, J.B. (ed.) 1972. Phytochemical Ecology. Academic Press, N.Y. 272 pp.

Harborne, J.B. 1973. Phytochemical methods. Chapman and Hall, London.

Harlow, W.M. and E.S. Harrar. 1941. Textbook of Dendrology. McGraw-Hill, New York. 542 pp.

Harring, C.M. and K. Mori. 1977. Pityokteines curvidens Germ. (Coleoptera: Scolytidae): aggregation in response to optically pure ipsenol. Z. Angew. Entomol. 82:327-329.

Harring, C.M. and J.P. Vite. 1975. "Ipsenol," der Populations lockstroff des krummzahnigen Tannenborken-kafers. Naturwiss. 62:488.

Harris, H. 1966. Enzyme polymorphisms in man. Proc. Royal Soc. (B) 164:298-310.

Hatchett, J.H. and R.L. Gallun. 1970. Genetics of the ability of the Hessian fly, Mayetiola destructor, to survive on wheats having different genes for resistance. Ann. Entomol. Soc. Amer. 63:1400-1407.

Hawker, L.E. 1957. The physiology of reproduction in fungi. Cambridge Univ. Press. 128 pp.

Hawker, L.E., and M.F. Madelin. 1976. The dormant spore. pp. 1-72 In D.J. Weber and W.M. Hess (eds.). The Fungal Spore. Wiley and Sons, New York.

Hawksworth. F.G. 1978a. Intermediate cuttings in mistletoe-infested lodgepole pine and southwestern ponderosa pine stands. pp. 86-99 In Dwarf mistletoe control through forest management--a symposium. U.S.D.A. For. Serv. Gen. Tech. Rep. PSW-31.

Hawksworth, F.G. 1978b. Biological factors of dwarf mistletoe in relation to control. In Dwarf mistletoe control through forest management--a symposium. U.S.D.A. For. Serv. Gen. Tech. Rep. PSW-31.

Hay, C.J. 1956. Experimental crossing of mountain pine beetle with Black Hills beetle. Ann. Entomol. Soc. Amer. 49:567-571.

Hedden, R., J.P. Vite and K. Mori. 1976. Synergistic effect of a pheromone and a kairomone on host selection and colonization by Ips avulsus. Nature (Lond.) 261:696-697.

Hedlin, A.F. 1974. Cone and seed insects of British Columbia. Environment Canada Forestry Serv., Victoria, B.C. 63 pp.

Heikkenen, H. 1977. Southern pine beetle: a hypothesis regarding its primary attractant. J. For. 75:404-413.

Heikkenen, H.J. and B.F. Hrutfiord. 1965. Dendroctonus pseudotsugae: A hypothesis regarding its primary attractant. Science 150:1457-1459.

Hemingway, R.W., G.W. McGraw, and S.J. Barras. 1977. Polyphenols in Ceratocystis minor-infected Pinus taeda: fungal metabolites, phloem and xylem phenols. J. Agric. and Food Chem. 25:717-722.

Hendry, L.B., B. Piatek, L.E. Browne, D.L. Wood, J.A. Byers, R.H. Rish and R.A. Hicks. 1980. In vivo conversion of a labeled host plant chemical to pheromones of the bark beetle Ips paraconfusus. Nature (Lond.) 284:485.

Henry, S.M. (ed.). 1966-1967. Symbiosis (2 Vols.). Academic Press, New York.

Henson, W.R. 1961. Laboratory studies on the adult behavior of Conophthorus coniperda (Schwarz) (Coleoptera: Scolytidae). I. Seasonal changes in the internal anatomy of the adult. Ann. Entomol. Soc. Amer. 54:698-701.

Henson, W.R. 1962. Laboratory studies on the adult behavior of Conophthorus coniperda (Schwarz) (Coleoptera: Scolytidae). III. Flight. Ann. Entomol. Soc. Amer. 55:524-530.

Hepting, G.H. 1971. Diseases of forest and shade trees of the United States. U.S.D.A Forest Serv. Agr. Hdbk. No. 386. 658 pp.

Hertert, H.D., D.L. Miller, and A.D. Partridge. 1975. Interaction of bark beetles (Coleoptera: Scolytidae) and root-rot pathogens in grand fir in northern Idaho. Can. Entomol. 107:899-904.

Hetrick, L.A. 1933. Some factors in natural control of the southern pine beetle, Dendroctonus brevicomis Zimm. J. Econ. Entomol. 33(3):554-556.

Hetrick, L.A. 1949. Some overlooked relationships of southern pine beetle. J. Econ. Entomol. 42:466-469.

Higby, P.K. and M.W. Stock. 1982. Genetic relationships between two sibling species of bark beetle (Coleoptera: Scolytidae), the Jeffrey pine beetle and the mountain pine beetle, in northern California. Ann. Entomol. Soc. Am. (in press).

Himes, W.E. and J.M. Skelly. 1972. An association of the black turpentine beetle Dendroctonus terebrans and Fomes annosus in loblolly pine. Phytopath. 62:670.

Hodges, J.D. and P.L. Lorio. 1975. Moisture stress and composition of xylem oleoresin in loblolly pine. For Sci. 22:283-290.

Hodges, J.D. and L.S. Prickard. 1971. Lightning in the ecology of the southern pine beetle, Dendroctonus frontalis (Coleoptera: Scolytidae). Can. Entomol. 103:44-52.

Hoffard, W.H. and J.E. Coster. 1976. Endoparasitic nematodes of Ips bark beetles in eastern Texas. Env. Entomol. 5:128-132.

Holst, E.C. 1936. Zygosaccharomyces pini a new species of yeast associated with bark beetles in pines. J. Ag. Res. [US] 53:513-518.

Holst, E.C. 1937. Asceptic rearing of bark beetles. J. Econ. Entomol. 30:676-677.

Holt, W.R. 1961. Metarrhizium anisopliae (Metchnikoff) Sorokin infecting larvae of the black terpentine beetle. J. Insect Path. 3:93.

Hood, L., J.H. Campbell, and S.C.R. Elgin. 1975. The organization, expression and evolution of antibody genes and other multigene families. Ann. Rev. Genetics 9:305-353.

Hooker, A.L. and W.A. Russell. 1962. Inheritance of resistance to Puccinia sorghi in six corn inbred lines. Phytopath. 52:122-128.

Hooper, R.G., H.S. Crawford, and R.F. Hadlow. 1973. Bird density and diversity as related to vegetation in forest recreational areas. J. For. 71:766-769.

Hopkins, A.D. 1899. Report on investigations to determine the cause of unhealthy conditions of spruce and pine from 1880-1893. Bull. W. Virg. Agr. Exp. Sta. 56, 461 pp.

Hopkins, A.D. 1902. Insect enemies of the pine in the Black Hills Forest Reserve. USDA Division of Entomology Bulletin 32. 24 pp.

Hopkins, A.D. 1909. Contributions toward a monograph of the scolytid beetles. I. The genus Dendroctonus. U.S.D.A. Tech. Ser. Bull. 17:1-164.

Hopkins, A.D. 1916. Economic investigations of the scolytid bark and timber beetles of North America. U.S. Dept. Agric. Program of Work, 1917.

Hopping, G.R. 1963. The natural groups of species in the genus Ips DeGeer (Coleoptera: Scolytidae) in North America. Can. Entomol. 95:508-516.

Hopping, G.R. 1963a. Generic characters in the tribe Ipini (Coleoptera:Scolytidae), with a new species, a new combination, and new synonymy. Can. Entomol. 95:61-68.

Hopping, G.R. 1963b. The North American species in Group I of Ips DeGeer (Coleoptera:Scolytidae). Can. Entomol. 95:1091-1096.

Hopping, G.R. 1963c. The North American species in Groups II and III of Ips DeGeer (Coleoptera:Scolytidae). Can. Entomol. 95:1202-1210.

Hopping, G.R. 1964. The North American species in Groups IV and V of Ips DeGeer (Coleoptera:Scolytidae). Can. Entomol. 96:970-978.

Hopping, G.R. 1965a. The North American species in Group VI of Ips DeGeer (Coleoptera:Scolytidae). Can. Entomol. 97:533-541.

Hopping, G.R. 1965b. The North American species in Group VII of Ips DeGeer (Coleoptera:Scolytidae). Can. Entomol. 97:533-541.

Hopping, G.R. 1965c. The North American species in Group VII of Ips DeGeer (Coleoptera:Scolytidae). Can. Entomol. 97:159-172.

Hopping, G.R. 1965d. The North American species in Group IX of Ips DeGeer (Coleoptera:Scolytidae). Can. Entomol. 97:422-434.

Hopping, G.R. 1965e. The North American species of Group X of Ips DeGeer (Coleoptera:Scolytidae). Can. Entomol. 97:803-809.

Houk, E.J. and G.W. Griffiths. 1980. Intracellular symbiotes of the Homoptera. Ann. Rev. Ent. 25:161-87.

Howe, V.K., A.D. Oberle, T.G. Keeth and W.J. Gordon. 1971. The role of microorganisms in the attractiveness of lightning-struck pines to southern pine beetle. Western Ill. Univ. Ser. Biol. Sci. 9:1-44.

Howe, W. 1949. Factors affecting the resistance of certain cucurbits to the squash borer. J. Econ. Entomol. 42:321-326.

Howe, W. 1950. Biology and host relationships of the squash vine borer. J. Econ. Entomol. 43:480-483.

Huettel, M.D. and G.L. Bush. 1973. Enzyme polymorphisms and the differentiation of sibling species. pp. 145-150 In P.H. Dunn (ed.). Proc. 2nd Intl. Symp. Biol. Control of Weeds. Commonwealth Inst. of Biol. Control Misc. Publ. No. 6.

Huffaker, C.B. (ed.). 1974. Biological Control. Plenum Press, New York.

Huffaker, C.B. (ed.). 1980. New Technology of Pest Control. John Wiley and Sons, New York. 500 pp.

Hughes, P.R. 1973a. Dendroctonus: production of pheromones and related compounds in response to host monoterpenes. Z. Angew. Entomol. 73:294-312.

Hughes, P.R. 1973b. Effect of a-pinene exposure on trans-verbenol synthesis in Dendroctonus ponderosae Hopk. Naturwiss. 60:261-262.

Hughes, P.R. 1974. Myrcene: a precursor of pheromones in Ips beetles. J. Insect Physiol. 20:1271-1275.

Hughes, P.R. 1975. Pheromones of Dendroctonus: origin of a-pinene oxidation products present in emergent adults. J. Insect Physiol. 21:687-691.

Hughes, P.R. 1976. Response of female southern pine beetles to the aggregation pheromone frontalin. Z. Angew. Entomol. 80:280-284.

Hughes, P.R. and J.A.A. Renwick. 1977a. Hormonal and host factors stimulating pheromone synthesis in female western pine beetles, Dendroctonus brevicomis. Physiol. Entomol. 2:289-292.

Hughes, P.R. and J.A.A. Renwick. 1977b. Neural and hormonal control of pheromone biosynthesis in the bark beetle, Ips paraconfusus. Physiol. Entomol. 2:117-123.

Hughes, P.R., J.A.A. Renwick and J.P. Vite. 1976. The identification and field bioassay of chemical attractants in the roundheaded pine beetle. Environ. Entomol. 5:1165-1168.

Hunt, D.J. and N.G.M. Hague. 1974. The distribution and abundance of Parasitaphelenchus oldhami, a nematode parasite of Scolytus scolytus and S. multistriatus, the bark beetle vectors of Dutch elm disease. Plant Pathology 23:133-135.

Hunt, J. 1956. Taxonomy of the genus Ceratocystis. Lloydia 19:1-58.

Hunt, R.S. and G.O. Poinar. 1971. Culture of Parasitorhabditis sp. (Rhabditida:Protorhabditinae) on a fungus. Nematologica 17:321-322.

Hunter, P.E. and R. Davis. 1963. Observations on Histiostoma gordius (Vitz.) (Anoetidae) and other mites associated with Ips beetles. Proceedings Entomol. Soc. Wash. 65:287-293.

Hynum, B.G. and A.A. Berryman. 1980. Dendroctonus ponderosae (Coleoptera:Scolytidae): Pre-aggregation landing and gallery initiation on lodgepole pine. Can. Entomol. 112:185-191.

Hynum, B.G. and A.A. Berryman. 1981. Dendroctonus ponderosae Hopkins (Coleoptera:Scolytidae): Gallery initiation on lodgepole pine during aggregation. Environ. Entomol. 10:842-846.

Ikeda, T., M. Fumio and D.M. Benjamin. 1977. Chemical basis for feeding adaptation of pine sawflies Neodiprion rugifrons and N. swainei. Science 197:497-499.

Integrated Pest Management. 1979. A program of research for the State Agricultural Experiment Stations and the Colleges of 1890. Intersociety Consortium for Plant Protection for the Experiment Station Committee on Organization and Policy. pp. 12-13.

Jackson, T.A., M.R. Lennartz, and R.G. Hooper. 1979. True age and cavity initiation by red-cockaded woodpecker. J. For. 77:102-103.

Jander, R. 1963. Insect orientation. Ann. Rev. Ent. 8:95-114.

Jantz, O.K. and J.A. Rudinsky. 1965. Laboratory and field methods for assaying olfactory responses of the Douglas-fir beetle, Dendroctonus pseudotsugae Hopkins. Can. Entomol. 97:935-941.

Jennings, D.T. and H.A. Pace, III. 1975. Spiders preying on Ips bark beetles. Southwestern Naturalist 20:225-229.

Johnson, P.C. and J.E. Coster. 1979. Techniques for evaluating the influence of behavioral chemicals on dispersion of the southern pine beetle within infestations. Pp.18-25 In T.E. Coster and J.L. Leary (eds.). Evaluating control tactics for the southern pine beetle--a symposium. USDA Forest Serv. Techn. Bull. 1613.

Johnson, P.C. and R. Denton. 1975. Outbreaks of the western spruce budworm in the American Northern Rocky Mountain area from 1922 through 1971. U.S.D.A.-For. Serv. Gen. Tech. Rep. INT-20.

Jorgensen, E. 1961. The formation of pinosylvin and its monomethyl ether in sapwood of Pinus resinosa. Ait. Can. J. Bot. 39:1765-1772.

Jouvenaz, D.P. and R.C. Wilkinson. 1970. Incidence of Serratia marcescens in wild Ips calligraphus populations in Florida. J. Invert. Path. 16:295-296.

Kalkstein, L. 1976. Effects of climatic stress upon outbreaks of the southern pine beetle. Env. Entomol. 5:653-658.

Kangas, von E., H. Oksanen and V. Perttunen. 1970. Responses of Blastophagus piniperda L. (Coleoptera: Scolytidae) to trans-verbenol, cis-verbenol, and verbenone, known to be population pheromones of some American bark beetles. Ann Entomol. Fenn. 36:75-83.

Kaston, B.J. 1936. The morphology of the elm bark beetle, Hylurgopinus (Hylastes) rufipes Eichh. Bull. Conn. Agric. Expt. Sta. 387:613-650.

Kendrick, W.B. and A.C. Molnar. 1965. A new Ceratocystis and its Verticicladiella imperfect state associated with the bark beetle Dryocoetes confusus on Abies lasiocarpa. Can. J. Bot. 43:39-43.

Kennedy, B.H. 1970. Dendrosoter protuberans (Hymenoptera: Braconidae), an introduced larval parasite of Scolytus multistriatus. Ann. Ent. Soc. Amer. 63:351-358.

Kennedy, B.H. 1979. The effect of multilure on parasites of the European elm bark beetle, Scolytus multistriatus. Bull. Ent. Soc. Amer. 25:116-118.

Kiehlmann, E., J.E. Conn and J.H. Borden. 1982. 7-Ethoxy-6-methoxy-2,2-dimethyl-2H-1-benzopyran. Org. Prep. and Proc. Int'l. (in press).

Kimmins, J.P. 1970. Probabilistic phototactic behavior in a bark beetle. Can. J. Zool. 48:919-923.

Kimney, J.W. and R.L. Furniss. 1943. Deterioration of fire-killed Douglas-fir. U.S.D.A. Tech. Bull. 851. 61 pp.

Kingsbury, J.M. 1964. Poisonous Plants of the United States and Canada. Prentice-Hall, New York. 626 pp.

Kinn, D.N. 1970. Acarine parasites and predators of the western pine beetle. pp. 128-131 In R.W. Stark and D.L. Dahlsten (eds.). Studies on the population dynamics of the western pine beetle, Dendroctonus brevicomis Leconte (Coleoptera: Scolytidae). Univ. of California, Div. of Agric. Sci.

Kinn, D.N. 1971. The life cycle of behavior of Cercoleipus coelonotus (Acarina: Mesostigmata), including a survey of phoretic mite associates of California Scolytidae. Univ. of California Publication in Entomology No. 65, 66 pp.

Kinzer, G.W., A.F. Fentiman, Jr., R.L. Foltz and J.A. Rudinsky. 1971. Bark beetle attractants: 3-methyl-2-cyclohexen-1-one isolated from Dendroctonus pseudotsugae. J. Econ. Entomol. 64:970-971.

Kinzer, G.W., A.F. Fentiman, Jr., T.F. Page, Jr., R.L. Foltz, J.P. Vite and G.B. Pitman. 1969. Bark beetle attractants: identification, synthesis and field bioassay of a new compound isolated from Dendroctonus. Nature (Lond.) 221:477-478.

Kishi, Y. 1970. Mimemodes japonus Reitter (Coleoptera: Rhizophagidae), an egg predator of the pine bark beetle, Cryphalus fulvus Niijima (Coleoptera: Ipidae). Kontyu 38:195-197.

Klein, W.H. 1978. Strategies and tactics for reducing losses in lodgepole pine to the mountain pine beetle by chemical and mechanical means. Pp. 148-158 In A.A. Berryman, G.D. Amman, R.W. Stark and D.L. Kibbee (eds.). Theory and practice of mountain pine beetle management in lodgepole pine forests--a symposium. April 25-27. College of Forest Resources, University of Idaho, Moscow.

Klein, W.H. 1979. Measuring damage to lodgepole pine caused by the mountain pine beetle. Pp. 35-42 In W.E. Waters (ed.). Current Topics in Forest Entomology. USDA Forest Serv. Gen. Tech. Rept. WO-8.

Klein, W.H., L.E. Stipe, and L.V. Frandsen. 1972. How damaging is a mountain pine beetle infestation? A case study. USDA Forest Serv. Region 4, Ogden, Utah. 12 pp.

Klimetzek, von D., J.P. Vite and K. Mori. 1980. Zur Wirkung und Formulierung des Populationslockstoffes des Nutzholzborkenkafers Trypodendron (=Xyloterus) lineatum. Z. Angew. Entomol. 89:57-63.

Kline, L.N., R.F. Schmitz, J.A. Rudinsky and M.M. Furniss. 1974. Repression of spruce beetle (Coleoptera) attraction by methylcyclohexenone in Idaho. Can. Entomol. 106:485-491.

Knell, J.D. and G.E. Allen. 1978. Morphology and ultrastructure of Unikaryon minutum sp.n. (Microsporidia:Protozoa), a parasite of the southern pine beetle, Dendroctonus frontalis. Acta Protozool. 17:271-278.

Knight, F.B. and F.M. Yasinski. 1956. Incidence of trees infested by the Black Hills beetle. USDA For. Serv. Res. Note RM-21. 4 pp.

Knipling, E.F. 1960. The eradication of the screwworm fly. Sci. Amer. 203:54-61.

Knipling, E.F. 1965. Some basic principles in insect population suppression. Bull. Ent. Soc. Amer. 12:7-15.

Koerber, T.W. (compiler) 1976. Lindane in forestry: a continuing controversy. U.S.D.A. For. Serv. Gen. Tech. Rep., PSW-14. 30 pp.

Koplin, J.R. 1969. The numerical response of woodpeckers to insect prey in a subalpine forest in Colorado. Condor 71:436-438.

Koplin, J. and P. Baldwin. 1970. Woodpecker predation on an endemic population of Engelmann spruce beetles. Amer. Midland Natur. 83:510-515.

Kosuge, T. 1969. The role of phenolics in host response to infection. Ann. Rev. Phytopathol. 7:195-222.

Kozak, V.T. 1976. The importance of natural enemies in reducing the numbers of bark-beetles in coniferous trees (In Russian). Zakhist Roslin 23:7-10.

Kozlowski, T.T. 1969. Tree physiology and forest pests. J. For. 67:118-123.

Krebs, C.J. 1978. Ecology: The Experimental Analysis of Distribution and Abundance. Harper and Row, New York. 678 pp.

Krisch, K. 1971. Carboxylic ester hydrolases. In P. Boyer (ed.). The Enzymes. Academic Press, New York.

Kroll, J.C. and R.R. Fleet. 1979. Impact of woodpecker predation on over-wintering within-tree populations of the southern pine beetle (Dendroctonus frontalis). pp. 269-281 In J.G. Dickson, R.N. Connor, R.R. Fleet, J.A. Jackson, and J.C. Kroll (eds.). The Role of Insectivorous Birds in Forest Ecosystems. Academic Press, New York.

Kulhavy, D.L., A.D. Partridge, and R.W. Stark. 1978. Mountain pine beetle and disease management in lodgepole pine stands: inseparable. Pp. 177-181 In A.A. Berryman, G.D. Amman, R.W. Stark, and D.L. Kibbee (eds.). Theory and practice of mountain pine beetle management in lodgepole pine forests--a symposium. Apr. 25-27. College of Forest Resources, University of Idaho, Moscow.

Kulman, H.M. 1964. Pitch defects in red pine associated with unsuccessful attacks by Ips spp. J. For. 62:322-325.

Kurashvili, B.E., P.S. Chanturishvili, A.O. Cholokava, G.A. Kakuliya, V.V. Odikadze, L.K. Maglakelidze, and Y.S. Dzhambazishvili. 1974. The results of experiments on the application of the white muscaridine fungus against the large spruce bark beetle (In Russian). Tbilisi, Georgian SSR, Metsniereba, 36 pp.

Lack, D. 1966. Population Studies of Birds. Clarendon Press, Oxford, 341 pp.

La Du, B.N. H.G. Mandel, and E.L. Way. 1971. Fundamentals of Drug Metabolism and Drug Disposition. The Williams and Wilkins Co., Baltimore. 615 pp.

Lanier, G.N. 1970. Biosystematics of the genus Ips (Coleoptera:Scolytidae) in North America. Hopping's Group IX. Can. Entomol. 102:1139-1163.

Lanier, G.N., M.C. Birch, R.F. Schmitz and M.M. Furniss. 1972. Pheromones of Ips pini (Coleoptera: Scolytidae): variation in response among three populations. Can. Entomol. 104:1917-1923.

Lanier, G.N. and W.E. Burkholder. 1974. Pheromones in speciation of Coleoptera. pp. 161-189 In M.C. Birch (ed.). Pheromones. North Holland, Amsterdam.

Lanier, G.N. and B.W. Burns. 1978. Barometric flux. Effects on the responsiveness of bark beetles to aggregation attractants. J. Chem. Ecol. 4:139-147.

Lanier, G.N., A. Claesson, T. Stewart, J.J. Piston and R.M. Silverstein. 1980. Ips pini: the basis for interpopulational differences in pheromone biology. J. Chem. Ecol. 6:677-687.

Lanier, G.N., W.E. Gore, G.T. Pearce, J.W. Peacock and R.M. Silverstein. 1977. Response of the European elm bark beetle, Scolytus multistriatus (Coleoptera: Scolytidae), to isomers and components of its pheromone. J. Chem. Ecol. 3:1-8.

Lanier, G.N. and D.L. Wood. 1968. Controlled mating, karyology, morphology, and sex-ratio in the Dendroctonus ponderosa complex. Ann. Entomol. Soc. Amer. 61:517-526.

Lawton, J., and S. McNeil. 1979. Between the devil and the deep blue sea: on the problem of being an herbivore. In R. Anderson, L. Taylor, B. Turner (eds.). Population Dynamics. Blackwell Scientific, Oxford.

Lawton, J.H. and D.R. Strong, Jr. 1981. Community patterns and competition in folivorous insects. Am. Nat. 118:317-338.

Leach, J.G. 1940. Insect Transmission of Plant Diseases. McGraw-Hill. New York. 615 pp.

Leach, J.G. L.W. Orr, C. Christianson. 1934. Interrelations of bark beetles and blue staining fungi in felled Norway pine timber. J. Agr. Res. 49:315-341.

LeConte, J.L. and G.H. Horn. 1878. The Rhynchophora of America north of Mexico. Amer. Philos. Soc. 15:341-391.

Lekander, B. 1968. Scandinavian bark beetle larvae, descriptions and classification. Royal College of Forestry Research Notes 4. 186 pp.

Leonard, T.J. and S. Dick. 1973. Induction of haploid fruiting by mechanical injury in Schizophyllum commune. Mycologia. 65:809-822.

Leuschner, W.A. 1979. Elements of a typical IPM system: The socio-economic and decision making model. pp. 263-267 In Integrated pest management for forest insects. Where do we stand today? Proc. Soc. Amer. Foresters (1978).

Leuschner, W.A. T.A. Max, G.D. Spittle, and H.W. Wisdom. 1978. Estimating southern pine beetle damages. Bull. Ent. Soc. Amer. 24:29-34.

Leuschner, W.A., D.G. Shore, and D.W. Smith. 1979. Estimating the southern pine beetles hydrologic impact. Bull. Ent. Soc. Amer. 25:147-150.

Leuschner, W.A., R.C. Thatcher, T.L. Payne, and P.E. Buffam. 1977. SBRAP--an integrated research and applications program. J. For. 75:478-480.

Lewontin, R.C., and J.L. Hubby. 1966. A molecular approach to the study of genic heterozygosity in natural populations. II. Amount of variation and degree of heterozygosity in natural populations of Drosophila pseudoobscura. Genetics 54:595-609.

Libbey, L.M., M.E. Morgan, T.B. Putman, and J.A. Rudinsky. 1976. Isomer of antiaggregative pheromone identified from male Douglas-fir beetle: 3-methylcyclohex-3-en-1-one. J. Insect Physiol. 22:871-873.

Light, D.M. and M.C. Birch. 1979. Inhibition of the attractive pheromone response in Ips paraconfusus by (R)-(-)-ipsdienol. Naturwiss. 66:159-160.

Lindgren, B.S. 1980. Pheromone based management of ambrosia beetles in dryland sorting areas on Vancouver Island. Unpublished Prog. Rept., Simon Fraser Univ., Burnaby, B.C., Canada.

Lindquist, E.E. 1964. Mites parasitizing eggs of bark beetles of the genus Ips. Can. Entomol. 96:125-126.

Lindquist, E.E. 1969. Review of Holarctic Tarsenemid mites (Acarina:Prostigmata) parasitizing eggs of Ipine bark beetle. Mem. Entomol. Soc. Can. 60. 111 pp.

Lindquist, E.E. 1969. Mites and the regulation of bark beetle populations. pp. 389-399 In G. Owen Evans (ed.). Proc. 2nd Internatl. Congr. Acarology, 1967.

Livingston, R.L. and A.A. Berryman. 1972. Fungus transport structures in the fir engraver, Scolytus ventralis (Coleoptera:Scolytidae). Can. Ent. 104:1793-1800.

Lomnicki, A. 1974. Evolution of the herbivore-plant, predator-prey, and parasite-host systems: a theoretical model. Amer. Natur. 108:167-180.

Loucks, O.L. 1970. Evolution of diversity, efficiency and community stability. Am. Zool. 10:17-25.

Lu, K.E., D.G. Allen, and W.B. Bollen. 1957. Association of yeasts with Douglas-fir beetle. For. Sci. 3:336-343.

Lysenko, O. 1959. Ecology of microorganisms in biological control of insects. Trans. Int. Conf. Insect Path. Biol. Control, I., Prague. pp. 109-133.

Mabry, T. and J. Gill. 1979. Sesquiterpene lactones and other terpenoids. pp. 502-537 In G. Rosenthal and D. Janzen (eds.). Herbivores: Their Interaction with Secondary Plant Metabolites. Academic Press, New York.

MacArthur, R.H. 1972. Geographical Ecology. Harper and Row, New York.

MacArthur, R.H., and E.O. Wilson. 1967. The Theory of Island Biogeography. Princeton Univ. Press, Princeton, New Jersey.

MacConnell, J.G., J.H. Borden, R.M. Silverstein and E. Stokkink. 1977. Isolation and tentative identification of lineatin, a pheromone from the frass of Trypodendron lineatum (Coleoptera: Scolytidae). J. Chem. Ecol. 3:549-561.

Mahon, R.J., C.A. Green, and R.H. Hunt. 1976. Diagnostic allozymes for routine identification of adults of the Anopheles gambiae group of species (Diptera:Culicidae). J. Med. Entomol. 15:297-299.

Mahoney, R.L. 1978. Lodgepole pine/mountain pine beetle risk classification methods and their application. pp. 106-113 In A.A. Berryman, G.D. Amman, R.W. Stark and D.L. Kibbee (eds.). Theory and Practice of Mountain Pine Beetle Management in Lodgepole Pine Forests. Univ. of Idaho, Moscow.

Malthus, T.R. 1798. An essay on the principle of population as it affects the future improvement of society. Johnson, London.

Margulis, L. 1974. Five kingdom classification and the origin and evolution of cells. Evol. Biol. 7:45-78.

Marler, J.E. and S.J. Barras. 1978. Identification of bacterial flora in galleries of the southern pine beetle, Dendroctonus frontalis Zimm. Final Rept. USFS, FS-SO-2203-1.29.

Mason, R. 1969. A simple technique for measuring oleoresin exudation flow in pines. For Sci. 15:56-57.

Massey, C.L. 1964. The nematode parasites and associates of the fir engraver beetle, Scolytus ventralis LeConte, in New Mexico. J. Insect Pathol. 6:133-155.

Massey, C.L. 1966. The influence of nematode parasites and associates on bark beetles in the United States. Bull. Ent. Soc. Amer. 12:384-386.

Massey, C.L. 1974. Biology and taxonomy of nematode parasites and associates of bark beetles in the United States. U.S.D.A. For. Serv. Agric. Handbook 446. Washington, D.C. 233 pp.

Massey, C.L. and N.D. Wygant. 1963. Woodpeckers: most important predators of the spruce beetle. Colorado Field Ornithologist 16:4-8.

Mathre, D.E. 1964. Survey of Ceratocystis spp. associated with bark beetles in California. Contrib. Boyce Thompson Inst. 22:353-362.

Mattson, W.J. and N.D. Addy. 1975. Phytophagous insects as regulators of forest primary production. Science 190:515-522.

Maxwell, F. 1972. Host plant resistance to insects--nutritional and pest management relationships. pp. 599-609 In J. Rodriquez (ed.). Insect and Mite Nutrition. North Holland Publishing Co., Amsterdam.

Maynard Smith, J. 1976. A comment on the red queen. Amer. Natur. 110:325-330.

Mayr, E. 1963. Animal species and evolution. Belknap Press, Cambridge.

Mayr, E. 1969. Principles of Systematic Zoology. McGraw-Hill, New York. 428 pp.

McCambridge, W.F. 1967. Nature of induced attacks by the Black Hills beetle, Dendroctonus ponderosae (Coleoptera: Scolytidae). Ann. Ent. Soc. Amer. 60:920-928.

McCambridge, W.F. 1969. Attraction of Black Hills beetles to ponderosa pine in the central Rocky Mountains. N. Cent. Br. Ent. Soc. Amer. Proc. 23:137-140.

McCambridge, W.F. 1971. Temperature limits of flight of the mountain pine beetle, Dendroctonus ponderosae. Ann. Entomol. Soc. Amer. 64:534-535.

McClelland, R. and S.S. Frissell. 1975. Identifying forest snags useful for hole-nesting birds. J. Forestry 73:414-417.

McCowan, J.C. and J.A. Rudinsky. 1958. Biological studies on the Douglas fir bark beetle. Weyerhaeuser Timber Co. For. Res. Note 11, 21 pp.

McDonald, G.I. 1979. Resistance of western white pine to blister rust: A foundation for integrated control. U.S.D.A. For. Serv. Res. Note INT-252. 5 pp.

McGraw, G.W. and R.W. Hemingway. 1977. 6, 8-dihydroxy-3-hydroxymethyl isocoumarin, and other phenolic metabolites of Ceratocystis minor. Phytochem. 6:1315-1316.

McGugan, B.M. and H.C. Coppel. 1962. A review of the biological control attempts against insects and weeds in Canada. Part II. pp. 35-216 In Biological Control of Forest Insects 1910-1958. Tech. Comm. No. 2. Commonwealth List of Biol. Control. Trinidad.

McKnight, R.C. 1979. Differences in response among populations of Dendroctonus ponderosae Hopkins to its pheromone complex. M.S. thesis, Univ. of Washington, Seattle. 77 pp.

McLean, J.A. and J.H. Borden. 1975. Survey for Gnathotricus sulcatus (Coleoptera: Scolytidae) in a commercial sawmill with the pheromone, sulcatol. Can. J. For. Res. 5:586-591.

McLean, J.A. and J.H. Borden. 1977. Suppression of Gnathotricus sulcatus with sulcatol-baited traps in a commercial sawmill and notes on the occurrence of G. retusus and Trypodendron lineatum. Can. J. For. Res. 7:348-356.

McLean, J.A. and J.H. Borden. 1978. An operational, pheromone-based suppression program for an ambrosia beetle, Gnathotricus sulcatus, in a commercial sawmill. J. Econ. Entomol. 72:165-170.

McMullen, L.H. and M.D. Atkins. 1961. Intraspecific competition as a factor in the natural control of the Douglas-fir beetle. For. Sci. 7:197-203.

McMurray, T. 1980. Effects of drought stress induced changes in host tree foliage quality on the growth of two forest insect pests. M.S. Thesis, Univ. of New Mexico.

McNeil, S. and T. Southwood. 1978. The role of nitrogen in the development of insect/plant relationships. pp. 77-98 In J. Harborne (ed.). Biochemical Aspects of Insect/Plant Interactions. Academic Press, New York.

Mergen, F., P. Hockstra, R. Echols. 1955. Genetic control of oleoresin yield and viscosity in slash pine. For. Sci. 1:19-30.

Meyer, H., and D. Norris. 1967. Behavioral responses of Scolytus multistriatus (Coleoptera: Scolytidae) to host (Ulmus) and beetle-associated chemotactic stimuli. Ann. Ent. Soc. Amer. 60:642-647.

Michalski, J. and S. Seniczak. 1974. Trichogramma semblidis (Chalcidoidea: Trichogrammatidae) as a parasite of bark beetle eggs (Coleoptera: Scolytidae). Entomophaga 19:237-242.

Michalson, E.L. and J. Findeis. 1979. Economic impact of mountain pine beetle on outdoor recreation. pp. 43-49 In Current Topics in Forest Entomology. Selected papers from the XVth Int. Congr. Ent. (1976). USDA For. Serv. Gen. Tech. Rep. WO-8.

Mickoleit, G. 1973. Uber den Ovipositor der Neuropteroidea und Coleoptera und seine phylogenetische Bedeutung (Insecta, Holometabola). Z. Morph. Okol. Tiere 74:37-64.

Miles, S.J. 1979. A biochemical key to adult members of the Anopheles gambiae group of species (Diptera:Culicidae). J. Med. Entomol. 15:297-299.

Miller, J.M. and F.P. Keen. 1960. Biology and control of the western pine beetle. U.S.D.A. For. Serv. Misc. Publ. 800. 381 pp.

Milligan, R.H. 1978. Hylastes ater (Paykull) (Coleoptera: Scolytidae)--black pine bark beetle. Forest and Timber Insects in New Zealand, No. 29. 8 pp.

Mirov, N. 1961. Composition of gum terpentines of pines. U.S. Forest Serv. Tech. Bul. No. 1239. 159 pp.

Mirov, N.T. 1953. Taxonomy and chemistry of the white pines. Madrono 12:81-89.

Mirov, N. and P. Iloff, Jr. 1958. Composition of gum terpentines of pines. XXIV. A report on Pinus ponderosa from five localities: central Idaho, central Montana, southeastern Wyoming, northwestern Nebraska, and central eastern Colorado. J. Amer. Pharm. Assoc., Sci. Ed. 47:404-409.

Mitter, C. and D.J. Futuyma. 1979. Population genetic consequences of feeding habits in some forest Lepidoptera. Genetics 92:1005-1021.

Mitter, C., D.J. Futuyma, J.C. Schneider, and J.D. Hare. 1979. Genetic variation and host plant relations in a parthenogenetic moth. Evolution 33:777-790.

Moeck, H.A. 1970. Ethanol as the primary attractant for the ambrosia beetle Trypodendron lineatum (Coleoptera: Scolytidae). Can. Entomol. 102:985-995.

Moeck, H.A. 1971. Field test of ethanol as a scolytid attractant. Canada Dept. Fish. and For., Bi-Mon. Res. Notes 27(2):11-12.

Moeck, H.A. 1975. Host selection behavior of bark beetles (Coleoptera: Scolytidae) attacking Pinus ponderosa with special emphasis on the western pine beetle, Dendroctonus brevicomis LeC. Ph.D. dissertation. Univ. Calif., Berkeley.

Moeck, H.A. 1980. Field test of Swedish "Drainpipe" pheromone trap with mountain pine beetle. Env. Canada. Bimonthly Res. Notes 36(1):2-3.

Moeck, H.A., D.L. Wood and K.Q. Lindahl, Jr. 1980. Host selection behavior of bark beetles (Coleoptera: Scolytidae) attacking Pinus ponderosa, with special emphasis on the western pine beetle, Dendroctonus brevicomis. J. Chem. Ecol. (In press).

Molnar, A.C. 1965. Pathogenic fungi associated with a bark beetle on alpine fir. Can. J. Bot. 43:563.

Moore, G.E. 1970. Isolating entomogenous fungi and bacteria, and tests of fungal isolates against the southern pine beetle. J. Econ. Entomol. 63:1702-1704.

Moore, G.E. 1971. Mortality factors caused by pathogenic bacteria and fungi of the southern pine beetle in Northern Carolina. J. Invert. Path. 17:28-37.

Moore, G.E. 1972. Southern pine beetle mortality in North Carolina caused by parasites and predators. Env. Entomol. 1:58-65.

Moore, G.E. 1972a. Microflora from the alimentary tract of healthy southern pine beetles, Dendroctonus frontalis (Coloptera: Scolytidae), and their possible relationship to pathogenicity. J. of Invert. Path. 19:72-75.

Moore, G.E. 1972b. Pathogenicity of ten strains of bacteria to larvae of the southern pine beetle. J. of Invert. Path. 20:41-45.

Moore, G.E. 1973. Pathogenicity of three entomogenous fungi to the southern pine beetle at various temperatures and humidities. Env. Entomol. 2:54-57.

Moore, G.E. and Thatcher, R.C. 1973. Epidemic and endemic populations of the southern pine beetle. U.S.D.A For. Serv. Res. Pap. SE-111.

Morgan, E.W. 1956. Silvical factors influencing resistance of ponderosa pine to Black Hills beetle attack. Soc. Am. For. Proc. 1:61-63.

Mori, K. 1976a. Absolute configuration of (+)-ipsdienol, the pheromone of Ips paraconfusus Lanier, as determined by the snythesis of its (R)-(-)-isomer. Tetrahedron Lett. 19:1609-1612.

Mori, K. 1976b. Synthesis of optically active forms of ipsenol, the pheromone of Ips bark beetles. Tetrahedron 32:1101-1106.

Mori, K. 1976c. Synthesis of (1S: 2R: 4S: 5R)-(-)-a-multistriatin, the pheromone in the smaller European elm bark beetle, Scolytus multistriatus. Tetrahedron 32:1979-1981.

Mori, K. 1977. Absolute configuration of (-)-4-methylheptan-3-ol, a pheromone of the smaller European elm bark beetle, as determined by the snythesis of its (3R, 4R)-(+)- and (3S, 4R)-(+)- isomers. Tetrahedron 33:289-294.

Mori, K., N. Mizumachi and M. Matsui. 1976. Synthesis of optically pure (1S, 4S, 5S)-2-pinen-4-ol (cis-verbenol) and its antipode, the pheromone of Ips bark beetles. Agr. Biol. Chem. 40:1611-1615.

Morimoto, K. 1976. Notes on the family characters of Apionidae and Brentidae (Coleoptera) with a key to related families. Kontyu 44:469-476.

Morris, O.N. and P. Olsen. 1970. Insect disease survey in British Columbia, 1965-1969. Canada, Fish and For., For. Res. Lab., CFS, Victoria. Info. Rep. BC-X-47.

Morrow, R.A. 1972. Dendroctonus ponderosae: The mountain pine beetle. Insect and Disease Laboratory, Colorado State Forest Serv., Colo. State Univ., Fort Collins.

Moser, J.C. 1975. Mite predators of the southern pine beetle. Ann. Entomol. Soc. Am. 68:1113-1116.

Moser, J.C. 1976. Surveying mites (Acarina) phoretic on the southern pine beetles in Allen Parish, Louisiana. Can. J. Ent. 103:1175-1198.

Moser, J.C., E.A. Cross, and L.M. Roton. 1971a. Biology of Pyemotes parviscolyti (Acari:Pyemotidas). Entomophaga 16:367-379.

Moser, J.C., B. Kielczewski, J. Wisniewski, and S. Balazy. 1978. Evaluating Pyemotot dryas (Vitzthum 1923) (Acari: Pyemotidae) as a parasite of the southern pine beetle. International Jour. Acarology 4:67-70.

Moser, J.C. and L.M. Roton. 1971. Mites associated with southern pine bark beetles in Allen Parish, Louisiana. Can. Entomol. 103:1775-1798.

Moser, J.C., R.C. Thatcher, and L.S. Pickard. 1971b. Relative abundance of southern pine beetle associates in east Texas. Ann. Entomol. Soc. Am. 64:72-77.

Moser, J.C., R.C. Wilkinson, and E.W. Clark. 1974. Mites associated with Dendroctonus frontalis Zimmerman (Scolytidae: Coleoptera) in Central America and Mexico. Turrialba 24(4):379-381.

Muesebeck, C.F.W. 1936. The genera of parasitic wasps of the braconid subfamily Euphorinae with a review of the Nearctic species. U.S.D.A. Misc. Publ. 241. 36 pp.

Mullick, D. 1969. Studies in periderm. II. Anthocyanidins in secondary periderm tissue of amabilis fir, grand fir, western hemlock, and western red cedar. Can. J. Botany 47:1419-1422.

Murray, R.A. and M.G. Solomon. 1978. A rapid technique for analysing diets of invertebrate predators by electrophoresis. Ann. Appl. Biol. 90:7-10.

Mustaparta, H., M.E. Angst and G.N. Lanier. 1977. Responses of single receptor cells in the pine engraver beetle, Ips pini (Say) (Coleoptera: Scolytidae) to its aggregation pheromone, ipsdienol, and its aggregation inhibitor, ipsenol. J. Comp. Physiol. 121:343-347.

Mustaparta, H., M.E. Angst and G.N. Lanier. 1979. Specialization of olfactory cells to insect-and host-produced volatiles in the bark beetle Ips pini (Say). J. Chem. Ecol. 5:109-123.

Nakashima, T. 1979. Several types of mycetangia found in (Scolytid and) platypodid ambrosia beetles. pp. 165-166 In Current Topics in Forest Entomology. Selected papers from the XVth Inter. Congr. Ent. (1976). U.S.D.A. For. Serv. Gen. Tech. Rep. WO-8.

Namkoong, G., T.H. Roberts, L.B. Nunnally, and H.A. Thomas. 1979. Isozyme variations in populations of southern pine beetles. Forest Sci. 25:197-203.

National Academy of Sciences. 1969. Principles of plant and animal pest control. 3. Insect-Pest Management and Control. Publ. 1695. NAS. Washington, D.C. 508 pp.

National Academy of Sciences. 1972. Pest Control Strategies for the Future. NAS. Washington, D.C. 376 pp.

National Academy of Sciences. 1975. Pest Control: An assessment of present and alternative technologies. In Forest Pest Control, Vol. 4, the Report of the Forest Study Team, National Research Council, Washington, D.C.

Nelson, R.M. and J.A. Beal. 1929. Experiments with bluestain fungi in southern pines. Phytopath. 19:1101-1106.

Nevo, E. 1978. Genetic variation in natural populations: patterns and theory. Theor. Pop. Biol. 13:121-177.

Nicholas, H. 1973. Terpenes. Recent Advances in Phytochemistry 2:254-309.

Nickle, W.R. 1962. The endoparasitic nematodes of California bark beetles with descriptions of Bovienema n.g. and Neoparasitylenchus n. subq. and with the presentation of new information on the life history of Contortylenchus elongatus n. comb. Ph.D. thesis. Univ. of California at Berkeley. 210 pp.

Nickle, W.R. 1963. Nematodes of bark beetles. Proc. Ent. Soc. Ontario. 93:131 (Abstr.).

Nickle, W.R. 1971. Behavior of the shothole borer, Scolytus regulosis altered by the nematode parasite Neoparasitylenchus regulasi. Ann. Ent. Soc. Amer. 64:751.

Nickle, W.R. 1978. Taxonomy of nematodes that parasitize insects and their use as biological control agents. pp. 37-51 In J.A. Romberger (ed.). Biosystematics in Agriculture. Beltsville Symposia in Agricultural Research No. 2, May 8-11, 1977. Allanheld, Osman and Co., Publ., Montclair, New Jersey.

Nienhaus, F. and R.A. Sikora. 1979. Mycoplasmas, spiroplasmas and rickettsia-like organisms as plant pathogens. Ann. Rev. Phytopath. 17:37-58.

Nijholt, W.W. and J.A. Chapman. 1968. A flight trap for collecting live insects. Can. Entomol. 100:1151-1153.

Nijholt, W.W. and J. Schonherr. 1976. Chemical response behavior of scolytids in West Germany and western Canada. Environ. Canada, Bi-mon. Res. Notes 32:31-32.

Nikitskii, N.B. 1976. Distribution of the tip bark beetle (Ips acuminatus) and its predators in the inhabited portion of the trunk (In Russian). Zoologicheskii Zhurnal 55:989-994.

Norris, D.M., J.M. Baker, and H.M. Chu. 1969. Symbiotic interrelationships between microbes and ambrosia beetles. III. Ergosterol as the source of sterol to the insect. Ann. Entomol. Soc. Am. 62:413-414.

Nuorteva, M. 1971. The bark beetles (Coleoptera, Scolytidae) and their insect enemies in Kuusamo commune, North Finland (In German). Annales Entomologici Fennici 37:65-72.

Nuorteva, M. and Laine L. 1968. On the possibility of insects as carriers of root fungus (Fomes annosus (Fr.) Cooke. Translation by Canada Dept. Fisheries and Forestry, Ottawa. Orig. source. Ann. Entomol. Fenn. 34:113-135.

Oester, P.T. and J.A. Rudinsky. 1975. Sound production in Scolytidae: stridulation by "silent" Ips bark beetles. Z. Ang. Ent. 79:421-427.

Oester, P.T. and J.A. Rudinsky. 1979. Acoustic behavior of three sympatric species of Ips (Coleoptera: Scolytidae) co-inhabiting Sitka spruce. Z. Angew. Entomol. 87:398-412.

Oester, P.T., L.C. Ryker and J.A. Rudinsky. 1978. Complex male premating stridulation of the bark beetle Hylurgops rufipennis (Mann.). Coleop. Bull. 32:93-98.

Oksanen, H., V. Perttunen and E. Kangas. 1970. Studies on the chemical factors involved in the olfactory orientation of Blastophagus piniperda (Coleoptera: Scolytidae). Contrib. Boyce Thompson Inst. 24:299-304.

Opler, P.A. 1974. Oaks as evolutionary islands for leaf mining insects. Am. Sci. 62:67-73.

Otvos, I.S. 1965. Studies on avian predators of Dendroctonus brevicomis LeConte (Coleoptera: Scolytidae) with special reference to Picidae. Can. Entomol. 97:1184-1199.

Otvos, I.S. 1969. Vertebrate predators of Dendroctonus brevicomis LeConte (Coleoptera: Scolytidae), with special reference to Aves. Ph.D. thesis, Univ. of California, Berkeley. 202 pp.

Otvos, I.S. 1970. Avian predation of the western pine beetle. pp. 119-127 In R.W. Stark and D.L. Dahlsten, (eds.). Studies on the Population Dynamics of the Western Pine Beetle, Dendroctonus brevicomis LeConte (Coleoptera: Scolytidae). Univ. of California, Div. of Agric. Sci.

Otvos, I.S. 1977. Observations on the food of three forest dwelling lizards in California. Herp. Review 8:6-7.

Overgaard, N.A. 1968. Insects associated with the southern pine beetle in Texas, Louisiana, and Mississippi. J. Econ. Entomol. 61:1197-1201.

Painter, R.H. 1951. Insect Resistance in Crop Plants. University Press of Kansas, Lawrence.

Parmeter, T.R., Jr., 1978. Forest stand dynamics and ecological factors in relation to dwarf mistletoe spread, impact, and control. pp. 16-30 In Dwarf mistletoe control through forest management--a symposium. U.S.D.A. For. Serv. Gen. Tech. Rep. 31.

Pathak, M. 1975. Utilization of insect-plant interactions in pest control. pp. 121-148 In D. Pimentel (ed.). Insects, Science, and Society. Academic Press, New York.

Payne, T.L. 1970. Electrophysiological investigations on response to pheromones in bark beetles. Contrib. Boyce Thompson Inst. 24:275-282.

Payne, T.L. 1971. Bark beetle olfaction. I. Electro-antennogram responses of the southern pine beetle (Coleoptera: Scolytidae) to its aggregation pheromone frontalin. Ann. Entomol. Soc. Amer. 64:266-268.

Payne, T.L. 1974a. Pheromone perception. pp. 35-61 In M.C. Birch (ed.). Pheromones. North-Holland, Amsterdam.

Payne, T.L. 1974b. Pheromone and host odor-stimulated potentials in Dendroctonus. Experientia 30:509.

Payne, T.L. 1975. Bark beetle olfaction. III. Antennal olfactory responsiveness of Dendroctonus frontalis Zimmerman and D. brevicomis LeConte (Coleoptera: Scolytidae) to aggregation pheromones and host tree terpene hydrocarbons. J. Chem. Ecol. 1:233-242.

Payne, T.L. 1979. Pheromone and host odor perception in bark beetles. pp. 27-57 In T. Narahashi (ed.). Neurotoxicology of Insecticides and Pheromones. Plenum Pub. Co., New York.

Payne, T.L., J.E. Coster, and P.C. Johnson. 1979. Development and evolution of synthetic inhibitors for use in southern pine beetle pest management. pp. 139-143 In W.E. Waters (ed.). Current Topics in Forest Entomology. Selected papers from the XVth Internt'l Congr. Ent. Washington, D.C. (1976). U.S.D.A. For. Serv. Tech. Rept. WO-8.

Payne, T.L., J.E. Coster, J.V. Richerson, L.J. Edson and E.R. Hart. 1978. Field response of the southern pine beetle to behavioral chemicals. Environ. Entomol. 7:578-582.

Payne, T.L. and J.C. Dickens. 1976. Adaptation to determine receptor system specificity in insect olfactory communication. J. Insect Physiol. 22:1569-1572.

Peacock, J.W. 1975. Research on chemical and biological controls for elm bark beetles. pp. 18-49 In Dutch Elm Disease. Proceedings of IUFRO Conference, Minneapolis-St. Paul, Minnesota, Sept. 1973. D.A. Burdekin and H.M. Heybroek, Compilers. U.S.D.A. For. Serv., North-eastern Forest Experimental Station, Upper Darby, PA. 94 pp.

Peacock, J.W., R.A. Cuthbert, W.E. Gore, G.N. Lanier, G.T. Pearce and 'R.M. Silverstein. 1975. Collection on Porapak Q of the aggregation pheromone of Scolytus multistriatus (Coleoptera: Scolytidae). J. Chem. Ecol. 1:149-160.

Peacock, J.W., A.C. Lincoln, J.B. Simeone and R.M. Silverstein. 1971. Attraction of Scolytus multistriatus (Coleoptera: Scolytidae) to a virgin-female-produced pheromone in the field. Ann. Entomol. Soc. Amer. 64:1143-1149.

Peacock, J.W., R.M. Silverstein, A.C. Lincoln and J.B. Simeone. 1973. Laboratory investigations of the frass of Scolytus multistriatus (Coleoptera: Scolytidae) as a source of pheromone. Environ. Entomol. 2:355-359.

Pearce, G.T., W.E. Gore and R.M. Silverstein. 1976. Synthesis and absolute configuration of multistriatin. J. Org. Chem. 41:2797-2803.

Pearce, G.T., W.E. Gore, R.M. Silverstein, J.W. Peacock, R. Cuthbert, G.N. Lanier and J.B. Simeone. 1975. Chemical attractants for the smaller European elm bark beetle, Scolytus multistriatus (Coleoptera: Scolytidae). J. Chem. Ecol. 1:115-124.

Person, H.L. 1931. Theory in explanation of the selection of certain trees by the western pine beetle. J. For. 31:696-699.

Person, H.L. 1940. The clerid, Thanisimus lecontei (Wolc.) as a factor in the control of the western pine beetle. J. For. 38:390-396.

Perttunen, V. and T. Boman. 1965. Laboratory experiments on the spontaneous take-off activity of Blastophagus piniperda (Coleoptera: Scolytidae) in relation to temperature and light intensity at different seasons of the year. Proc. 12th Int. Congr. Entomol. (1964) 5:344-345.

Perttunen, V., O. Oksanen and E. Kangas. 1970. Aspects of the external and internal factors affecting the olfactory orientation of Blastophagus piniperda (Coleoptera: Scolytidae). Contrib. Boyce Thompson Inst. 24:293-297.

Peterman, R.M. 1977. An evaluation of the fungal inoculation method of determining the resistance of lodgepole pine to mountain pine beetle (Coleoptera: Scolytidae) attacks. Can. Entomol. 109:443-448.

Peterman, R.M. 1978. The ecological role of mountain pine beetle in lodgepole pine forests. 224 pp. In A.A. Berryman, G.D. Amman, R.W. Stark and D.L. Kibbee (eds.). Theory and practice of mountain pine beetle management in lodgepole pine forests. Forest, Wildlife and Range Expt. Sta., Univ. Idaho, Moscow.

Pettersen, H. 1976a. Chalcid-flies (Hymenoptera: Chalcidoidea) reared from Ips typographus L. and Pityogenes chalcographus L. at some Norwegian localities. Norw. J. Entomol. 23:47-50.

Pettersen, H. 1976b. Parasites (Hymenoptera: Chalcidoidea) associated with bark beetles in Norway. Norw. J. Entomol. 23:75-77.

Pfister, R.D. and R. Daubenmire. 1975. Ecology of lodgepole pine Pinus contorta Dougl. pp. 27-46 In D.M. Baumgartner (ed.). Management of lodgepole pine ecosystems. Washington State University Coop. Ext. Serv., Pullman.

Pimentel, D. 1961. Animal population regulation by the genetic feedback mechanism. Amer. Natur. 95:65-79.

Pimentel, D. 1968. Population regulation and genetic feedback. Science 159:1432-1437.

Pimentel, D. 1971. Population regulation and genetic feedback. 195 pp. In I.A. McLaren (ed.). Natural Regulation of Animal Populations. Atherton Press, New York.

Pimentel, D. and A.C. Bellotti. 1976. Parasite-host population systems and genetic stability. Amer. Natur. 110:877-888.

Pitman, G.B. 1971. Trans-verbenol and alpha-pinene: their utility in manipulation of the mountain pine beetle. J. Econ. Entomol. 64:426-430.

Pitman, G.B. 1973. Further observations on Douglure in a Dendroctonus pseudotsugae management system. Environ. Entomol. 2:109-112.

Pitman, G.B., R.L. Hedden and R.I. Gara. 1975. Synergistic effects of ethyl alcohol on the aggregation of Dendroctonus pseudotsugae (Coloptera: Scolytidae) in response to pheromones. Z. Angew. Entomol. 78:203-208.

Pitman, G.B., R.A. Kliefoth and J.P. Vite. 1965. Studies on the pheromone of Ips confusus (LeConte). II. Further observations on the site of production. Contrib. Boyce Thompson Inst. 23:13-18.

Pitman, G.B., M.W. Stock and R.C. McKnight. 1978. Pheromone application in mountain pine beetle--lodgepole pine management: theory and practice. pp. 165-173 In A.A. Berryman, G.D. Amman and R.W. Stark (eds.). Theory and practice of mountain pine beetle management in lodgepole pine forests. College of Forest Resources, Univ. of Idaho, Moscow.

Pitman, G.B. and J.P. Vite. 1969. Aggregation behavior of Dendroctonus ponderosae (Coleoptera: Scolytidae) in response to chemical messengers. Can. Entomol. 101:143-149.

Pitman, G.B. and J.P. Vite. 1970. Field response of Dendroctonus pseudotsugae (Coleoptera: Scolytidae) to synthetic frontalin. Ann. Entomol. Soc. Amer. 63:661-664.

Pitman, G.B. and J.P. Vite. 1971. Predator-prey response to western pine beetle attractants. J. Econ. Entomol. 64:402-404.

Pitman, G.B. and J.P. Vite. 1975. Biosynthesis of methylcyclohexenone by male Douglas-fir beetle. Environ. Entomol. 3:886-887.

Pitman, G.B., J.P. Vite, G.W. Kinzer and A.F. Fentiman, Jr. 1968. Bark beetle attractants: trans-verbenol isolated from Dendroctonus. Nature (Lond.) 218:168-169.

Pitman, G.B., J.P. Vite, G.W. Kinzer and A.F. Fentiman, Jr. 1969. Specificity of population-aggregating pheromones in Dendroctonus. J. Insect Physiol. 15:363-366.

Plummer, E.L., T.E. Stewart, K. Byrne, G.T. Pearce and R.M. Silverstein. 1976. Determination of the enantiomeric composition of several insect pheromone alcohols. J. Chem. Ecol. 2:307-331.

Poinar, C.O., Jr. 1970. Nematode parasites and associates of the western pine beetle. pp. 132-133 In R.W. Stark and D.L. Dahlsten (eds.). Studies on the population dynamics of the western pine beetle Dendroctonus brevicomis LeConte (Coleoptera: Scolytidae). Univ. Calif. Div. of Agr. Sci.

Poinar, G.O. 1975. Entomogenous Nematodes. E.J. Brille, Holland. 317 pp.

Poinar, G.O. and J.N. Caylor. 1974. Neoparasitylenchus amvtocercus sp. N. (Tylenchida: Nematodea) from Conophthorus monophyllae (Scolytidae: Coleoptera) in California with a synopsis of the nematode genera found in bark beetles. J. Invert. Pathol. 24:112-119.

Polozhentsev. P.A. and V.F. Kozlov. 1975. Insect enemies of bark beetles. Zashchita Rastenii 1:41-44.

Ponomarenko, A.G. 1969. Istoricheskoe razvitie zhestkokrylykharkhostemat. Trudy Paleontologicheskogo Inst. AN SSSR, Tom 125:1-240.

Powell, J.A. and R.A. Mackie. 1966. Biological interrelationships of moths and Yucca whipplei. Univ. Calif. Publ. Entomol. 42:1-59.

Powell, J.M. 1966. Distribution and outbreaks of Dendroctonus ponderosae Hopk. in forests of western Canada. Dept. For., Forest Research Lab. Calgary, Alberta Info. Rept. A-X-2. 19 pp.

Powell, J.R. 1978. The founder-flush speciation theory: an experimental approach. Evolution 32:465-474.

Price, P.W. 1975. Insect Ecology. John Wiley and Sons, New York. 514 pp.

Price, P.W. 1980. Evolutionary Biology of Parasites. Princeton Univ. Press, Princeton.

Prokopy, R.J. 1981. Epideictic pheromones influencing spacing patterns of phytophagous insects. pp. 181-213 In D.A. Nordlund, R.L. Jones and W.J. Lewis (eds.). Semiochemicals: Their Role in Pest Control. John Wiley and Sons, New York.

Puritch, G.S. 1977. Distribution and phenolic composition of sapwood and heartwood in Abies grandis and effects of the balsam woolly aphid. Can. J. For. Res. 7:59-61.

Purrini, K. 1975. Contribution to the knowledge of the diseases of the large elm bark beetle, Scolytus scolytus F. in the Kosova district, Yugoslavia (in German). Anzeiger fur Schadlingskunde, Pflanzenschutz, Umweltschutz 48:154-156.

Purrini, K. 1978a. On Malameba locustae King and Taylor (Protozoa, Rhizopoda, Amoebidae) in the tufted spruce bark beetle, Dryocoetes autographus Katz. (Coleoptera: Scolytidae) (in German). Anzeiger fur Schadlingskunde, Pflanzenschutz, Umweltschutz 51:139-141.

Purrini, K. 1978b. Protozoa as pathogens of some species of bark beetles (Coleoptera: Scolytidae) in the Konigssee district, Upper Bavaria (in German). Anzeiger fur Schalingskunde Pflanzenschutz Umweltschutz 51:171-175.

Rabb, R.L. and F.E. Guthrie (Eds.). 1972. Concepts of Pest Management. North Carolina State Univ., Raleigh, N.C. 242 pp.

Raffa, K.F. 1980. The role of host resistance in the colonization behavior, ecology and evolution of bark beetles. Ph.D. Thesis, Washington St. Univ., Pullman. 137 pp.

Raffa, K.F. and A.A. Berryman. 1980. Flight responses and host selection by bark beetles. 278 pp. In A.A. Berryman and L. Safranyik (eds.). Dispersal of Forest Insects: Evaluation, Theory and Management Implications. Coop. Ext. Serv., Wash. St. Univ., Pullman.

Ramirez, B.W. 1970. Host specificity of fig wasps (Agaonidae). Evolution 24:680-691.

Read, C. 1970. Parasitism and Symbiology, an Introductory Text. Ronald Press Co. 316 pp.

Reece, C.D. 1979. Evolution of pest management. In R.D. Gale (ed.). Integrated Pest Management Colloquium. U.S.D.A. For. Ser. Gen. Tech. Rept. WO-14.

Reeve, R.J., J.E. Coster and P.C. Johnson. 1980. Spatial distribution of flying southern pine beetle (Coleoptera: Scolytidae) and the predator Thanasimus dubius (Coleoptera: Cleridae). Environ. Entomol. 9:113-118.

Reid, R.W. 1957. The bark beetle complex associated with lodgepole pine slash in Alberta. Part II. Notes on the biologies of several Hymenopterous parasites. Can. Entomol. 89:5-8.

Reid, R.W. 1958. Internal changes in the female mountain pine beetle, Dendroctonus monticolae Hopk., associated with egg laying and flight. Can. Entomol. 90:464-468.

Reid, R.W. 1958. Nematodes associated with the mountain pine beetle. Canada Dept. Agr. Div. For. Biol. Bimonthly Progr. Rep. 14:3.

Reid, R.W. 1962. Biology of the mountain pine beetle Dendroctonus monticolae Hopkins in the East Kootenay Region of British Columbia. II. Behavior in the host fecundity and internal changes in the female. Can. Ent. 94:605-613.

Reid, R.W. 1963. Biology of the mountain pine beetle, Dendroctonus monticolae Hopkins, in the East Kootenay Region of British Columbia. III. Interaction between the beetle and its host, with emphasis on brood mortality and survival. Can. Entomol. 95:225-238.

Reid, R.W. and H. Gates. 1970. Effect of temperature and resin on hatch of eggs of the mountain pine beetle (Dendroctonus ponderosae). Can. Entomol. 102:617-622.

Reid, R.W., H.S. Whitney and J.A. Watson. 1967. Reactions of lodgepole pine to attack by Dendroctonus ponderosae Hopkins and blue stain fungi. Can. J. Bot. 45:1115-1125.

Reisch, J. 1975. Forest protection from the point of view of protection of the environment (in German). pp. 61-65 In Fortschritte im integrierten Pflanzenschutz. Band 1. Aktuelle Probleme in integrierten Pflanzenschutz. Dietrich Steinkopff Verlag.

Renwick, J.A.A. 1967. Identification of two oxygenated terpenes from the bark beetles Dendroctonus frontalis and Dendroctonus brevicomis. Contrib. Boyce Thompson Inst. 23:355-360.

Renwick, J.A.A., P.R. Hughes and I.S. Krull. 1976b. Selective production of cis- and trans-verbenol from (-)- and (+)-alpha-pinene by a bark beetle. Science 191:199-201.

Renwick, J.A.A., P.R. Hughes and J.P. Vite. 1975. The aggregation pheromone system of a Dendroctonus bark beetle in Guatemala. J. Insect Physiol. 21:1097-1100.

Renwick, J.A.A. and G.B. Pitman. 1979. An attractant isolated from female Jeffrey pine beetles, Dendroctonus jeffreyi. Environ. Entomol. 8:40-41.

Renwick, J.A.A., G.B. Pitman and J.P. Vite. 1976a. 2-Phenylethanol isolated from bark beetles. Naturwiss. 63:198.

Renwick, J.A.A. and J.P. Vite. 1968. Isolation of the population aggregating pheromone of the southern pine beetle. Contrib. Boyce Thompson Inst. 24:65-68.

Renwick, J.A.A. and J.P. Vite. 1969. Bark beetle attractants: mechanism of colonization by Dendroctonus frontalis. Nature (Lond.) 224:1222-1223.

Renwick, J.A.A. and J.P. Vite. 1970. Systems of chemical communication in Dendroctonus. Contrib. Boyce Thompson Int. 24:283-292.

Renwick, J.A.A. and J.P. Vite. 1972. Pheromones and host volatiles that govern aggregation of the six-spined engraver beetle, Ips calligraphus. J. Insect Physiol. 18:1215-1219.

Renwick, J.A.A., J.P. Vite and R.F. Billings. 1977. Aggregation pheromones in the ambrosia beetle, Platypus flavicornis. Naturwiss. 64:226.

Rhoades, D. 1979. Evolution of plant chemical defenses against herbivores. pp. 3-54 In G. Rosenthal and D. Janzen (eds.). Herbivores: Their Interaction with Secondary Plant Metabolites. Academic Press, New York.

Rhoades, D.F. 1977. The anti-herbivore defenses of Larrea. pp. 135-175 In T.J. Mabry, J. Hunziker and D.R. DiFeo, Jr. (eds.). The Biology and Chemistry of the Creosotebush (Larrea) in the New World Deserts. Dowden, Hutchinson and Ross, Inc., Stroudsburg, PA.

Rhoades, D.F. and R.G. Cates. 1976. Toward a general theory of plant antiherbivore chemistry. In J.W. Wallace and R.L. Mansell (eds.). Biochemical Interaction Between Plants and Insects.

Ribereau-Gayon, P. 1972. Plant Phenolics. Oliver and Boyd, Edinburgh. 254 pp.

Rice, E. 1974. Allelopathy. Academic Press, New York. 353 pp.

Rice, E. 1979. Allelopathy--an update. Bot. Rev. 45:17-109.

Rice, R.E. 1969. Response of some predators and parasites of Ips confusus (LeC.) (Coleoptera: Scolytidae) to olfactory attractants. Contrib. Boyce Thompson Inst. 24:189-194.

Richards, G.A. and M.A. Brooks. 1958. Internal symbiosis in insects. Ann. Rev. Entomol. 3:37-56.

Richerson, J.V. and J.H. Borden. 1972. Host finding by heat perception in Coeloides brunneri (Hymenoptera: Braconidae). Can. Entomol. 104:1877-1881.

Richerson, J.V. and T.L. Payne. 1979. Effects of bark beetle inhibitors on landing and attack behavior of the southern pine beetle and beetle associates. Environ. Entomol. 8:360-364.

Ringold, G.B., P.J. Gravelle, D. Miller, M.M. Furniss, and M.D. McGregor. 1975. Characteristics of Douglas-fir beetle infestation in northern Idaho resulting from treatment with Douglure. U.S.D.A. For. Serv. Res. Note INT-189. 10 pp.

Roberts, D.W., and A.S. Campbell. 1977. Stability of entomopathogenic fungi. pp. 19-76 In Environmental Stability of Microbial Insecticides. Ent. Soc. of Amer., Maryland. p. 119.

Robinson, R.C. 1962. Blue stain fungi in lodgepole pine (Pinus contorta Dougl. var. latifolia Engelm.) infested by the mountain pine beetle (Dendroctonus monticolae Hopk.). Can. J. Bot. 40:609-614.

Robinson-Jeffrey, R.C. and R.W. Davidson. 1968. Three new Europhium species with Verticicladiella imperfect states on blue-stained pine. Can. J. Bot. 46:1523-1527.

Robinson-Jeffrey, R.C. and A.H.H. Grinchenko. 1964. A new fungus in the genus Ceratocystis occurring on bluestained lodgepole pine attacked by bark beetles. Can. J. Bot. 42:527-532.

Roe, A.L. and G.D. Amman. 1970. The mountain pine beetle in lodgepole pine forests. U.S.D.A. For. Serv. Res. Pap. INT-71, 23 pp.

Romano, A.H. 1966. Dimorphism. pp. 181-209 In G.C. Ainsworth and A.S. Sussman (eds.). The Fungi, Vol. 2. Academic Press, New York.

Rosenthal, G. and D. Janzen (eds.). 1979. Herbivores: Their Interaction with Secondary Plant Metabolites. Academic Press, New York. 718 pp.

Rosenzweig, M.L. 1973. Evolution of the predator isocline. Evolution 27:84-94.

Ross, H.H. 1965. A Textbook of Entomology. John Wiley and Sons, New York.

Roth, L.M. and T. Eisner. 1962. Chemical defenses of arthropods. Ann. Rev. Entomol. 7:107-136.

Roughgarden, J. 1975. Evolution of marine symbiosis--a simple cost-benefit model. Ecology 56:1201-1208.

Rudinsky, J.A. 1961. Factors affecting the population density of bark beetles. Proc. 13th IUFRO Cong. Vol. 1. Sec. 24-11. 13 pp.

Rudinsky, J.A. 1962. Ecology of Scolytidae. Ann. Rev. Entomol. 7:327-348.

Rudinsky, J.A. 1966. Scolytid beetles associated with Douglas-fir: response to terpenes. Science 152:218-219.

Rudinsky, J.A. 1968. Pheromone-mask by the female Dendroctonus pseudotsugae Hopk., an attraction regulator. Pan-Pac. Entomol. 44:248-250.

Rudinsky, J.A. 1969. Masking of the aggregating pheromone in Dendroctonus pseudotsugae Hopk. Science 166:884-885.

Rudinsky, J.A. 1973a. Multiple functions of the southern pine beetle pheromone, verbenone. Environ. Entomol. 2:511-514.

Rudinsky, J.A. 1973b. Multiple functions of the Douglas fir beetle pheromone, 3-methyl-2-cyclohexen-1-one. Environ. Entomol. 2:579-585.

Rudinsky, J.A. 1979. Chemoacoustically induced behavior of Ips typographus (Coleoptera: Scolytidae). Z. Angew. Entomol. 88:537-541.

Rudinsky, J.A. and G.E. Daterman. 1964. Field studies on flight patterns and olfactory responses of ambrosia beetles in Douglas-fir forests of western Oregon. Can. Entomol. 96:1339-1352.

Rudinsky, J.A., G.W. Kinzer, A.F. Fentiman, Jr. and R.L. Foltz. 1972a. Trans-verbenol isolated from Douglas-fir beetle: laboratory and field bioassay in Oregon. Environ. Entomol. 1:485-488.

Rudinsky, J.A., M.M. Furniss, L.N. Kline, and R.F. Schmitz. 1972b. Attraction and repression of Dendroctonus pseudotsugae (Coleoptera: Scolytidae) by three synthetic pheromones in traps in Oregon and Idaho. Can. Entomol. 104:815-822.

Rudinsky, J.A. and R.R. Michael. 1973. Sound production in Scolytidae: stridulation by female Dendroctonus beetles. J. Insect Physiol. 19:689-705.

Rudinsky, J.A. and R.R. Michael. 1974. Sound production in Scolytidae: "rivalry" behavior of male Dendroctonus beetles. J. Insect Physiol. 20:1219-1230.

Rudinsky, J.A., M.E. Morgan, L.M. Libbey and R.R. Michael. 1973. Sound production in Scolytidae: 3-methyl-2-cyclohexen-1-one released by the female Douglas-fir beetle in response to male sonic signal. Environ. Entomol. 2:505-509.

Rudinsky, J.A., M.E. Morgan, L.M. Libbey and T.B. Putnam. 1974a. Antiaggregative-rivalry pheromone of the mountain pine beetle, and a new arrestant of the southern pine beetle. Environ. Entomol. 3:90-98.

Rudinsky, J.A., M.E. Morgan, L.M. Libbey and T.B. Putnam. 1974b. Additional components of the Douglas fir beetle (Coleoptera: Scolytidae) aggregative pheromone and their possible utility in pest control. Z. Angew. Entomol. 76:65-77.

Rudinsky, J.A., M.E. Morgan, L.M. Libbey and T.B. Putnam. 1976. Release of frontalin by male Douglas-fir beetle. Z. Angew. Entomol. 81:267-269.

Rudinsky, J.A., V. Novak and P. Svihra. 1971. Attraction of the bark beetle Ips typographus L. to terpenes and a male-produced pheromone. Z. Angew. Ent. 67:179-188.

Rudinsky, J.A., P.T. Oester and L.C. Ryker. 1978. Gallery initiation and male stridulation of the polygamous spruce bark beetle Polygraphus rufipennis. Ann. Entomol. Soc. Amer. 71:317-321.

Rudinsky, J.A. and L.C. Ryker. 1976. Sound production in Scolytidae: rivalry and premating stridulation of male Douglas-fir beetle. J. Insect Physiol. 22:997-1003.

Rudinsky, J.A. and L.C. Ryker. 1977. Olfactory and auditory signals mediating behavioral patterns of bark beetles. pp. 195-209 In V. Lebeyrie (ed.). Comportment des Insectes et Trophique Milieu, Coll. Int. C.N.R.S. No. 265, Tours, France.

Rudinsky, J.A. and L.C. Ryker. 1979. Field bioassay of male Douglas-fir beetle compound 3-methylcyclohex-3-en-1-one. Experientia 35:1302.

Rudinsky, J.A. and L.C. Ryker. 1980. Multifunctionality of Douglas-fir beetle pheromone 3,2-MCH confirmed with solvent dibutyl phthalate. J. Chem. Ecol. 6:193-201.

Rudinsky, J.A., L.C. Ryker, R.R. Michael, L.M. Libbey, and M.E. Morgan. 1976. Sound production in Scolytidae: female sonic stimulus of male pheromone release in two Dendroctonus beetles. J. Insect Physiol. 22:1675-1681.

Rudinsky, J.A. and J.P. Vite. 1956. Effects of temperature upon the activity and the behavior of the Douglas-fir beetle. For. Sci. 2:258-267.

Rumbold, C.T. 1931. Two blue-staining fungi associated with beetle infestation of pines. J. Agric. Res. 43:847-873.

Rumbold, C.T. 1936. Three blue staining fungi, including two new species, associated with bark beetles. J. Agric. Res. 52:419-437.

Rumbold, C.T. 1941. A blue stain fungus Ceratospomella montium n. sp. and some yeast associated with two species of Dendroctonus. J. Agric. Research 62:589-601.

Russell, C.E. and A.A. Berryman. 1976. Host resistance in the fir engraver beetle. 1. Monoterpine composition of Abies grandis pitch blisters and fungus-infected wounds. Can. J. Bot. 54:14-18.

Ryan, C.A. 1979. Proteinase inhibitors. pp. 599-618 In G. Rosenthal and D. Janzen (eds.). Herbivores: Their Interaction with Secondary Plant Metabolites. Academic Press, New York.

Ryan, R.B. 1959. Termination of diapause in the Douglas-fir beetle, Dendroctonus pseudotsugae Hopkins (Coleoptera: Scolytidae), as an aid to continuous laboratory rearing. Can. Entomol. 91:520-525.

Ryan, R.B. 1961. A biological and developmental study of Coeloides brunneri Vier., a parasite of the Douglas-fir beetle, Dendroctonus pseudotsugae Hopk. Ph.D. Thesis. Oregon State Univ. Corvallis. 172 pp.

Ryan, R.B. and J.A. Rudinsky. 1962. Biology and habits of the Douglas-fir beetle parasite, Coeloides brunneri Viereck in western Oregon. Can. Entomol. 94:748-763.

Ryker, L.C., L.M. Libbey and J.A. Rudinsky. 1979. Comparison of volatile compounds and stridulation emitted by the Douglas-fir beetle from Idaho and western Oregon populations. Environ. Entomol. 8:789-798.

Ryker, L.C. and J.A. Rudinsky. 1976. Sound production in Scolytidae: aggressive and mating behavior of the mountain pine beetle. Ann. Entomol. Soc. Amer. 69:677-680.

Safranyik, L. 1978. Effects of climate and weather on mountain pine beetle populations. pp. 77-84 In A.A. Berryman, G.D. Amman, R.W. Stark and D.L. Kibbee (eds.). Theory and practice of mountain pine beetle management in lodgepole pine forests. Forest, Wildlife and Range Expt. Stn., Univ. Idaho, Moscow. 224 pp.

Safranyik, L., D.M. Shrimpton and H.S. Whitney. 1974. Management of lodgepole pine to reduce losses from the mountain pine beetle. Environ. Canada For. Serv., For. Tech. Rept. 1. 24 pp.

Safranyik, L., D.M. Shrimpton, and H.S. Whitney. 1975. An interpretation of the interaction between lodgepole pine, the mountain pine beetle and its associated blue stain fungi in western Canada. pp. 406-428 In D. Baumgartner (ed.). Management of Lodgepole Pine Ecosystems. Wash. State Univ. Coop. Ext. Service.

Sartwell, C. and R.E. Stevens. 1975. Mountain pine beetle in ponderosa pine--prospects for silvicultural control in second-growth stands. J. For. 73:136-140.

Sasakawa, M., F. Ohta and T. Negishi. 1976. Relationship between response to the aggregation pheromone trap of the minute pine bark beetle, Taenioglyptes fulvus (Niijima) (Coleoptera: Scolytidae) and regulation of the entrance hole density. Bull. Kyoto Prefectural Univ. Forests 20:42-48.

Sasamoto, K. 1958. Studies on the relation between silica content in the rice plant and the insect pests, VI. On the injury of silicated rice plant caused by the rice stem borer and its feeding behavior. Japan. J. Appl. Entomol. Zool. 2:88-92.

Saunders, J.L. 1967. Diurnal emergence of Xyleborus ferrugineus (Coleoptera: Scolytidae) from cacao trunks in Ecuador and Costa Rica. Ann. Entomol. Soc. Amer. 60:1094-1096.

Savely, Jr., H.E. 1939. Ecological relations of certain animals in dead pine and oak logs. Ecol. Monogr. 9:321-385.

Saxena, K.M.S. and A.L. Hooker. 1968. On the structure of a gene for disease resistance in maize. Proc. Nat. Acad. Sci. 61:1300-1305.

Scharpf, R.F. and J.R. Parmeter, Jr. 1978. Proceedings of the symposium on dwarf mistletoe control through forest management. U.S.D.A. For. Serv. Gen. Tech. Rep. PSW-31. 190 pp.

Schedl, K.E. 1931. Morphology of the bark-beetles of the genus Gnathotrichus Eichh. Smithsonian Misc. Coll. 82. 88 pp.

Schedl, K.E. 1978. Evolutionszentren bei den Scolytoidea (Coleoptera). Ent. Abh. Mus. Tierk. Dresden 41(9):311-323.

Schmid, J.M. 1972. Emergence, attack, densities, and seasonal trends of mountain pine beetle (Dendroctonus ponderosae) in the Black Hills. U.S.D.A. For. Ser. Res. Note RM-211. 7 pp.

Schmid, J.M. and T.E. Hinds. 1974. Development of spruce-fir stands following spruce beetle outbreaks. U.S.D.A. For. Serv. Res. Pap. RM-131. 16 pp.

Schmid, J.M. and R.H. Frye. 1977. Spruce beetle in the Rockies. U.S.D.A. For. Serv. Gen. Tech. Rept. RM-49. 38 pp.

Schmitz, R.F. and A.R. Taylor. 1969. An instance of lightning damage and infestation of ponderosa pines by the pine engraver beetle in Montana. U.S.D.A. For. Serv. Res. Note INT-88.

Schonherr, J. 1976. Mountain pine beetle: Visual behavior related to integrated control. pp. 449-452 In Proc. XVIth IUFRO World Congr. Div. II. Oslo, Norway.

Schowalter, T.D., R.N. Coulson, and D.A. Crossley, Jr. 1981. The role of southern pine beetle and fire in maintenance of structure and function of the southeastern coniferous forest. Environ. Entomol. 10:821-825.

Schroeder, D. 1974. Possibilities of biological control of elm bark beetles (Scolytidae) as a means of limiting Dutch elm disease (in German). Z. Angew. Ent. 76:150-159.

Schwerdtfeger, F. 1973. Forest entomology. pp. 361-386 In R.F. Smith et al. (eds.). History of Entomology. Annual Reviews Inc., Palo Alto, California.

Scott, V.E. 1978. Characteristics of ponderosa pine snags used by cavity-nesting birds in Arizona. J. For. 76:26-28.

Scott, V.E., K.E. Evans, D.R. Patton, C.D. Stone, and A. Singer. 1977. Cavity-nesting birds of North American forests. U.S.D.A. For. Serv. Agr. Hdbk. 511. 112 pp.

Scott, V.E., T.A. Whelan, and R.R. Alexander. 1978. Dead trees used by cavity-nesting birds on the Fraser Experimental Forest: a case history. U.S.D.A. Forest Serv. Res. Note RM-360. 4 pp.

Scudder, G.G.E. and S.S. Duffey. 1972. Cardiac glycosides in the Lygaeinae (Hemiptera:Lygaeidae). Can. J. Zool. 50:35-42.

Seifert, R.P. 1975. Clumps of Heliconia inflorescences as ecological islands. Ecology 56:1416-1422.

Seigler, D. and P. Price. 1976. Secondary compounds in plants: primary functions. Am. Nat. 110:101-105.

Selander, R.K. 1976. Genetic variation in natural populations. pp. 21-45 In F.J. Ayala (ed.). Molecular Evolution. Sinauer Associates, Sunderland.

Selander, R.K. and W.E. Johnson. 1973. Genetic variation among vertebrate species. Ann. Rev. Ecol. Syst. 4:75-91.

Shain, L. 1967. Resistance of sapwood in stems of loblolly pine to infection by Fomes annosus. Phytopathol. 57:1034-1045.

Shain, L. and W.E. Hillis. 1971. Phenolic extractives in Norway spruce and their effects on Fomes annosus. Phytopathol. 61:841-845.

Shelford, V.E. 1913. Animal communities in temperate America. Bull. Georgr. Soc. Chicago No. 5, 362 pp.

Shepherd, R.F. 1965. Distribution of attacks by Dendroctonus ponderosae Hopk. on Pinus contorta Dougl. var. latifolia Engelm. Can. Ent. 97:207-215.

Shepherd, R.F. 1966. Factors influencing the orientation and rates of activity of Dendroctonus ponderosae Hopkins (Coleoptera:Scolytidae) Can. Ent. 98:507-518.

Sheppard, P.M. 1962. Some aspects of the geography, genetics, and taxonomy of a butterfly. In D. Nichols (ed.). Taxonomy and Geography. Syst. Assoc. Publ. 4:1-32.

Shifrine, M. and H.J. Phaff. 1956. The association of yeasts with certain bark beetles. Mycologia 48:41-55.

Shook, R.S. and P.H. Baldwin. 1970. Woodpecker predation on bark beetles in Engelmann spruce logs as related to stand density. Can. Ent. 102:1345-1354.

Shorey, H.H. 1977. Interaction of insects with their chemical environment. pp. 1-5 In H.H. Shorey and J.J. McKelvey, Jr. (eds.). Chemical Control of Insect Behavior. John Wiley and Sons, New York.

Shrimpton, D.M. 1973a. Extractives associated with wound response of lodgepole pine attacked by the mountain pine beetle and associated microorganisms. Can. J. Bot. 51:527-535.

Shrimpton, D.M. 1973b. Age and size--related response of lodgepole pine to inoculation with Europhium clavigerum. Can. J. Bot. 51:1155-1160.

Shrimpton, D.M. 1978. Resistance of lodgepole pine to mountain pine beetle infestations. pp. 64-76 In A.A. Berryman, G.D. Amman, R.W. Stark, and D.L. Kibbee (eds.). Theory and practice of mountain pine beetle management in lodgepole pine forests--a symposium. Apr. 25-27. College of Forest Resources, Univ. of Idaho, Moscow.

Shrimpton, D.M. and H.S. Whitney. 1968. Inhibition of growth of blue stain fungi by wood extractives. Can. J. Bot. 46:757-761.

Sikorowski, P.P., G.S. Pabst and O. Tomson. 1979. The impact of diseases on southern pine beetle in Mississippi. Miss. Agric. Forestry Exp. Sta. Tech. Bull. 99. 9 pp.

Silverstein, R.M. 1970. Methodology for isolation and identification of insect pheromones--examples from Coleoptera. pp. 285-299 In D.L. Wood, R.M. Silverstein and M. Nakajima (eds.). Control of Insect Behavior by Natural Products. Academic Press, New York.

Silverstein, R.M. and J.C. Young. 1976. Insects generally use multicomponent pheromones. pp. 1-29 In M. Beroza (ed). Pest Management with Insect Sex Attractants and Other Behavior-controlling Chemicals. Amer. Chem. Soc., Washington, D.C.

Silverstein, R.M., J.O. Rodin and D.L. Wood. 1966. Sex attractants in frass produced by male Ips confusus in ponderosa pine. Science 150:509-510.

Silverstein, R.M., J.O. Rodin and D.L. Wood. 1967. Methodology for isolation and identification of insect pheromones with reference to studies on California five-spined ips. J. Econ. Entomol. 60:944-949.

Silverstein, R.M., R.G. Brownlee, T.E. Bellas, D.L. Wood and L.E. Browne. 1968. Brevicomin: principal sex attractant in the frass of the female western pine beetle. Science 159:889-890.

Simpson, G.G. 1961. Principles of Animal Taxonomy. Columbia University Press, New York. 247 pp.

Sinclair, G.D. and D.K. Dymond. 1973. The distribution and composition of extractives in Jack pine trees. Can. J. For. Res. 3:516-521.

Slatkin, M. and J.M. Smith. 1979. Models of coevolution. Quart. Rev. Biol. 54:233-263.

Slessor, K.N., A.C. Oehlschlager, B.D. Johnston, H.D. Pierce, Jr., S.K. Grewal and L.K.G. Wickremesinghe. 1980. Lineatin: regioselective synthesis and resolution leading to the chiral pheromone of Trypodendron lineatum. J. Org. Chem. 45:2290-2297.

Smart, J., and N.F. Hughes. 1973. The insect and the plant: progressive paleoecological integration. pp. 143-155 In H.F. Van Emden (ed.). Insect/Plant Relationships. Blackwell Scientific Publ., Oxford.

Smith, R.H. 1961. The fumigant toxicity of three pine resins to Dendroctonus brevicomis and D. jeffreyi. J. Econ. Entomol. 54:365-369.

Smith, R.H. 1963. Toxicity of pine resin vapors to three species of Dendroctonus bark beetles. J. Econ. Entomol. 56:827-831.

Smith, R.H. 1964. Variation in the monoterpenes of Pinus ponderosa Laws. Science 143:1337-1338.

Smith, R.H. 1964. The monoterpenes of lodgepole pine oleoresin. Phytochem. 3:259-262.

Smith, R.H. 1965. A physiological difference among beetles of Dendroctonus ponderosae (=D. monticolae) and D. ponderosae (D. jeffreyi). Ann. Entomol. Soc. Amer. 58:440-442.

Smith, R.H. 1965. Effect of monoterpene vapors on the western pine beetle. J. Econ. Entomol. 58(3):509-510.

Smith, R.H. 1966a. Forcing attacks of western pine beetles to test resistance of pines. U.S.D.A. For. Serv. Res. Note PSW-119. 119 pp.

Smith, R.H. 1966b. Resin quality as a factor in the resistance of pines to bark beetles. pp. 189-196 In H. Gerhold, R. McDermott, E. Schreiner, and J. Winieski (eds.). Breeding Pest-Resistant Trees. Pergamon Press, Oxford.

Smith, R.H. 1967. Variations in the monoterpene composition of the wood resin of Jeffrey, Washoe, Coulter, and lodgepole pines. For. Sci. 13:246-252.

Smith, R. 1969. Xylem resin as a factor in the resistance of pines to forced attacks by bark beetles. pp. 1-13 In Second World Consultation on Forest Tree Breeding, Washington. FAO International Union of Forestry Research Organizations.

Smith, R.H. 1972. Xylem resin in the resistance of the Pinaceae to bark beetles. U.S.D.A. For. Serv. Gen. Tech. Rep. PSW-1. 7 pp.

Smith, R.H. 1975. Formula for describing effect of insect and host tree factors on resistance to western pine beetle attack. Jour. Econ. Entomol. 68:841-844.

Smith, R.H. 1977. Monoterpenes of ponderosa pine xylem resin in western United States. U.S.D.A. For. Serv. Tech. Bull. No. 1532. 48 pp.

Smith, R.H. and R.C. van Borstel. 1972. Genetic control of insect populations. Science 178:1164-1174.

Snellgrove, T.A. and T.D. Fahey. 1977. Market values and problems associated with utilization of dead timber. For. Products J. 27:74-79.

Society of American Foresters. 1958. "Forest Terminology." Soc. Amer. For., Washington, D.C. 84 pp.

Southwood, T.R.E. 1960. The evolution of the insect-host tree relationship--a new approach. Proc. XI Int. Cong. Entomol. 1:651-655.

Southwood, T.R.E. 1961. The number of species of insect associated with various trees. J. Anim. Ecol. 30:1-8.

Southwood, T.R.E. 1973. The insect/plant relationship--an evolutionary perspective. In H.F. Van Emden (ed.), Insect/Plant Relationships. Blackwell Scientific Publ., Oxford.

Squillace, A.E. 1976. Analyses of monoterpenes of conifers by gas-liquid chromatography. pp. 120-139 In J.P. Miksche (ed.). Modern Methods in Forest Genetics. Springer-Verlag, N.Y.

Squillace, A.E. 1971. Inheritance of monoterpene composition in cortical oleoresin of slash pine. For. Sci. 17:381-389.

Srivastava, L. 1963. Secondary Phloem in the Pinaceae. University of California, Berkeley, Calif. 141 pp.

Stage, A.R. 1973. Prognosis model for stand development. U.S.D.A. For. Serv. Res. Pap. INT-137. 32 pp.

Stage, A.R. 1975. Forest stand prognosis in the presence of
pests. pp. 233-245 In D.M. Baumgartner (ed.). Management
of lodgepole pine ecosystems. Washington State Univ.
Ext. Serv. Pullman.

Stallcup, P.L. 1963. A method for investigating avian
predation on the adult black hills beetle. M.S. thesis,
Colorado State University, Fort Collins. 60 pp.

Stark, R. 1965. Recent trends in forest entomology. Ann.
Rev. Entom. 10:303-324.

Stark, R.W. 1976. The concept of impact in integrated pest
management. pp. 110-116 In Current Topics in Forest
Entomology. Selected papers from the XVth Inter. Congr. of
Ent., Washington, D.C. U.S.D.A. For. Serv. Gen. Tech.
Rept. WO-8.

Stark, R.W. 1977. Integrated pest management in forest
practice. J. For. 75:251-254.

Stark, R.W. 1979. Forest management and integrated pest
management. pp. 34-42 In R.D. Gale (ed.). Integrated Pest
Management Colloquium. Symposium Proceedings Oct.
17-18, 1978. Millford, PA. U.S.D.A. For. Serv. Gen. Tech.
Rept. WO-14.

Stark, R.W. and J.H. Borden. 1965. Observations on mortality
factors of the fir engraver beetle, Scolytus ventralis
(Coleoptera: Scolytidae). J. Econ. Entomol. 58:1162-1163.

Stark, R.W. and F.W. Cobb, Jr. 1969. Smog injury, root
diseases and bark beetle damage in ponderosa pine. Calif.
Agric. 23:13-15.

Stark, R.W. and D.L. Dahlsten (eds.). 1970. Studies on the
population dynamics of the western pine beetle,
Dendroctonus brevicomis LeConte. Univ. Calif. Div. Agric.
Sci. Publ. 174 pp.

Stark, R.W. and A.R. Gittins. 1973. Pest Management in the 21st Century. Proc. Symp. Oct. 13-14, 1972. Moscow, Idaho. Idaho Res. Foundation Inc. Natural Res. Series No. 2. 102 pp.

Stark, R.W., P.R. Miller, F.W. Cobb, D.L. Wood and J.R. Parmeter. 1968. Photochemical oxidant injury and bark beetle (Coleoptera: Scolytidae) infestation of ponderosa pine, I. Incidence of bark beetle infestation in injured trees. Hilgardia 39:121-126.

Starks, K., R. Muniappan, and R. Eikenbary. 1972. Interaction between plant resistance and parasitism against the greenbug on barley and sorghum. Ann. Ent. Soc. Amer. 65:650-655.

Starr, M.P. 1975. A generalized scheme for classifying organismic associations. Symp. Soc. Exp. Biol. 29:1-21.

Stebbins, G.L. and F.J. Ayala. 1981. Is a new evolutionary synthesis necessary? Science 213:967-971.

Steinhaus, E.A. 1949. Principles of insect pathology. McGraw, New York. 757 pp.

Steinhaus, E.A. 1963. Insect Pathology, an Advanced Treatise. Academic Press, New York.

Steinhaus E.A., and G.A. Marsh. 1962. Report of diagnoses of diseased insects. Hilgardia 33:349-490.

Stephen, F.M. and D.L. Dahlsten 1976. The arrival sequence of the arthropod complex following attack by Dendroctonus brevicomis (Coleoptera: Scolytidae) in ponderosa pine. Can. Entomol. 108:283-304.

Stevens, R., C. Myers, W. McCambridge, G. Downing, and J. Laut. 1975. Mountain pine beetle in front range ponderosa pine: what it's doing and how to control it. U.S.D.A. For. Serv. Gen. Tech. Rep. RM-7.

Stewart, T.E., E.L. Plummer, L.L. McCandless, J.R. West and R.M. Silverstein. 1977. Determination of enantiomer composition of several bicyclic ketal insect pheromone components. J. Chem. Ecol. 3:27-43.

Stock, M.W. and G.D. Amman. 1980. Genetic differentiation among mountin pine beetle populations from lodgepole pine and ponderosa pine in northeast Utah. Ann. Entomol. Soc. Amer. 73:472-478.

Stock, M.W. and J.D. Guenther. 1979. Isozyme variation among mountain pine beetle (Dendroctonus ponderosae) populations in the Pacific Northwest. Environ. Entomol. 8:889-893.

Stock, M.W., J.D. Guenther, and G.B. Pitman. 1978. Implications of genetic differences between mountain pine beetle populations to integrated pest management. pp. 197-204 In A.A. Berryman, G.D. Amman, R.W. Stark, and D.L. Kibbee (eds.). Theory and practice of mountain pine beetle management in lodgepole pine forests--a symposium. Apr. 25-27. College of Forest Resources, University of Idaho, Moscow.

Stock, M.W., G.B. Pitman and J.D. Guenther. 1979. Genetic differences between Douglas-fir beetles (Dendroctonus pseudotsugae) from Idaho and coastal Oregon. Ann. Entomol. Soc. Amer. 72:394-397.

Stoszek, K.J. 1978. Forest management considerations. In M.H. Brookes, R.W. Stark, and R.W. Campbell (eds.). The Douglas-fir Tussock Moth: A Synthesis. U.S.D.A. For. Serv. Tech. Bull. 1585. 331 pp.

Stoszek, K., P. Mika, and H. Osborne. 1977. Comparative studies on the physiological environmental indices of grand fir stands located on high, moderate, and low DFTM hazard sites in northern Idaho. Final Report, For., Wildlife and Range Exp. Stn., Univ. of Idaho, Moscow.

Strobel, G.A. and G.N. Lanier. 1981. Dutch elm disease. Sci. Amer. 245:56-66.

Stroh, R., and H. Gerhold. 1965. Eastern white pine characteristics related to weevil feeding. Silv. Genet. 14:160-169.

Strong, D.R., Jr. 1974a. Nonasymptotic species richness models and the insects of British trees. P.N.A.S. 71:2766-2769.

Strong, D.R., Jr. 1974b. Rapid asymptotic species accumulation on phytophagous insect communities: the pests of Cacao. Science 185:1064-1066.

Strong, D.R. and D.A. Levin. 1979. Species richness of plant parasites and growth form of their host. Am. Nat. 114:1-22.

Strong, D.R., Jr., E.D. McCoy, and J.R. Rey. 1977. Time and the number of herbivore species: the pests of sugarcane. Ecology 58:167-175.

Sturgeon, K.B. 1979. Monoterpene variation in ponderosa pine xylem resin related to western pine beetle predation. Evolution 33:803-814.

Sturgeon, K.B. 1980. Evolutionary interactions between the mountain pine beetle, Dendroctonus ponderosae Hopkins, and its host trees in the Colorado Rocky Mountains. Ph.D. dissertation, University of Colorado, Boulder. 160 pp.

Swaine, J.M. 1907. Practical and popular entomology. 21. The Scolytidae or engraver beetles. Can. Entomol. 39:191-195.

Swaine, J.M. 1909. Catalogue of the described Scolytidae of America north of Mexico. pp. 76-159 In New York State Mus. Bull. 134, Report of State Entomology for 1908, Appendix B.

Swaine, J.M. 1918. Canadian bark beetles. Part 2. A preliminary classification, with an account of the habits and means of control. Dom. Canada Dept. Agriculture, Entomol. Branch Tech. Bull. 14(2). 143 pp.

Swain, T. 1979. Tannins and lignins. pp. 657-682 In G. Rosenthal and D. Janzen (eds.). Herbivores: Their Interactions with Secondary Plant Metabolites. Academic Press, New York.

Tartof, K.D. 1975. Redundant genes. Ann. Rev. Ecol. Sys. 9:355-385.

Templeton, A. and M.A. Rankin. 1978. Genetic revolutions and control of insect populations. In: R.H. Richardson (ed.) The Screwworm Problem. Univ. of Texas Press.

Templeton, A.R. 1980a. The theory of speciation via the founder principle. Genetics 94:1011-1038.

Templeton, A.R. 1980b. Modes of speciation and inferences based on genetic distances. Evolution 34:719-729.

Tepedino, V.J. and N.L. Stanton. 1976. Cushion plants as islands. Oecologia 25:243-256.

Thalenhorst, W. 1956. Grundzuge der populationsdynamik des grossen Fitchenborkenkafers Ips typographus L. Shriftenreihe Forstl. Fakultat Univ. Gottingen. No. 21.

Thatcher, R.C. 1960. Bark beetles affecting southern pines: A review of current knowledge. U.S.D.A. For. Serv. Occas. Pap. 180. 25 pp.

Thatcher, R.C. and L.S. Pickard. 1966. The clerid beetle, Thanasimus dubias, as a predator of the southern pine beetle. J. Econ. Entomol. 59:955-957.

Thomas, G.M. 1974. Diagnostic techniques. pp. 1-48 In G.E. Cantwell (ed.). Insect diseases. Marcel Dekker, Inc. New York. 595 pp.

Thomas, G.M. and G.O. Poinar Jr. 1973. Report of diagnoses of diseased insects 1962-1972. Hilgardia 42:261-360.

Thomas, J.B. 1957. The use of larval anatomy in the study of bark beetles (Coleoptera:Scolytidae). Can. Entomol. Supplement 5. 45 pp.

Thomas, J.B. 1960. The immature stages of Scolytidae: the tribe Xyloterini. Can. Entomol. 92:410-419.

Thomas, J.B. 1965. The immature stages of Scolytidae: the genus Dendroctonus Erichson. Can. Entomol. 97:374-400.

Thomas, T.W., G.L. Crouch, R.S. Bumstead, and L.D. Bryant. 1975. Silvicultural options and habitat values in coniferous forests. pp. 272-287 In D.R. Smith (Tech. Coor.). Proceedings of Symposium on Managing Forest and Range Habitats for non-game birds. U.S.D.A. For. Serv. Gen. Tech. Rep. WO-1.

Thompson, S.N. and R.B. Bennett. 1971. Oxidation of fat during flight of male Douglas-fir beetles, Dendroctonus pseudotsugae. J. Insect Physiol. 17:1555-1563.

Thorsteinson, A.J. 1960. Host selection in phytophagous insects. Ann. Rev. Entomol. 5:193-218.

Tilden, P.E., W.D. Bedard, D.L. Wood, K.Q. Lindahl and P.A. Rauch. 1979. Trapping the western pine beetle at and near a source of synthetic attractive pheromone. Effects of trap size and position. J. Chem. Ecol. 5:519-531.

Tillyard, R.J. 1924. Upper Permian Coleoptera and a new order from the Behmont beds. Proc. Linnaean Soc. N.S. Wales 49:429-435.

Tinsley, T.W. and K.A. Harrap. 1978. Viruses of invertebrates. pp. 1-101 In H. Fraenkel-Conrat and R.R. Wagner (eds.). Comprehensive Virology, Vol. 12. Plenum, New York.

Torgerson, T.R. 1978. Parasites and hyperparasites of the Douglas-fir tussock moth. In M.H. Brooks, R.W. Stark, and R.W. Campbell (eds.). The Douglas-Fir Tussock Moth: a synthesis. U.S.D.A. For. Serv. Tech. Bull. 1585. 331 pp.

Trager, W. 1970. Symbiosis. Van Nostrand Reinhold, New York. 100 pp.

Turnock, W.T., K.L. Taylor, D. Schroder, and D.L. Dahlsten. 1976. Biological control of pests of coniferous forests. pp. 289-311 In C.B. Huffaker and P.S. Messenger (eds.). Theory and practice of biological control. Academic Press, New York.

Unnithan, G.C. and K.K. Nair. 1977. Ultrastructure of juvenile hormone-induced degenerating flight muscles in a bark beetle, Ips paraconfusus. Cell Tiss. Res. 185:481-490.

Unnithan, G.C., K.K. Nair and W.S. Bowers. 1977. Precocene-induced degeneration of the corpus allatum of adult females of the bug Oncopeltus fasciatus. J. Insect Physiol. 23:1081-1094.

Upadhyay, H.P. and W.B. Kendrick. 1975. Prodromus for a revison of Ceratocystis (Microascales, Ascomycetes) and its conidial states. Mycologia 67:798-805.

U.S.D.A. Forest Service. 1972. Report of the workshop on impacts of insects and diseases on uses, values, and productivity of forest resources. Marana Air Park, Arizona. February 7-11. Proc. Rept. 64 pp.

van Emden, H.F. (ed.). 1973. Insect/Plant Relationships. John Wiley and Sons, Inc., New York. 215 pp.

van Emden, H. and M. Way. 1973. Host plants in the population dynamics of insects. pp. 181-199 In H. van Emden (ed.). Insect/Plant Relationships, John Wiley and Sons, New York.

Van Valen, L. 1973. A new evolutionary law. Evolutionary Theory 1:1-30.

Vaughn, J.L. 1974. Virus and rickettsial diseases, pp. 49-84 In G.E. Cantwell (ed.). Insect Diseases, Vol. 1. Marcel Dekker, New York. 300 pp.

Vite, J.P. 1961. The influence of water supply on oleoresin exudation pressure and resistance to bark beetle attack in Pinus ponderosa. Contr. Boyce Thompson Inst. 21:37-66.

Vite, J.P. and A. Bakke. 1979. Synergism between chemical and physical stimuli in host colonization by an ambrosia beetle. Naturwiss. 66:528-529.

Vite, J.P., A. Bakke and P.R. Hughes. 1974. Ein populations--lockstoff des zwolfsahnigen Kiefernborkenkafers Ips sexdentatus. Naturwiss. 61:365-366.

Vite, J.P., A. Bakke and J.A.A. Renwick. 1972b. Pheromones in Ips (Coleoptera; Scolytidae): occurrence and production. Can. Entomol. 104:1967-1975.

Vite, J.P. and R.G. Crozier. 1968. Studies on the attack behavior of the southern pine beetle. IV. Influence of host condition on aggregation pattern. Contrib. Boyce Thompson Inst. 24:87-94.

Vite, J.P. and R.I. Gara. 1962. Volatile attractants from ponderosa pine attacked by bark beetles (Coleoptera: Scolytidae). Contrib. Boyce Thomspon Inst. 21:251-254.

Vite, J.P., R.I. Gara and R.A. Kliefoth. 1963. Collection and bioassay of a volatile fraction attractive to Ips confusus (LeC.) (Coleoptera: Scolytidae). Contrib. Boyce Thompson Inst. 22:39-50.

Vite, J.P., R.I. Gara and H.D. von Scheller. 1964. Field observations on the response to attractants of bark beetles infesting southern pines. Contrib. Boyce Thompson Inst. 22:461-470.

Vite, J.P., R. Hedden and K. Mori. 1976b. Ips grandicollis: field response to the optically pure pheromone. Naturwiss. 63:43-44.

Vite, J.P., F. Islas S., J.A.A. Renwick, P.R. Hughes, and R.A. Kleifoth. 1974. Biochemical and biological variation of southern pine beetle populations in North and Central America. Z. Angew. Ent. 75:422-435.

Vite, J.P., D. Klimetzek, G. Loskant, R. Hedden and K. Mori. 1976a. Chirality of insect pheromones: response interruption by active antipodes. Naturwiss. 63:582-583.

Vite, J.P., G. Ohloff, and R.F. Billings. 1978. Pheromonal chirality and integrity of aggregation response in southern species of the bark beetle Ips sp. Nature (Lond.) 272:817-818.

Vite, J.P. and G.B. Pitman. 1969. Aggregation behavior of Dendroctonus brevicomis in response to synthetic pheromones. J. Insect Physiol. 15:1617-1622.

Vite, J.P., G.B. Pitman, A.F. Fentiman, Jr., and G.W. Kinzer. 1972a. 3-Methyl-2-cyclohexen-1-ol isolated from Dendroctonus. Naturwiss. 59:469.

Vite, J.P. and J.A.A. Renwick. 1971a. Inhibition of Dendroctonus frontalis response to frontalin by isomers of brevicomin. Naturwiss. 58:418.

Vite, J.P. and J.A.A. Renwick. 1971b. Population aggregating pheromone in the bark beetle, Ips grandicollis. J. Insect. Physiol. 17:1699-1704.

Vite, J.P. and D.L. Williamson. 1970. Thanasimus dubius: prey perception. J. Insect Physiol. 16:233-239.

von Rudloff, E. and G. Rehfeldt. 1980. Chemosystematic studies in the genus Pseudotsuga. IV. Inheritance and geographical variation in the leaf oil terpenes of Douglas-fir from the Pacific Northwest. Can. J. Bot. (In press).

Von Schrenk, H. 1903. The "Bluing" and the "Red Rot" of the western pine, with special reference to the Black Hills Forest Reserve. USDA. Bur. Plant Indus. Bull. 36.

Vosylyte, B. 1978. Biological characteristics of the nematode Contortylenchus pseudodiplogaster Slankis, (Sphaerulariidae)-- a parasite of the engraver beetle Ips sexdentatus (in Russian). Acta Parasitologica Lituanica 16:99-106.

Wagner, T.L., J.A. Gagne, P. Doraiswamy, R.N. Coulson, and K.W. Brown. 1980. Tree moisture and xylem water potential in relation to southern pine beetle development and mortality. Environ. Entomol. 8:1129-1138.

Walker, M.V. 1938. Evidence of Triassic insects in the Petrified Forest National Monument, Arizona. Proc. U.S. Nat. Mus. 85(3033):137-141.

Wallace, J. and R. Mansell (eds.). 1976. Biochemical Interaction Between Plants and Insects. Plenum, New York. 425 pp.

Walsh, B.D. 1864. On phytophagic varieties and phytophagous species. Proc. Ent. Soc. Phil. 3:403-430.

Waring, R.H. and G.B. Pitman. 1980. A simple model of host resistance to bark beetles. Oregon St. Univ., School of Forestry, Res. Note 65, 2 pp.

Waters, W.E. 1974. Systems approach to managing pine bark beetles. pp. 12-14 In T.L. Payne, R.N. Coulson, and R.C. Thatcher (eds.). Southern Pine Beetle Symposium Proceedings. Texas Agr. Exp. Stn.

Waters, W.E. and R.W. Stark. 1980. Forest pest management: concept and reality. Ann. Rev. Entomol. 25:479-509.

Webb, J.W. and R.T. Franklin. 1978. Influence of phloem moisture on brood development of the southern pine beetle (Coleoptera: Scolytidae). Environ. Entomology 7:405-410.

Weijman, A.C.M. and G.S. de Hoog. 1975. On the subdivision of the genus Ceratocystis. Ant. Van Leeuwenhoek 41:353-360.

Weiser, J. 1970. Three new pathogens of the Douglas fir beetle, Dendroctonus pseudotsugae: Nosema dendroctoni n. sp., Ophryocystis dendroctoni n. sp. and Chytridiopsis typographi n. comb. Jr. Invert. Pathol. 16:436-441.

Wellner, C.A. 1978. Management problems resulting from mountain pine beetles in lodgepole pine forests. pp. 9-18 In A.A. Berryman, G.D. Amman, R.W. Stark, and D.L. Kibbee (eds.). Theory and practice of mountain pine beetle management in lodgepole pine forests--a symposium. Apr. 25-27. College of Forest Resources, University of Idaho, Moscow.

Werner, R.A. 1972. Response of the beetle, Ips grandicollis to combinations of host and insect-produced attractants. J. Insect Physiol. 18:1403-1412.

White, G.K. 1976. The impact of the pine beetle on recreational values. Ph.D. dissertation. Washington State Univ., Pullman. 87 pp.

White, M.J.D. 1978. Modes of Speciation. W.H. Freeman and Co., San Francisco.

White, T.C.R. 1969. An index to measure weather-induced stress of trees associated with outbreaks of psyllids in Australia. Ecol. 50:905-909.

White, T.C.R. 1974. A hypothesis to explain outbreaks of looper caterpillars with special reference to populations of Selidosema suavis in a plantation of Pinus radiata in New Zealand. Oecologia 16:279-301.

White, T.C.R. 1976. Weather, food, and plagues of locusts. Oecologia 22:119-134.

Whitney, H.S. 1971. Association of Dendroctonus ponderosae (Coleoptera: Scolytidae) with blue stain fungi and yeasts during brood development in lodgepole pine. Can. Entomol. 103:1495-1503.

Whitney, H.S. and R.A. Blauel. 1972. Ascospore dispersion in Ceratocystis spp. and Europhium clavigerum in conifer resin. Mycologia 64:410-414.

Whitney, H.S. and F.W. Cobb, Jr. 1972. Non-staining fungi associated with the bark beetle Dendroctonus brevicomis (Coleoptera:Scolytidae) on Pinus ponderosa. Can. J. Bot. 50:1943-1945.

Whitney, H.S. and S.H. Farris. 1970. Maxillary mycangium in the mountain pine beetle. Science 167:54-55.

Whitney, H.S. and A. Funk. 1977. Pezizella chapmanii n. sp., a discomycete associated with bark beetle galleries in western conifers. Can. J. Bot. 55:888-891.

Whitney, H.S., L. Safranyik, S.J. Muraro and E.D.A. Dyer. 1978. In defense of the concept of direct control of mountain pine beetle populations in lodgepole pine: some modern approaches. pp. 159-164 In A.A. Berryman, G.D. Amman, and R.W. Stark (eds.). Theory and Practice of mountain pine beetle management in lodgepole pine forests; symposium proceedings, Apr. 25-27, 1978, Washington State University, Pullman, Wash. University of Idaho, Moscow.

Whitney, H.S., R.D. Whitney and N.J. Whitney. 1979. Significance of wood stains in living conifers. Abstr. Proc. Ann. Meeting Can. Phytopath. Soc. Lethbridge, Alta. June 1979. Proc. Can. Phytopath. Soc. 46 June 25-27, 1979.

Whittaker, R.H. 1969. Evolution of diversity in plant communities. Brookhaven Symp. Biol. 22:178-196.

Wickerham, L.J. 1960. Hansenula holstii, a new yeast important in the early evolution of the heterothallic species of its genus. Mycologia 52:171-183.

Wickerham, L.J. and K. Burton. 1961. Phylogeny of phosphomannan--producing yeasts, I. The Genera. J. Bacteriol. 82:265-268.

Wickman, B. 1978. Tree injury. pp. 66-77 In M. Brookes, R. Stark, and R. Campbell (eds.). The Douglas-Fir Tussock Moth: A Synthesis. U.S.D.A. Tech. Bull. 1585. 338 pp.

Wilkenson, R.C. 1968. Reproduction and diet in three species of Ips bark beetles. In E. Hodgson (ed.). Florida Inst. Food and Agric. Ser. Ann. Res. Rep., Univ. Fla., Gainesville.

Wilkinson, C. 1968. In E. Hodgson (ed.). Enzymatic Oxidation of Toxicants. North Carolina State University Press, Raleigh. 113 pp.

Williamson, D.L. 1971. Olfactory discernment of prey by Medetera bistriata (Diptera: Dolicopodidae). Ann. Entomol. Soc. Amer. 64:586-589.

Williamson, D.L. and S.P. Vite. 1971. Impact of insecticidal control on the southern pine beetle population in Eastern Texas. J. Econ. Entomol. 64:1440-1444.

Wilson, A.C. 1975. Evolutionary importance of gene regulation. Stadler Genet. Symp. 7:117-134.

Wilson, D.S. 1980. The Natural Selection of Populations and Communities. The Benjamin/Cummins Publ. Co., Menlo Park, Calif.

Wilson, L.F. 1977. A guide to insect injury of conifers in the Lake States. pp. 67-72 In U.S.D.A. For. Serv. Agr. Hdbk. 501.

Wong, B.L. and A.A. Berryman. 1977. Host resistance to the fir engraver beetle. 3. Lesion development and containment of infection by resistant Abies grandis inoculated with Trichosporium symbioticum. Can. J. Bot. 55:2358-2365.

Wood, D.L. 1961. The occurrence of Serratia marcescens Bizio in laboratory populations of Ips confusus (LeConte) (Coleoptera:Scolytidae). J. Insect Path. 3:330-331.

Wood, D.L. 1972. Selection and colonization of ponderosa pine by bark beetles. pp. 101-117 In H.F. van Emden (ed.). Insect/Plant Relationships. Blackwell Scientific, London.

Wood, D.L. 1976. Host selection by bark beetles in the mixed-conifer forests of California. In T. Jermy (ed.). The Host-Plant in Relation to Insect Behavior and Reproduction. Plenum Press, New York.

Wood, D.L. 1979. Development of behavior modifying chemicals for use in forest pest management in the USA. pp. 261-279 In F.T. Ritter (ed.). Chemical ecology: About Communication in Animals. Elseveer, North Holland Biomedical Press.

Wood, D.L. (collator). 1980. Approach to research and forest pest management for the western pine beetle control. In C.B. Huffaker (ed.). New Technology of Pest Control. J. Wiley and Sons. New York. 500 pp.

Wood, D.L. and W.D. Bedard. 1976. The role of pheromones in the population dynamics of the western pine beetle. Proc. XV Int. Congr. Entomol., Washington, D.C.

Wood, D.L., L.E. Browne, W.D. Bedard, P.E. Tilden, R.M. Silverstein and J.O. Rodin. 1968. Response of Ips confusus to synthetic sex pheromones in nature. Science 159:1373-1374.

Wood, D.L., L.E. Browne, B. Ewing, K. Lindahl, W.D. Bedard, P.E. Tilden, K. Mori, G.B. Pitman and P.R. Hughes. 1976. Western pine beetle: specificity among enantiomers of male and female components of an attractant pheromone. Science 192:896-898.

Wood, D.L., L.E. Browne, R.M. Silverstein and J.O. Rodin. 1966. Sex pheromones of bark beetles--I. Mass production, bio-assay, source, and isolation of the sex pheromone of Ips confusus (LeC.). J. Insect Physiol. 12:523-536.

Wood, D.L. and R.W. Bushing. 1963. The olfactory response of Ips confusus (LeConte) (Coleoptera: Scolytidae) to the secondary attraction in the laboratory. Can. Entomol. 95:1066-1078.

Wood, D.L., F.W. Cobb, Jr., D.J. Cohen, L.E. Browne, H.A. Moeck and R.W. Stark. 1974. Host selection by bark beetles (Coleoptera:Scolytidae) in the mixed conifer forests of California. Symp. on "The Host Plant in Relation to Insect Behavior and Reproduction," June 11-14, 1974, Tihany, Hungary.

Wood, S.L. 1954a. A revision of the North American Cryphalini (Scolytidae, Coleoptera). Univ. Kansas Sci. Bull. 36(2), No. 15:959-1089.

Wood, S.L. 1954b. Bark beetles of the genus Carphoborus Eichhoff (Coleoptera:Scolytidae) in North America. Can. Entomol. 86:502-526.

Wood, S.L. 1957. Ambrosia beetles of the tribe Xyloterini (Coleoptera:Scolytidae) in North America. Can. Entomol. 89:337-354.

Wood, S.L. 1958. Bark beetles of the genus Pityoborus Blackman (Coleoptera: Scolytidae). Great Basin Nat. 18:46-56.

Wood, S.L. 1963. A revision of the bark beetle genus Dendroctonus Erichson (Coleoptera: Scolytidae). Great Basin Natur. 23:1-117.

Wood, S.L. 1972. New synonymy in the bark beetle tribe Cryphalini (Coleoptera: Scolytidae). Great Basin Nat. 32:40-54.

Wood, S.L. 1978. A reclassification of the subfamilies and tribes of Scolytidae (Coleoptera). Ann. Soc. Entomol. France (N.S.) 14:95-122.

Wood, S.L. 1982. The bark and ambrosia beetles of North and Central America (Coleoptera: Scolytidae), a taxonomic monograph. Great Basin Nat. Mem. 6:1-1326 (+ index).

Wood, T.K. 1980. Divergence in the Enchenopa binotata Say complex (Homoptera: Membracidae) effected by host plant adaptation. Evolution 34:147-160.

Wool, D., H.F. van Emden and S.W. Bunting. 1978. Electrophoretic detection of the internal parasite, Aphidius matricariae in Myzus persicae. Ann. Appl. Biol. 90:21-26.

Wright, E.J. 1935. Trichosporium symbioticum n. sp. a wood staining fungus associated with Scolytus ventralis. J. Agr. Res. 50:525-538.

Wright, L.C., A.A. Berryman, and S. Gurusiddaiah. 1979. Host resistance to the fir engraver beetle, Scolytus ventralis (Coleoptera: Scolytidae), 4. Effect of defoliation on wound monoterpene and inner bark carbohydrate concentrations. Can. Entomol. 111:1255-1262.

Wright, R.H., D.L. Chambers, and I. Keiser. 1971. Insect attractants, antiattractants, and repellents. Can. Entomol. 103:627-630.

Index